Securing Delay-Tolerant Networks with BPSec

Securing Delay-Tolerant Networks with BPSec

Dr Edward J. Birrane III, Sarah Heiner, and Ken McKeever

This edition first published 2023
© 2023 John Wiley & Sons, Inc.

The right of Dr Edward J. Birrane III, Sarah Heiner, and Ken McKeever to be identified as the authors of this work has been asserted in accordance with law.

Registered Office
John Wiley & Sons, Inc., 111 River Street, Hoboken, NJ 07030, USA

For details of our global editorial offices, customer services, and more information about Wiley products visit us at www.wiley.com.

Wiley also publishes its books in a variety of electronic formats and by print-on-demand. Some content that appears in standard print versions of this book may not be available in other formats.

Library of Congress Cataloging-in-Publication Data Applied for:
ISBN: 9781119823476 (hardback)

Cover Design: Wiley
Cover Image: © Sergey Panychev/Shutterstock

Set in 9.5/12.5pt STIXTwoText by Straive, Chennai, India

This book is dedicated to the memory of Ralph E. Piersanti – educator, athlete, actor, friend, and (for large parts of my childhood) an eminently convincing Batman.

Dr Edward J. Birrane III

This book is dedicated to my family. Thank you for seeing, and being, the magic.

Sarah Heiner

This book is dedicated to Gillian McKeever, for her support, patience, and most of all understanding of the multitude of endeavors (like this one) that seem to keep me otherwise busier than I should rightly be.

Ken McKeever

Contents

Acronyms

AAD	Additional Authenticated Data
ABE	Attribute-Based Encryption
AEAD	Authenticated Encryption with Associated Data
AES	Advanced Encryption Standard
AH	Authentication Header
API	Application Programming Interface
ARPAnet	Advanced Research Projects Agency Network
ASB	Abstract Security Block
BAB	Bundle Authentication Block
BCB	Block Confidentiality Block
BIB	Block Integrity Block
BPA	Bundle Protocol Agent
BPCF	Block Processing Control Flags
BPSec	Bundle Protocol Security
BP	Bundle Protocol
BPv6	Bundle Protocol version 6
BPv7	Bundle Protocol version 7
BSP	Bundle Security Protocol
CA	Certificate Authority
CBOR	Concise Binary Object Representation
CCSDS	Consultative Committee for Space Data Systems
CDDL	Concise Data Definition Language
CIA	Confidentiality, Integrity, and Availability
CLA	Convergence-Layer Adapter
CMS	Cryptographic Message Syntax
CONOPS	Concept of Operations
COSE	CBOR Object Signing and Encrypting
COVID-19	Coronavirus Disease 2019
CRC	Cyclic Redundancy Check
DES	Data Encryption Standard
DNSSEC	Domain Name System Security Extensions
DTLS	Datagram Transport Layer Security
DTN	Delay Tolerant Networking
ECDHE	Elliptic Curve Diffie–Hellman Exchange
EID	Endpoint Identifier
ESB	Extension Security Block

ESP	Encapsulating Security Payload
FTP	File Transfer Protocol
GCM	Galois Counter Mode
GPS	Global Positioning System
HMAC	Hashed Message Authentication Code
HTTPS	Hypertext Transfer Protocol Secure
IANA	Internet Assigned Numbers Authority
ID	Identifier
IETF	Internet Engineering Task Force
IKE	Internet Key Exchange
ION	Interplanetary Overlay Network
IPN	Interplanetary Network
IPSec	Internet Protocol Security
IP	Internet Protocol
IPPT	Integrity-Protected Plaintext
IRTF	Internet Research Task Force
ISP	Internet Service Provider
IV	Initialization Vector
JPL	Jet Propulsion Laboratory
JSON	JavaScript Object Notation
LAN	Local Area Network
LTP	Licklider Transport Protocol
MAC	Message Authentication Code
MET	Mission Elapsed Time
MILnet	Military Network
MTMR	Multiple-Target Multiple-Result
MTSR	Multiple-Target Single-Result
MTU	Maximum Transmission Unit
NASA	National Aeronautics and Space Administration
NCP	Network Control Program
NIST	National Institute of Standards and Technology
NPA	Near-Path Attacker
NP	Network-layer Protocol
NSFnet	National Science Foundation Network
NTP	Network Time Protocol
OPA	On-Path Attacker
OP	Operation
OpenSSL	Open Secure Socket Layer
OSB	Other Security Block
PCB	Payload Confidentiality Block
PDU	Protocol Data Unit
PIB	Payload Integrity Block
PKI	Public Key Infrastructure
PNB	Previous Node Block
PSK	Pre-Shared Key
RAM	Random Access Memory
RC4	Rivest Cipher 4
RFC	Request for Comments

RSA	Rivest-Shamir–Adleman
SBSP	Streamlined Bundle Security Protocol
SCPS	Space Communications Protocol Suite
SHA	Secure Hash Algorithm
SOp	Security Operation
SSI	Solar System Internet
SSL	Secure Socket Layer
STMR	Single-Target Multiple-Result
STSR	Single-Target Single-Result
TCP	Transmission Control Protocol
TC	Telecommand
TLS	Transport Layer Security
TM	Telemetry
TTP	Trusted Third Parties
UDP	User Datagram Protocol
URI	Uniform Resource Identifier
UTC	Coordinated Universal Time
VPN	Virtual Private Network
WLAN	Wireless Local Area Network
WoT	Web of Trust

About the Authors

Dr Edward J. Birrane III

Dr Birrane is a computer scientist and embedded software engineer who focuses on the adaption of computer networking protocols for use in non-traditional environments. He has supported a variety of embedded software engineering efforts, to include the NASA New Horizons mission to explore the Pluto-Charon system and the NASA Parker Solar Probe mission to observe the outer corona of our sun.

He works with industry, government, and academia on the design and development of protocols to implement the Delay-Tolerant Networking (DTN) architecture. He co-chairs the DTN working group within the Internet Engineering Task Force (IETF) where he is a coauthor of BPv7 (RFC 9171), BPSec (RFC 9172), and the default security contexts for BPSec (RFC 9173). He is a member of the principal professional staff at the Johns Hopkins University Applied Physics Laboratory as the Chief Engineer for Space Constellation Networking. He is also an adjunct professor of computer science at both the University of Maryland, Baltimore County, and The Johns Hopkins University.

Sarah Heiner

Sarah Heiner is part of a passionate community of inventors working on a set of emerging networking protocols. She has contributed to the body of DTN work in several areas, with a primary focus on Bundle Protocol Security (BPSec). Involved with both the Internet Engineering Task Force and the Consultative Committee for Space Data Systems, Sarah is a coauthor of RFC 9173 BPSec Default Security Contexts, the Bundle Protocol Security Protocol Red Book, and the Asynchronous Management Architecture, among other documents. She received a bachelor's degree in computer science from the University of Maryland, Baltimore County, and is currently pursuing a master's degree in computer science at The Johns Hopkins University Applied Physics Laboratory.

Ken McKeever

Despite early aspirations of a career as a mad scientist, Ken McKeever is now a communication systems engineer and researcher. He received a bachelor's degree in electrical engineering from the Pennsylvania State University and a master's degree in electrical and computer engineering at The Johns Hopkins University Applied Physics Laboratory. His current focus is on the design of resilient networked communication systems operating in constrained environments and is the coauthor of BPSec (RFC 9172). His other research interests include wired and wireless communications security, protocol design, and test and evaluation of communication systems.

Foreword

I have been involved with computer and network security since 1975 after I received my under-graduate degree in electrical and computer engineering. Back then computer security was a fledgling profession primarily concerned with processing data of multiple classification levels simultaneously (i.e. multilevel security). Likewise, network security, in the ARPAnet days with the few attached computer systems running the Network Control Program (NCP) prior to the deployment of TCP/IP, was primarily concerned with the ability to ensure data confidentiality from sender to receiver. As the ARPAnet begat the MILnet which in turn begat the NSFnet and TCP/IP replaced NCP, the worldwide Internet explosion took place and networking evolved to what we know it to be today. I became involved with security for space with an early effort to use Internet protocols (e.g. TCP, IP, FTP) for space missions. An outgrowth of the Internet Protocols (IPs) in space was the Interplanetary Internet (IPN) and what we now know as the Bundle Protocol (BP) and its associated Bundle Protocol Security (BPSec). As one of the most exciting projects in my career, I worked on the Jet Propulsion Laboratory (JPL) team that architected the IPN/DTN and specifically, the security architecture and the security protocol.

In the early 1990s, while the Internet was beginning its world-wide growth using a standard protocol stack, the space communications community was still using decades old link layer proto-col standards such as Telemetry (TM), and Telecommand (TC). When upper layer protocols were needed for space missions, there were no standards, and protocols would be (re)invented for each mission. This was costly, time consuming, and inefficient.

With the success of the Internet's Transmission Control Protocol/Internet Protocol (TCP/IP) standards, the thought at the JPL was why not adapt those protocols for use in space missions. The result was a series of protocols known as the Space Communications Protocols Suite (SCPS) which consisted of a space-specific network layer protocol (NP), a transport protocol (TP) which was an adaptation of TCP for use in long delay-bandwidth environments, a space-specific "thin" security protocol (SP), and an adaptation of the Internet file transfer protocol (FP). Even in the 1990s we knew that security had to be included in the protocol stack from the get-go.

It was determined that the SCPS protocols would be excellent upper layer protocols for space missions in near-earth orbit and maybe even at lunar distances. However, there were too many issues which prevented SCPS protocols to work beyond lunar distances.

NASA's next bold adventure was the unmanned exploration of Mars. It was anticipated that a large array of Mars orbiters, landers, and rovers would be launched to Mars over the decades. Lan-ders and rovers might have direct-to-earth links, but more than likely they would make use of orbiting relays to minimize the size and weight of the landers. To the networking community, this looked like the beginnings of a network at Mars – both space-based around Mars and surface-based on Mars.

We knew that the SCPS protocols would work on the surface of Mars and near Mars but that between Earth and Mars, we needed an entirely new networking paradigm. It had to support orbital mechanic obscuration, loss of connectivity, long delays, ability to operate through relays, and it needed to be secure.

As a result, the IPN was "born." The intent was to design a set of protocols that could provide secure Internet-like services at long distances, while plagued with intermittent connectivity and the inherent bandwidth delays. As with SCPS, security was baked in from the outset. Too often, security is an afterthought (if at all considered) and then "bolted on" resulting in spectacular security failures. We decided that we were not going to go down that road once again. Over time, the IPN morphed to become the Delay Tolerant Internet and the Disruption Tolerant Internet (DTN). While the original vision was a secure network paradigm for deep-space networking, other non-space networking communities realized that the same architecture could be used to enable their needs. This included terrestrial sensor networks and battlefield networks – both suffering from intermittent connectivity resulting in long delivery delays under which TCP would not operate efficiently. DTN could also provide an architectural solution for other resource constrained environments, i.e. Internet connectivity in rural locations over narrow bandwidth and long-delay connectivity.

For deep space, it was believed that the IPN/DTN would have a finite (i.e. small) amount of bandwidth available and therefore it was of paramount need to assure that only those who needed access were provided with the ability to use the network (i.e. access control). Likewise, data integrity/authentication and data confidentiality were paramount DTN security requirements. Like space, non-space environments might also suffer from intermittent connectivity, narrow bandwidths, and long delivery delays.

The BP is the fundamental DTN protocol. BP provides the means to efficiently transport data across an intermittently connected network with widely differing bandwidth links. While TCP was designed to operate in fully connected, low-delay environments, BP was designed to operate in intermittently connected, long delay environments. However, BP does not address the DTN security requirements, nor should it. Rather, in a layered approach, a separate security protocol was developed. Akin to the Internet Protocol Security (IPSec) Protocol providing security for the IP, the Bundle Security Protocol (BSP) was designed to provide standard mechanisms to provide integrity, authentication, and confidentiality services for BP.

As is the case with IPSec, the BSP security services are optional to use. However, users determined that the original version of BSP was a complicated protocol to implement and even more so, complicated to ensure its security. As a result, a "streamlined" version of the Bundle Security Protocol was developed appropriately called the Streamlined Bundle Security Protocol (SBSP). SBSP was "lighter" and "simpler" than its BSP predecessor. SBSP was paired as the security protocol through Bundle Protocol version 6 (BPv6). But with major changes between BPv6 and BPv7, yet another set of changes resulted in the evolvement to a newer version of the bundle security protocol which we now know as BPSec. This book will provide the technical details of BPSec, its layering on BP, its security services and mechanisms, and how it enables a secure DTN architecture.

While the protocols to implement a secure Delay Tolerant Network now exist, there are still many security issues that need to be researched and solved. Among them are security management and the associated cryptographic key management. The Internet, with its fully connected and short delay connectivity can use asymmetric key agreement and certificate authorities. However, in resource constrained environments and intermittently connected environments, asymmetric key agreement may not work. In small networks, traditional pre-placed cached key material might suffice but will not scale with network growth. These constraints will probably result in the need for multiple standards to support the various DTN environments. More work is still needed to arrive at a set of security management standards. But while more research and work remain, the basis of DTN security has been developed and is ready for use to provide secure services now.

We've come a long way from the days of TM and TC providing point-to-point ground-to-space connectivity. We now have networks on the ground and in space, with many more to come both on Earth and interplanetary. The secure DTN architecture has provided the network technology to enable the design, implementation, and secure operation of those future interplanetary networks.

Howard Weiss
Parsons, Inc.
Principal Project Manager

Preface

A preface must answer (at least) one important question:

What should I be learning from this book?

Lurking in that query are existential questions about the nature of the book and, at times, the nature of the reader. So let me answer that question – *the question* – as straightforwardly as I can.

Over the next hundreds of pages, of course, you will find technical details of how to apply a particular kind of security to a particular kind of networking environment. You will see details about why design decisions were made and the implications of all of this on the architects, engineers, and operators making real networks work.

This book is meant to be a little more than just all that.

Allow me to make an observation about the nature of new ideas: sometimes the difference between a very big idea and a very small idea is the person looking at that idea.

When creating the Bundle Protocol, version 7 (BPv7) there were two very small ideas – expressive headers and stored messages. Yet…in the right hands they became very big ideas about connecting our planet and our solar system. Similarly, BPSec has a small idea – secure message parts individually. Yet…in your hands that can become a very big idea about secure networks.

A long-established tradition in the education of computer scientists and software engineers is to periodically repeat the phrase *"include security from the start."* It makes good sense and there are certainly cautionary tales involving applying security too late. Mostly, this adage is hard to translate into preventative action. What does it mean to "include security" and at the start of *what*?

BPv7 and BPSec are a chance to build security into the foundations of our messaging from that first byte of header to that last bit of payload. This very small idea might grow into very big ideas about the tools and techniques we use to build networks. About rethinking encapsulation. About securing data at rest. That very small idea could change what we think of when we say *include security from the start*.

Therein lies the challenge to the reader and the thing you should learn from this book. Do not just understand the technical implementation of these small ideas; uncover and discover the big ideas hiding right behind them.

So, enjoy this book with an open mind and a creative imagination.

And happy hunting.

Dr Edward J. Birrane III

About the Companion Website

This book is accompanied by a companion website:

www.wiley.com/go/birrane/securingdelay-tolerantnetworks

This website material contains the CCDL code and its notes as software resources.

1

Introduction

Delay-Tolerant Networking (DTN) refers to both a networking architecture and a set of protocols and practices used to instantiate that architecture. One unique characteristic of this architecture is that it combines the concepts of store-and-forward and network overlay to federate otherwise disparate networks. In doing so, the DTN architecture provides resilience to challenging conditions that would break more commonplace network architectures.

This chapter discusses the design, implementation, and configuration of security for stressed networking environments using the protocols and extensions built to secure the DTN environment. This discussion includes the rationale for the DTN architecture, how solutions to that architecture can apply to other stressed environments, and why securing this architecture requires new security mechanisms.

This chapter provides three different but equally important introductions to this material.

1. **Why a DTN Architecture**: Expectations for what a computer network should do, and the way in which it should operate, stem from experiences with existing networks. There is a growing desire to more pervasively instrument, monitor, and interact with our environment – both on and off planet Earth. Doing so requires expanding our concept of what a computer network is and how it should operate.
2. **What Motivated This Book**: This book is written to answer questions relating to the design, implementation, and anticipated usage of DTN security mechanisms. This includes observations on the challenges inherent in a DTN environment and the unique capabilities of DTN transport protocols.
3. **Where is Information Located**: The content of this book is structured to enable rapid referencing of materials as a function of the role of the reader. Early material discusses aspects related to DTN network security, middle material explains existing protocol capabilities and behaviors, and later materials explain how to combine and configure these protocols in unique networking deployments.

1.1 A Pervasively Networked World

The scientific, social, and commercial benefits of easy access to communication services are apparent to anyone who has used the Internet since the turn of the century. Since its inception in the 1980s, the Internet has continued to grow in number of users, data transport speeds, and overall amount of data communicated. Recent increases in Internet use is driven by increases in user communities (to include social media), larger data volumes (to include higher resolution multimedia), and the rise of machine-to-machine connections.

Securing Delay-Tolerant Networks with BPSec, First Edition. Edward J. Birrane III, Sarah Heiner, and Ken McKeever.
© 2023 John Wiley & Sons, Inc. Published 2023 by John Wiley & Sons, Inc.
Companion website: www.wiley.com/go/birrane/securingdelay-tolerantnetworks

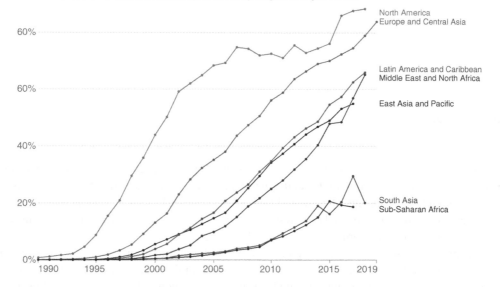

Share of the population using the Internet
All individuals who have used the Internet in the last three months are counted as Internet users. The Internet can be used via a computer, mobile phone, personal digital assistant, gaming device, digital TV, etc.

Figure 1.1 Internet usage is growing (unevenly) worldwide. Source: International Telecommunication Union/OurWorld in Data/CC BY 4.0. The adoption of Internet usage is growing, albeit unevenly, worldwide. This shows both the existing scale, and therefore the future work, needed to complete the instrumentation of our world.

Consider the share of the population accessing the Internet, illustrated in Figure 1.1. This listing of selected regions shows three clusters of use and adoption: rapid adoption, fast-following, and emerging. Data such as these show both the tremendous scale of the existing Internet and the tremendous growth remaining before usage is truly saturated.

Coincident with the increase in Internet users, access methods for the network are diversifying. Figure 1.2 shows, within the United States during the data reporting period, the rise in mobile access to the Internet to dominate the number of hours users spend producing and consuming network data.

In well-resourced environments, the Internet continues to demonstrate reliability and resilience, even in the presence of significantly different types of traffic utilization. A particularly motivating analysis of the behavior of the network in abnormal conditions can be performed on 2020 trends associated with the COVID 19 pandemic.

Through the pandemic, intuition would suggest growth in Internet usage driven by more access to entertainment, video conferencing, and less in-person communication. Initial analyses indicate that, as expected, network usage was increased during the pandemic, although this increase did not have the expected magnitude. Internet usage grew by approximately 15% with a more modest 5% increase in peak traffic volumes [1]. Specific applications (such as video teleconferencing) did experience higher increases in traffic usage, but this was offset by reductions in the use of other applications.

Person-to-person communication is the most recognized use of the Internet. However, this type of communication no longer represents the bulk of network traffic. This is why changes in how people use the Internet, such as during a pandemic, did not have the same impact on peak network traffic as intuition would otherwise imply.

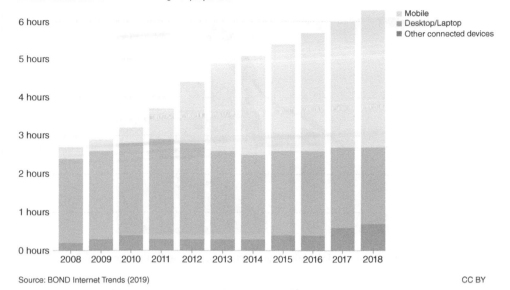

Figure 1.2   Digital media is predominantly access via mobile devices.
Source: OurWorldInData.org/internet.

In particular, machine-to-machine connections are growing with the number of online devices. The use of the Internet for machine-to-machine communications – termed the Internet of Things (IoT) – is becoming the dominant source of networking traffic. It is estimated that upwards of a trillion such devices will be online by 2035 [2]. See **Focus: The Internet of Not People** on next page.

The rise of IoT as a source of network traffic has implications beyond just the need to increase data speeds and data throughput. IoT represents the aspiration of extending the instrumentation and observation of our environment to increasingly remote locations on Earth, in near-space, and across our solar system. As exemplified by the extreme case of deep-space robotic spacecraft, networked devices can be placed in environments where humans cannot go.

The desire to collect information about remote parts of planet Earth and the solar system is outpacing the ability to densely populate these areas with resources. Adapting the Internet model to very challenged environments requires building an extensive infrastructure that cost of which rapidly reaches a point of diminishing returns.

Rather than making the solar system look like a densely populated city, new networking algorithms and protocols are being developed to operate in areas that have less access to, and less reliance on, a dedicated communications infrastructure.

> Social networking, telerobotics, telemedicine, distance learning, and even self-driving cars are examples of how more and more rapid communications change everyday aspects of our lives…[t]o sustain the rate of progress in this area, new approaches must be adopted, such as … highly mobile network nodes to provide coverage in challenged environments.
>
> The Path to Space-Terrestrial Internetworking [4]

Focus: The Internet of Not People

Bios Robotlab Writing Robot. Source: [3], Mirko Tobias Schaefer, Wikimedia Commons, CC BY 2.0.

The IoT refers to the growing segment of Internet use for machine-to-machine communication, such as sensor data transfer, notifications, and machine health and status information.

IoT communications benefit from more reliability and lower latency than human-in-the-loop communications. This is because devices can process data faster than humans but cannot intelligently adapt to abnormal circumstances.

Further attempts to distinguish traffic patterns have led to derivative notations such as the "Internet of Moving Things" to represent vehicular networks and the "Internet of Space" to represent space-terrestrial internetworks.

1.1.1 A New Networking Approach

DTN protocols represent a standardized approach to store-and-forward data distribution when timely, end-to-end connectivity cannot be guaranteed. These protocols were originally developed to enable deep-space, packetized networks impacted by long signal propagation delays and link disruptions.

DTN approaches also provide benefits for nearer-to-Earth networks that might experience disruptions resulting in a delay of messages from a sender to a receiver. These disruptions come from both physical issues (such as available transmitter power or antenna pointing) as well as logical issues (such as administrative permissions and traffic prioritization schemes).

Near-Earth spacecraft constellations, "IoT" networks, undersea networks, and vastly distributed sensing systems all experience these physical and logical disruptions. Across these environments, and more, DTN protocols can be used to build federated networks unlike any built before.

Figure 1.3 illustrates this concept of federating vastly disparate networks that cross environments with extreme communications challenges. Constructs like this cannot reasonably require that all devices stay in persistent, powered, timely contact with a centrally organized and managed network.

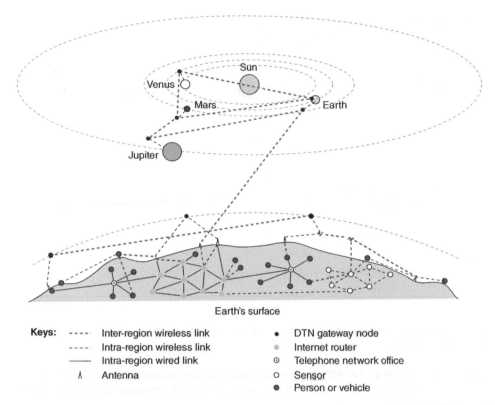

Keys:
- - - - - Inter-region wireless link   • DTN gateway node
- - - - - Intra-region wireless link   • Internet router
———— Intra-region wired link   ⊙ Telephone network office
⋏ Antenna   ○ Sensor
    ● Person or vehicle

Figure 1.3 DTN protocols federate vastly disparate networks. Source: Forrest Warthman and NASA.

The beneficial use cases for building these challenged networks are many. They can better the understanding and utilization of our planet by allowing sensor placement in increasingly remote and challenges areas of the world. They enable more ubiquitous and cost-effective telecommunications by relaxing real-time bandwidth requirements on networks for non-real-time data. They can provide graceful network degradation and more rapid network healing in times of disaster and war [4].

1.1.2 A New Transport Mechanism

NASA, in coordination with other space agencies, industry, and academia, has standardized the Bundle Protocol (BPv7) [5] as an alternative to the Internet Protocol (IP) in cases where networks are challenged by significant signal propagation delays or frequent link disruptions.

BPv7 agents implement store-and-forward behaviors and the BPv7 protocol data unit, the bundle, is designed to carry end-to-end session information and other annotations. By carrying additional information, bundles can be processed by challenged network nodes that cannot independently maintain end-to-end session state.

As is the case with many networking protocols, BPv7 should be expected to operate in deployments where network nodes (and the links amongst them) cannot always be trusted. BPv7 is not, inherently, secure. Just like the Transmission Control Protocol (TCP) and IP protocols prevalent on the Internet, BPv7 must operate within a security ecosystem to ensure proper delivery of bundles.

Figure 1.4 illustrates some of the security protections used over the Internet to provide a secure ecosystem for the exchange of TCP and IP messages. Individual protections, their configuration,

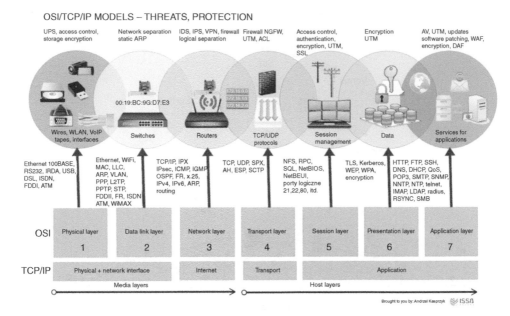

OSI/TCP/IP MODELS – THREATS, PROTECTION

Figure 1.4   Common network threats and protection. Source: [6], Andrzej Kasprzyk, Wikimedia Commons, CC BY 3.0.

and best practices for their use will evolve over time as capabilities and threats evolve. The value of this view of network security is that it showcases how multiple elements must work together to enable security.

The potentially stressed nature of a BPv7 operating environment imposes unique conditions where usual security mechanisms may not be sufficient. For example, the store-carry-forward nature of the network may require protecting data at rest, preventing unauthorized consumption of critical resources such as storage space, and operating without regular contact with a centralized security oracle (such as a certificate authority). An end-to-end security service is needed that operates in all the environments where BPv7 operates.

Securing Challenged Networks

The deployment of BPv7 leads to a fundamental question: how can we secure store-and-forward networks? How can we secure BPv7 bundles?

1.1.3   A New Security Mechanism

Security protocols used to secure the transport layer, the network layer, or both must operate within the set of policies, assumptions, constraints, and performance requirements of the networks in which they operate.

For example, the set of security protocols used over the Internet may differ from similar protocols used within cellular networks, near-Earth or deep-space spacecraft constellations, and very resource-constrained IoT devices. In some cases, these deployments may utilize similar algorithms for generating and processing cryptographic materials and only differ in how these algorithms are configured and applied. In other cases, new algorithms may need to be created to properly function.

Because BPv7 may operate in any one of a myriad of challenged networking domains, the security of bundles might not be able to reuse the familiar transport and network security mechanisms used on the Internet today. For example, the capabilities illustrated in Figure 1.4 might not be applicable to spacecraft constellations or sensors deployed in remote areas.

Security ecosystems will continue to develop and mature to adapt to the characteristics of new deployment scenarios. To the extent that a network conveys BPv7 bundles, a security mechanism built specifically for BPv7 must be a part of that security ecosystem. The Bundle Protocol Security (BPSec) specification defines a set of security features built specifically to secure BPv7 bundles using unique BPv7 features.

> **BPSec**
>
> The BPSec specification uses BPv7 features to secure BPv7 bundles.

1.2 Motivation For This Book

The motivation for this book is to answer questions related to DTN architectures and how they might effectively be secured. Some of these questions address how existing elements, such as emerging protocols, should be used to achieve security results. Others address how these networks should be architected for (and policies built around) emerging protocols.

Specifically, this book answers the following questions.

Can challenged networks effectively be secured? Network designers and application programmers cannot download and use traditional security packages (as-is) in challenged networks. However, this does not mean that network layer security is impossible in stressed environments. This book discusses the mechanisms available to secure even the most disrupted networks.

How can I make my security infrastructure more resilient to attacks? Operational networks often reduce their security posture if the performance of the network degraded. This invites malicious perturbation of network communications as a common network attack. If the performance of the network can be degraded enough that less security is applied (or verified at each hop) then the overall security posture of the network is reduced. Network engineers and application developers must consider ways of operating securely in degraded scenarios rather than waiting out periods of instability. This book describes the separation of cryptographic materials and the protocols that carry that material and how this distinction can be used to increase the resilience of security protocols when networks become degraded.

How can I prevent my network architecture from being over-constrained by traditional security approaches? Network security engineering often starts by building networks that can support the assumptions/requirements of traditional security protocols. When adopting security protocols from a well-resourced environment into a less-resourced environment, network engineers would need to over-engineer a challenged network deployment, making it more expensive to build and operate. Some environments are so challenged there is no way to over-engineer the network to make it behave like the Internet. This book demonstrates how networks can be architected differently if a delay-tolerant security approach is adopted, because such an approach places fewer assumptions/requirements on the networking layer.

Can DTN security continue to protect data when it is not actively in the network? Network designers and application developers typically differentiate data security within a network and the security of the data when it is no longer in the network. The DTN architecture blurs these boundaries by allowing data to be stored on nodes that are not the destination of the data. This book discusses how the security of data-at-rest can be achieved using BPScc without the need for application-specific mechanisms.

1.3   Conventions

There are a few conventions employed throughout this book to bring special attention to information. Understanding what mechanisms are used to highlight what information is helpful in understanding how to quickly find and review information.

1.3.1   Focus Studies

Focus studies refer to a relatively short, focused set of information providing some context for an expressed point of view. These focuses are offset from the main book text to call attention to them, usually with the addition of a representative illustration.

Focus: The Use of Focus Studies

Source: Marek/Pexels.

Where helpful, focus studies call special attention to ideas, concepts, or examples helpful to understanding the context, implementation, or benefits of some element of DTN security. While this book presumes basic knowledge of networking concepts, some experiences or technologies have been more influential in the design of DTN solutions than others.

   These studies can highlight subtle connections between DTN features and existing capabilities. In doing so, they provide hints for further reading or otherwise help to frame the context for the DTN conversation.

1.3.2   Summary Boxes

Many of the security concepts discussed in this book are derived as a series of evolutions starting with observations of an existing capability and progressing through logical modifications leading up to a DTN capability. To assist the reader in understanding where important conclusions are drawn, summary boxes are used at the end of some discussions to highlight a particular conclusion.

> **Summary Boxes**
>
> Information provided in a summary box provides an important summary of information provided in a more verbose form earlier in the text.

These boxes assist with both forward and backward looking content. When reading through a set of text looking forward to the summary can provide a good context on where a particular set of reasoning is going. Alternatively, readers might read summary boxes first to identify the text that is of most interest for reading.

1.3.3   Margin Notes

Margin notes serve as very short reminders of important concepts covered in the text of a chapter. They are placed partially in the margins to indicate that they are not part of the regular flow of information but, instead, are annotative in nature.

> **Margin Notes**
>
> These notes contain terse content reminders.

These notes are used for a variety of purposes. They emphasize important information in a section of text near the location of the note. They remind readers of information defined elsewhere in a chapter, or in some other chapter, that is important to placing existing text into context. They also identify the implications of something being discussed in the text.

1.3.4   Extract Quotes

A significant goal of this book is to explain the design decisions (and normative processing) of DTN security specifications. To that end, major design and processing directives related to BPv7 and BPSec are addressed. When doing so, the text of the relative specifications is quoted.

> Similar to cipher suites, security policies are based on the nature and capabilities of individual networks and network operational concepts.
>
> BPSec, RFC 9172 §1.2, [7]

This convention provides a straightforward mapping of discussion concepts and normative behavior. Additionally, the quoted text is indexed in the back of this book by the specification and section to help readers rapidly find information of interest.

1.3.5   Definitions

Custom definitions are used to ensure that terms with unique meanings in the context of DTN are not conflated with similar terms from other networking disciplines.

Definition 1.1   *(Definitions):*   Unique terminology that is either highlighted in specifications related to DTN security, or otherwise as uniquely introduced in this book.

1.4   Organization

This book is organized into three sections: (i) security considerations for delay-tolerant networks, (ii) the design, implementation, and customization of the BPSec security extensions, and (iii) how this protocol can be applied, combined with other security protocols, and deployed in emerging network environments.

The first section presents the challenges of securing disrupted networks. The section starts with an observation that network architectures are over-constrained by how we provide security services and that there is an important distinction between cryptography as a discipline and the security protocols used to exchange cryptographic materials. This is followed by a discussion of the unique stressors found in delay-tolerant networks and how these stressing conditions break many of the protocols used to secure the Internet. Finally, this section discusses the characteristics of a security architecture for delay-tolerant networks and the unique security considerations that must be considered when defining a delay-tolerant security protocol.

The second section focuses on the history and development of BPv7 and the BPSec security extensions, starting with their unique design principles and those practices borrowed from more traditional security protocols such as those comprising the Internet Protocol Security (IPSec) and the Transport Layer Security (TLS) protocol suites. Next, the mechanisms used to represent cryptographic materials in bundles are discussed along with rules used for the processing of BPSec materials. This section concludes with a discussion of two unique and enabling characteristics of the BPSec protocol: security context definitions and out-of-band policy configurations.

The third section discusses pragmatic considerations for deploying BPSec in both regular and delay-tolerant networks. This section discusses how the protocol can be extended to encounter a variety of methods for defining and communicating cryptographic material. This is followed by a discussion of design patterns associated with the use of BPSec security services and some network-level use cases that build upon these patterns. Finally, this section concludes with a discussion of what can go wrong along the way when deploying BPSec in a challenged network.

1.5 Summary

The desire to more pervasively instrument both our world and our celestial neighborhood has challenged computer networking researchers to develop new algorithms and protocols. These new approaches to data communication must operate in conditions fundamentally dissimilar to today's Internet.

However, the design of network protocols are built around the capabilities of the underlying networking architecture. Therefore, if these new challenged networks cannot provide the same operational guarantees as the Internet, then IPs might not behave the same when used in these new environments.

This book addresses fundamental questions related to how data communications can be secured in challenged networks, with a focus on networks that use BPv7 for transport services and BPSec extensions for securing BPv7 bundles. In doing so, the motivation, design, and behavior of the design of the BPSec are explained.

References

1 Feldmann, A., Gasser, O., Lichtblau, F., Pujol, E., Poese, I., Dietzel, C., Wagner, D., Wichtlhuber, M., Tapiador, J., Vallina-Rodriguez, N., Hohlfeld, O. and Smaragdakis, G. [2021]. Implications of the COVID-19 pandemic on the internet traffic, *Broadband Coverage in Germany; 15th ITG-Symposium*, pp. 1–5.
2 Sparks, P. [2017]. The route to a trillion devices, *White Paper, ARM*.

3 Schaefer, M. T. [2008]. Bios robotlab writing robot.jpg. License: CC BY 2.0, https://creativecommons.org/licenses/by/2.0viaWikimediaCommons. **URL:** https://commons.wikimedia.org/wiki/File:Bios_robotlab_writing_robot.jpg.

4 Birrane, E. J., Copeland, D. J. and Ryschkewitsch, M. G. [2017]. The path to space-terrestrial internetworking, *2017 IEEE International Conference on Wireless for Space and Extreme Environments (WiSEE)*, pp. 134–139.

5 Burleigh, S., Fall, K. and Birrane, E. J. [2022]. Bundle Protocol Version 7, RFC 9171. **URL:** https://www.rfc-editor.org/info/rfc9171.

6 Kasprzyk, A. [2019]. Iso-model-threats-network-protection.png. License: CC BY 3.0, https://creativecommons.org/licenses/by/3.0viaWikimediaCommons. **URL:** https://commons.wikimedia.org/wiki/File:ISO-model-threats-network-protection.png.

7 Birrane, E. J. and McKeever, K. [2022]. Bundle Protocol Security (BPSec), RFC 9172. **URL:** https://www.rfc-editor.org/info/rfc9172.

2

Network Design Considerations

More and more unique network architectures are being developed to extend data communications into increasingly remote areas on Earth, in near-Earth space, and throughout the solar system. Projects that seek to build computer networks in new environments representing new domains must rethink both architecture and design activities to uncover unique design constraints. This chapter discusses how constraints and network architectural patterns impact design activities and why new transport and security mechanisms might be needed in these new and exotic deployments.

After reading this chapter you will be able to:

- Summarize how constraints impact network design and implementation.
- Describe the security implications of a layered networking architecture.
- Distinguish between the roles of cryptography and network security.
- Compare different strategies for configuring cryptographic functions when used in a network.

2.1 Designing for Challenged Networks

Successful design requires both creative and precise thinking. Creativity determines how to achieve design goals outside of existing patterns or other pre-supposed solutions. Precision ensures designs meet their requirements within imposed constraints and assumptions about the operating environment. Yet, the design of any complex system cannot be optimal for every user in every circumstance. Good designers address competing needs by formulating joint optimizations across a variety of user needs and circumstances.

The design of any complex system is a blend of artistry and engineering – discovering constraints and synthesizing solutions within those constraints. The difficulty inherent in creating joint optimizations for complex systems has generated many design-related *bon mots*, such as *"Make the common case fast"* and *"Premature optimization is the root of all evil"* [1].

The design of network architecture, protocols, and procedures is no exception to these rules. The pervasiveness of existing computer networks does not imply that future network design can be done without creativity or without a focus on unique constraints. The increasing desire to connect our world (and other worlds in the solar system) requires significant creativity as we connect highly mobile devices, intermittently connected sensors, and even deep-space spacecraft and planetary landers.

Securing Delay-Tolerant Networks with BPSec, First Edition. Edward J. Birrane III, Sarah Heiner, and Ken McKeever.
© 2023 John Wiley & Sons, Inc. Published 2023 by John Wiley & Sons, Inc.
Companion website: www.wiley.com/go/birrane/securingdelay-tolerantnetworks

Even the relatively straightforward design of Wireless, Local Area Networks (WLANs) must contend with the joint optimizations around the number and placement of wireless access points, frequency selection, and user capacity needs [2]. Incorporating increasingly exotic devices into a network design increases this optimization complexity. This can create problems when

> **New Network Designs**
>
> Our networked future requires new and creative ways to transport and secure data.

constraints associated with such devices preclude the use of existing networking building blocks.

Therefore, building usable, new networks (or incorporating new devices into an existing network) must start with an understanding of how new constraints impact network design and limit the ability to reuse existing networking building blocks.

2.1.1 Network Design Constraints

The elaboration of a design as a systems engineering discipline is guided by the generation of assumptions, requirements, and constraints. Assumptions represent ways to reason about uncertainty early in a design process and might eventually be retired as the understanding of the system and its use increases. Requirements represent what capabilities must be enabled by the design and subsequent implementation of a system. Together, assumptions and requirements are often settled prior to beginning a design activity.

Design constraints differ from assumptions and requirements in that they address elements of *how* the system will operate whereas assumptions and requirements address *what* the system must accomplish. Design activities undertaken prior to analyzing potential constraints risk producing incorrect or inefficient designs.

> **Designing with Constraints**
>
> Design activities must start by understanding applicable design constraints.

Definition 2.1 *(Network Design Constraint):* Any condition that must be accommodated in the design of a network. A design cannot be validated unless its implementation will function in the presence of all identified design constraints.

Any individual design constraint can state either that the design *must* include some element or that the design *must not* include some element. Therefore, constraints can be categorized based on whether they include or exclude design elements.

Inclusion Constraint: A constraint that identifies a condition that must be included in a design. These types of constraints should be kept to a minimum as they limit the ability of network designers to apply creative ways to solve network design problems. Some examples of these kinds of constraints include:
  - Interfaces that must be supported to communicate with existing systems.
  - Protocols that must be included as a requirement for some operational certification.
  - The use of specific vendors or community tools or libraries.

Exclusion Constraint: A constraint that identifies a condition that must be excluded in a design. Usually, these constraints represent conditions that violate assumptions on how the network will behave and, thus, prohibit the use of items that need those assumptions to function. These types of constraints are more friendly to design activities; by only identifying things to avoid, designers retain more freedom to solve design problems in a more creative way. Some examples of these kinds of constraints include:

- Deprecated security mechanisms or certain deprecated modes of security mechanisms.
- Transport protocols that require instantaneous, end-to-end connectivity.

As they are identified, design constraints impact how data might be represented and transported in a network – to include how data in the network might be secured. Understanding the security challenges inherent in a new networking design means that constraints must be analyzed to assess whether they require a change to established network security solutions.

This is particularly important because network security solutions often make assumptions about the capabilities of the network in which they are deployed. For example, Internet security solutions assume that security sessions can be negotiated end-to-end and that there will be third-party oracles available to verify node identities. If exclusion constraints are levied on a new network design that prevent assuming contemporaneous end-to-end connectivity or trusted third parties in a network, then existing security solutions might also be precluded from the network design.

Network Constraints Become Security Design Challenges

When constraints prevent the use of existing security protocols, new security mechanisms may need to be developed. This requires rethinking overall security design in the context of these constraints.

2.1.2 Finding Constraints

One way to increase the likelihood that no constraints have been missed is to identify common sources and types of constraints and analyze each of them accordingly. Reviewing constraint sources for each constraint type can result in a more comprehensive understanding of the true needs of the network.

2.1.2.1 Constraint Sources

In the context of network design, constraints can be generated either by stakeholders of the network or by the characteristics of the environment in which the network must be deployed. Additionally, the act of codifying design choices may introduce new constraints in the system, making the elicitation of constraints an iterative part of the design process.

Network users represent the largest set of entities that have some stake in the successful build and operation of the network. Common sources of constraints include operators, developers, users, vendors, the deployment environment, existing infrastructure, and existing networks.

The set of **operators** include those tasked with the day-to-day upkeep of the network and the management and configuration of nodes across the network. Networks must be secure and managed in order to be deployed and considered fit for any operational use. The policies, procedures, and required accreditations that determine fit-for-use must be supported by the design of the network.

Application developers provide the applications that run on the network being designed and, ultimately are the reason for networks to exist in the first place. The requirements placed on these applications may become constraints on the network itself. For example, real-time interactive applications require low-latency networks – placing a real-time application over a Delay Tolerant Networking (DTN) architecture would cause the application to fail [3].

Vendors build the products from which networks and networked devices are built. Vendors manufacture products that generate value for the vendor and their own stakeholders. This value is sometimes achieved by the direct profitability of a product, the projected sales volume of a product, or the reputation or sponsor impact of their product. The products that are supported by the vendor community can lead to constraints on the network design when products do not implement needed features.

The **deployment environment** describes the resources available for networking devices to operate and the challenges to signals into and out of such devices. As networks are deployed into increasingly remote areas, the physical ability to communicate data may be limited in ways that affect network design. This can be particularly impactful to devices that harvest energy from their environment, as they may not be constantly powered or have enough power for wireless network transmission. Similarly, devices that support narrow-beam pointing may break links regularly as part of device mobility.

Existing infrastructure represents the overlap of vendor capabilities in the physical environment. The hardware capabilities of a networked device – such as processing power and available storage – may be limited as a function of the cost of the product, the power available to operate the product, or the processing that a product must do to operate in a harsh environment.[1] The on-board storage and processing capabilities of a device may limit the data rates and store-and-forward features of a node and, across a network design, this has multiple implications relating to congestion management, routing, and security at rest.

Existing networks are often federated into a new network design to assist with data delivery. Custom networks interface with existing networks either as a gateway to general users or as infrastructure between custom users. For example, a corporate intranet might use an Internet Service Provider (ISP) to transport data between sites. Similarly, cellular networks carry traffic from cellular devices back to a packet network gateway, such as the Internet. In these cases, custom network design must support the protocols, interfaces, and policies needed to exchange information with existing infrastructure. Additionally, the level of trust of existing networks may place additional constraints relating to security configurations and the type of data that may be passed over those networks.

2.1.2.2 Constraint Types

Requirements can be decomposed into various types (functional, performance, configuration, interface, and error handling requirements), and this is done to help ensure that no aspect of required behavior of a system is overlooked. The same logic can be applied to constraints; decomposing constraints into various types can increase the quality and completeness of identified constraints.

There are numerous ways in which constraints could be decomposed. We propose three major types of constraints as engineering, interoperability, and behavior.

Engineering constraints refer to the physical capabilities of the network, its devices, and its applications. These limits represent what can be achieved within the cost, schedule, and technology that can be included in the design. Consider the task of architecting an interplanetary network between Earth and Mars. Such an endeavor would need to adjust for the engineering constraint associated with the signal propagation delay, as signals take between 5 and 20 minutes to travel the distance between these planets.

Interoperability constraints refer to elements of the design that must be in place to make the system operate with other, existing systems. These constraints are typically expressed in the languages, algorithms, protocols, or other software comprising external interfaces into and out of the system. Supporting the Internet Protocol (IP) naming and addressing is a likely interoperability constraint for network architectures communicating to, and through, the Internet.

---

1 Such as building deep space avionics that must be radiation tolerant.

Behavioral constraints affect the design of a network by describing how the network as a whole, or individual networked devices, would need to behave. Often networks are built from multiple layers of capability and multiple vendor offerings federating otherwise disparate systems. Common reactions to common events across this hardware and software diversity are needed to ensure that the network appears to behave as a cohesive entity. Additionally, identified security threats and associated best practices to counter those threats may place behavioral constraints on a network.

Potential relationships amongst constraint sources and types is illustrated in Table 2.1. In this table, engineering constraints are driven by developers and the operating environment, with some consideration for what vendors support and the infrastructure available to the network. Interoperability constraints tend to be driven by vendor offerings and existing networks, with input from operators. Finally, behavior constraints are driven by the operators entrusted with the behavior of the network, as impacted by the entire construction of the network.

2.1.3 Identifying Security Challenges

When a network design constraint prevents the use of a common network security tool, protocol, or configuration then the design of the network must creatively adapt to using alternative security mechanisms. These kinds of adaptations are not unprecedented even in long-lived networks as security threats and new use cases continue to change the threat landscape. Relatively recent examples of addressing large-scale security challenges include the development of the Domain Name System Security Extensions (DNSSEC) [4] or the practice of security through Hypertext Transfer Protocol Secure (HTTPS) requests [5].

However, the design of network security protocols is, itself, based on its own sense of engineering and interoperability limits and assumptions on the behavior of the network. Unless there is a *clear distinction* between security requirements and security assumptions a circular dependency can be created: a network may be given an interoperability constraint based on security tools that, themselves, assume network function that violates an engineering constraint.

> **Security Protocols**
>
> New security mechanisms are created when existing approaches are no longer sufficient.

Table 2.1 Sources of constraint types.

	Engineering constraints	Interoperability constraints	Behavior constraints
Operators	○	◐	●
Developers	●	○	◐
Vendors	◐	●	◐
Environment	●	○	◐
Infrastructure	◐	○	◐
Networks	○	●	◐

The level to which various constraint types are imposed by various constraint sources can be categorized as small (○), medium (◐), and large (●). This magnitude implies the extent to which a particular constraint source might define an impactful constraint type.

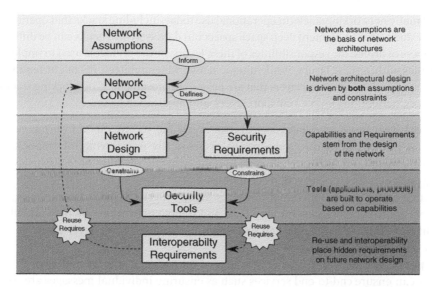

Figure 2.1 **Tooling generates interoperability requirements.** Network CONOPS and security requirements are often subject to constraints from interoperability requirements levied by tool selection. Identifying and accepting these interoperability requirements is a key part of determining a networking architecture.

Consider the design of some custom sensor network. Figure 2.1 illustrates some of the activities associated with the design of such a network. Most network design activities start with a focus on data flows through the network (the purpose of sensor networks being to move data). Network assumptions discuss the ability of the network to pass data and inform the overall network Concept of Operations (CONOPs). CONOPs include use cases that define what is needed in the network design. These CONOPs also address security threats within an environment and, thus, define security requirements for the network. This activity constrains the implementation (and adaptation) of a variety of tools implementing network functions, to include security tools. However, this process is not linear. Tool selection may impose interoperability requirements that, themselves, alter the original CONOPs. Just as those defining CONOPs must consider the impact on security tool selection, those performing security tool selections must consider the impact on CONOPs definitions.

2.2 Layered Network Architectures

There are multiple operations that must occur to successfully transmit user data along a network path. These operations can be interpreted as a series of responsibilities implemented by various steps of a network communications workflow. This workflow can be decomposed into a series of macro-steps that handle large data transformation and related operations with each macro-step further decomposed into a series of smaller steps, thus forming a hierarchical model of actions taken on user data. This conceptual model is a popular way to separate the functionality of the process needed to transform user data into network traffic. The set of layers used by a particular network for this purpose is called a network stack.

Definition 2.2 *(Network Stack):* A hierarchical set of protocols (and their implementations) that build information starting from the acceptance of user data to the modulation of information over some physical medium. The hierarchies of protocols comprising the stack typically abstract information from "higher" layers (closer to applications) to "lower" layers (closer to the physical medium).

There are many networking stacks in operational use today, including stacks that operate over the Internet and stacks that implement deep space spacecraft data systems. Stacks can be differentiated by the numbers of layers, the responsibilities of these layers, or the protocols used to implement the responsibilities within these layers. While unique networking architectures might require unique networking stacks, there are some layers that are present in most modern networking stacks: physical link layers, networking layers, transport layers, and application layers.

1. A **physical link layer** modulates logical data onto (and receives data from) some physical medium. All networks require some means of physically transferring information between nodes. This includes a fundamental data framing scheme, medium access capabilities, and possibly error detection or correction capabilities.
2. A **network layer** guides individual messages along a network path. Protocols in this layer implement route computations through local and remote networks. Any network where traffic might take different paths through nodes requires some network layer to manage routing through the network.
3. A **transport layer** transmits messages end-to-end through one or more networks. Protocols in this layer can ensure end-to-end services such as ensuring individual messages are received at their destination and that messages are delivered in order when they are part of a larger data set.
4. An **application layer** places data into the network for transmission and receives data from the network as part of message delivery. Because networks exist to communicate data between and among applications, this layer is fundamental to endpoints in a network.

Figure 2.2 illustrates the ways in which different layers might (or might not) be implemented on devices in a network. This figure illustrates five sample devices ((a) through (e)). Endpoint devices (a) and (e) implement a full network stack handling physical data, local networking, end-to-end services, and the applications sending and receiving data. Local area networking and repeater devices (b) such as hubs, switches and bent-pipe spacecraft amplify physical link data without interpreting it. Routing devices (c) route data based on local information without concerns for end-to-end guarantees. Other networking devices (d), such as VPNs or firewalls, might implement end-to-end services but not be the destination of a particular message.

The layered nature of networking stacks means that securing a message at one layer does not guarantee security for other layers of the same stack. This leads to an interesting design dilemma: individual networking layers are designed and implemented as stand-alone capabilities, but they must function seamlessly as a secure networking stack.

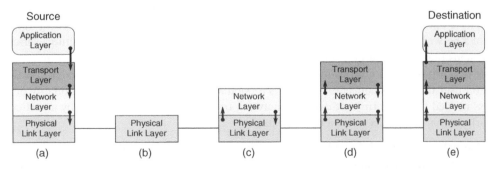

Figure 2.2 Network devices implement different networking stacks. Nodes representing the source (a) and destination (e) of a message path might traverse hubs that only look at physical link traffic (b), switches that examine individual network data units (c), or other full stack devices (d).

Pragmatically, this leads network designers to work through a small set of well-understood networking stacks and to use those stacks in the design of networks. Put another way, network architecture is often constrained by the best practices and implementations of both individual network layers and the best practices for combining them in a networking stack.

Just as each of these layers provides unique services, they require security specific to the implementation of that layer. This stack-based architecture imposes constraints on the design of the network itself. These constraints include keeping layer data insulated (with encapsulation), ensuring the timely exchange of messaging (delay/disruption intolerance), and securing higher-layer data using an all-or-nothing approach (course-grained security) [6].

> **Networking Stacks**
>
> Network layers are implemented individually but must work together to implement networking workflows.

The remainder of this section describes each of these constraints.

2.2.1 Encapsulation

The layered approach of a network stack requires some mechanism for both easy access to layer-specific information and information hiding across layers. This is frequently accomplished through the definition of a customized header-and-payload message structure at each layer and the use of encapsulation to describe how messages from a "higher" layer are represented in the messages of "lower" layers.

This concept of encapsulation is illustrated in Figure 2.3. In this figure, a *higher layer* message consists of headers and a message payload. The entirety of this message is encapsulated as the message payload of a middle-layer message that, in turn, is encapsulated as the message payload of some lower layer message.

This implies that per-layer protocols never interact with each other because the responsibilities of each layer are distinct. Routing decisions made as part of a network layer do not need to understand the end-to-end, in-order-delivery algorithms used by a transport layer. Rather, a transport layer protocol message may be encapsulated as the payload of a network protocol message.

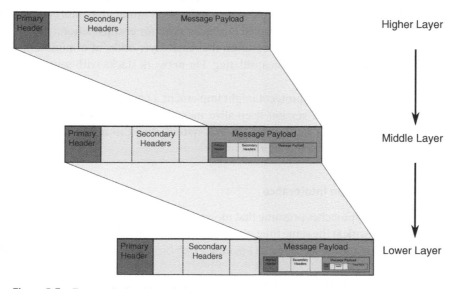

Figure 2.3   Encapsulation hides information from other network layers.

Because there is little coordination between layers, each "layer protocol" may require its own security. Security at one layer does not imply security for all layers. A secured transport layer protocol message does not mean that the underlying network layer protocol is also secure because the network layer protocol adds additional information (such as a message header) that is unseen by the transport layer.

2.2.1.1 Design Benefits

A layered network architecture has the motivating benefits that it makes individual protocol design simpler and provides a way to mix-and-match implementations.

Some design benefits of a layered network architecture include:

- **Information Hiding**: Encapsulation enforces an isolation of responsibilities. Individual protocols need only handle a small set of conditions and functional responsibilities, allowing for simpler protocol designs. Simpler protocols handle fewer special cases and are more likely to result in efficient, high-performance implementations.
- **Interchangeable Implementations**: Layer isolation allows for mixing and matching different implementations from different vendors as part of network architecture and deployment, which increases reuse. Vendors might sell network devices that implement only a physical link layer and require higher layer implementations to handle other responsibilities such as routing and retransmission. This level of interchangeability helps reduce the impact of vendor and infrastructure constraints on network designs.

2.2.1.2 Challenges

Strict information hiding, while beneficial for the implementation of software for a particular layer, introduces challenges relating to how these layers function together in a network stack to implement a networking workflow. Network layers must function as a seamless whole, which is difficult to do with limited cross-layer information.

Common challenges include the following:

- **Information Hiding**: This encapsulation benefit is also a challenge. Hiding information from lower layers of a networking stack might preclude optimizations on how data are treated. For this reason, networking stacks often implement some type of annotative information that can accompany a network message as it is passed to a lower layer for encapsulation.[2]
- **Increased Data Size**: Encapsulation presumes that each layer adds, at a minimum, headers to the higher-layer information it is encapsulating. For network stacks with several layers, this increases individual message size.
- **Nested Loops**: Each "per-layer" protocol might implement a control loop associated with data timeouts, retransmissions, and session keep-alive messaging. When these control loops are implemented independently at multiple layers in a system, they can interfere with each other causing unexpected behaviors in the network.

2.2.2 Delay and Disruption Intolerance

Many network security approaches presume that messaging endpoints (the source and the destination) co-exist in the network at the same time. For human interactions this is a sensible assumption. Consider the relatively common workflow of opening a web browser to access a web server. In this

---

2 Sometimes, annotative data are specific to a vendor implementation and provided as part of the value-added efficiency of a specific vendor ecosystem.

use case, both the user at the browser and the web server being accessed must be connected to the network. If the connection is disrupted the connection is terminated and must be restarted. Similarly, if the round-trip data exchange takes too long, secure sessions to the web server might time out, also requiring that a connection be terminated and that the user try again.

In the web server example, the network is not tolerant of delays and disruption. If these occur, the network is allowed to cancel a connection with the assumption that the user (or the application layer) will start over again.

Definition 2.3 *(Network Disruption):* The loss of end-to-end connectivity between two nodes in a network that prevents the exchange of messages. Disruption is usually discussed in the context of loss of connectivity between a message source and a message destination. Connectivity loss can result from the loss of a single link between two adjacent nodes in a network, or loss of multiple nodes in a network.

This concept of disruption is illustrated in Figure 2.4. When all nodes are connected and a path **P1** exists as shown in Figure 2.4a through the network, the network is not disrupted. Similarly, if certain links (such as links from node **A** to nodes **B** and **C**) are lost, but a path **P2** still exists, as shown in Figure 2.4b, the network is again not disrupted. If additional links are lost (such as links

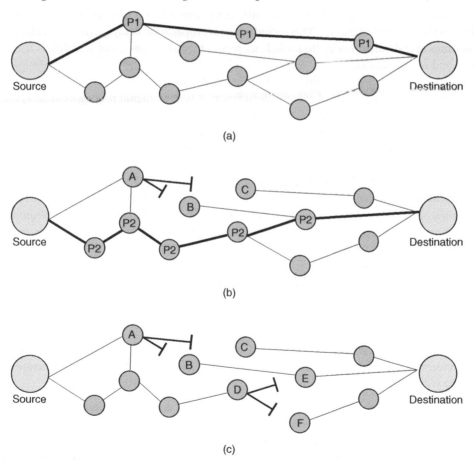

(a)

(b)

(c)

Figure 2.4 Disruption prevents end-to-end message exchange. (a) End-to-end connectivity through path P1. (b) End-to-end connectivity through path P2. (c) No end-to-end connectivity.

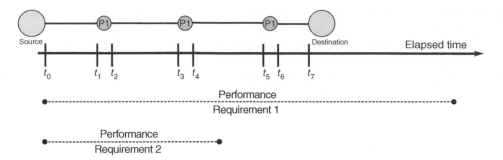

Figure 2.5 **Delay can only be assessed relative to some performance requirement.**

from node **D** to nodes **E** and **F**) as shown in Figure 2.4 then there is no viable end-to-end path and the network is disrupted.

Definition 2.4 *(Network Delay):* The failure of a message to be communicated end-to-end within a required timeframe. While network disruption or latency can be defined in an absolute way, delays are defined relative to a performance requirement.

This definition of delay accounts for the entire time from when a message is first transmitted until it is received at its destination. This includes signal propagation delays across individual links, process times at individual nodes (to include store-and-forward times), and any delays related to timeouts and retransmission.

This concept of overall delay is illustrated in Figure 2.5. In this illustration a single message is transmitted at time t_0 and received at some destination at time t_7. Signal propagation delays occur in the intervals $[(t_0 - t_1), (t_2 - t_3), (t_4 - t_5),$ and $(t_6 - t_7)]$. Node processing delays occur in the intervals $[(t_1 - t_2), (t_3 - t_4),$ and $(t_5 - t_6)]$. The determination of whether the total elapsed time is considered *delayed* can only occur in the context of some performance requirement.

Regardless of the total elapsed time, the determination of whether a network is a delayed network occurs relative to some performance measurement [7]. From Figure 2.5, a message in a network with *Performance Requirement 1* would not be considered a delayed network because the required delivery time occurs after time t_7. However, a message in a network with *Performance Requirement 2* would be considered a delayed network because the required maximum elapsed time would be after the delivery time at time t_7.

The presumption of timely access to nodes in a network has several benefits from the standpoint of algorithm design and implementation. This presumption can be a challenging one when attempting to re-use algorithms and implementations in challenged networking scenarios.

2.2.2.1 Design Benefits

Omnipresent, low-latency communication is a clear benefit that can be discussed as enabling for two reasons: it enables closing control loops over a network and allows for measurements (to include security monitoring and responses) to occur over the network.

- **Closed-Loop Control**: Without delays and disruptions, distributed algorithms can be run between nodes in a network to produce a coordinated result. This closed-loop operation can be used for any kind of negotiation, such as establishing session keys between security protocols, or checking certificates over a network. This type of control is only a benefit in low-latency networks because when loops are closing, applications are waiting.

- **Measurement and Reaction**: When network nodes are rapidly reachable over a network, characteristics of the network between nodes can be measured and reported in a timely fashion allowing for rapid responses to changes in network state, such as the application of quality-of-service provisions in response to increasing congestion.

2.2.2.2 Challenges

Presuming low-latency, omnipresent communications have the immediate challenge of graceful degradation when communications are delayed or disrupted, as can occur when nodes drop out of a network, when quality-of-service deprioritizes some message traffic, or when network links become congested.

In addition to the fact that no message data might be communicated during these times, algorithms that try to close control loops and perform measurements over delayed and disrupted links can result in making the delay/disruption problem worse, for two reasons.

- **Dropped Data**: When network timers do not account for delays and disruptions, or when local node storage is very small, data will be dropped by network nodes. In cases where data are unique or require more timely delivery when the network comes back, having dropped data may actually result in a longer time to delivery than if the data were stored locally by a node along the data's path.
- **More Network Chatter**: Protocols that attempt to perform real-time measurement and closed-loop control generate large amounts of traffic in a network. When timers expire, this level of chatter increases as negotiation protocols attempt to re-establish connections. In cases where delays/disruptions reduce the overall amount of data through a network, this level of chatter reduces the likelihood that actual user data is delivered.

2.2.3 Coarse-Grained Security

Layered network design simplifies algorithm design by focusing protocol contents on fewer responsibilities. This implies that the contents of a given protocol data unit represent similar kinds of information and, thus, should be secured using a common approach. This type of coarse-grained security means that there is no need to apply different security mechanisms to different portions of a message.

2.2.3.1 Design Benefits

Applying security services to entirely encapsulated chunks of data avoids the need to process those chunks or otherwise differentiate data provided by "higher" layers of a networking stack. The natural benefit of ignoring this complexity is that security can be optimized for speed and with less special cases or other complexities.

- **Simpler**: While the implementation, configuration, and policy associated with security services is not necessarily simpler on a protocol-by-protocol basis, having a regular scope to security operations means less choices have to be made by the security implementation. The IPSec Encapsulating Security Payload (ESP) mechanism [8] operating in tunnel mode encrypts entire packets whereas the same mechanism operating in transport mode encrypts entire payloads and, in the case of IPv6, some headers.
- **Faster**: Fewer choices for how to construct cipher suite inputs lead to faster data processing and more chances for optimization. When protocol data units are of fixed size, firmware-based security implementations can more easily (and rapidly) apply security.

2.2.3.2 Challenges

The challenge of an all-or-nothing approach to applying security to a protocol message is that it presupposes all applicable portions of a message should have the same level of security applied. In cases where different headers might include different types of information, a single security service using the same user configurations might not be appropriate for all data being encapsulated.

- **Over-encryption**: In cases where protocol message headers might have some headers that can be transmitted in the open and other headers that needs to be encrypted, a coarse-grained security approach might struggle to apply this. This could lead to encrypting all headers (and the payload) when not all information is required to be encrypted. This approach complicates the processing of data downstream and can lead to early exhaustion of encryption keys.
- **Over-sharing**: In cases where different parts of a message might need to be accessed by different recipients, a coarse-grained strategy requires that all recipients be able to process the same, single security service applied to messages. This could result in the use of group keys or other mechanisms where secrets are shared among a larger group of networking nodes. Generally, this approach to pervasively shared secrets reduces the confidence in the security services applied in the network.

2.2.4 Impact on Protocol Design

A layered networking architecture built around common, well-understood networking stacks provides reliable, scalable communications under the conditions those stacks were designed to operate. Variations of this approach have resulted in successful deployments across boutique sensor networks, the Internet, corporate intranetworks, and others. The most notable benefits of this approach, discussed in the prior sections, can be summarized as follows:

1. Hiding information between layers simplifies protocol reuse, design, and implementation.
2. Vendors can build focused implementations that users can mix-and-match.
3. Distributed control loops can be used in timely, high-rate networks.
4. Network management can be built from responsive monitoring of measured link state.
5. Hiding information also simplifies applying security at network layers.

When faced with building a new network there is an understandable preference to base the design of that network on network stacks that exist, perform well, and for which there exists broad vendor support. This desire usually means that the activity of designing a network becomes the activity of making the environment support the presumptions of pre-existing network stacks and protocols.

Existing Implementations Drive Network Design

The activity of designing a network becomes the activity of making the environment support the presumptions of pre-existing network stacks.

However, the Internet does not function without significant, constant investment, maintenance, and improvement. For this reason, these common Internet services are only available in areas that provide ready access to infrastructure. The cost to extend this level of service to remote regions of the world with few (or no) users far outweighs the potential revenue or other benefits to network operators. This means that networks that need to operate in these challenged regions will not be built with the same assumptions and guarantees made over the Internet.

As we have discussed, there exists a relationship between the capabilities of a network and the protocols operating over that network. Networking architectures with different assumptions and requirements will not fully benefit from the reuse of existing networking protocols. The rise of challenged networks that behave differently from the Internet has caused a re-evaluation of how network protocols – including security protocols – might operate in these new environments.

2.3 Cryptography and Network Security

Another important design consideration for a network is how the algorithms that generate and process cryptographic materials interface with the protocols that transport these materials and configure these algorithms. Networks cannot be secured without cryptographic algorithms, but these algorithms cannot be used unless the network elements can provide them the proper inputs. This couples the design of the network to the use of particular cryptographic algorithms that would be used to secure the network.

Analyzing the relationship between cryptography and security should be done at each layer of a network stack because each layer might present its own security assumptions and constraints. When reasoning about layer-by-layer security, network designers must understand three aspects of cryptosystem implementation.

1. **Algorithm Capabilities**: The capabilities and limitations of cryptographic algorithms form the basis of any security ecosystem. Network layers cannot be secured without cryptographic algorithms.
2. **Configurations**: Cryptographic algorithms only produce usable cryptographic materials when properly configured. This involves following implementation and processing recommendations by the cryptosystem implementer. Network designs must address how unique inputs to algorithms can be provided.
3. **Packaging and Transport**: The way in which network protocols scope security operations for messages impacts how cryptographic materials are generated and placed in network messages. Network designers must explain how message structure both identifies information requiring security and the ways in which related security materials are represented in messages.

2.3.1 Cryptographic Algorithm Capabilities

The role of cryptographic algorithms in network security cannot be overstated; the generation and exchange of cryptographic materials is fundamental to the concept of secure, trusted communications. Yet there are often significant limits associated with their successful application in a network.

Beyond the inherent limitations of some algorithms to certain types of attacks, otherwise secure algorithms can be applied in insecure ways as a function of their parameterization, packaging, and implementation. The ways in which cryptographic algorithms are applied in a network is as important as the underlying strength of the algorithm itself. One way to ensure that individual cryptographic algorithms are used appropriately together is to associate multiple algorithms together as part of a *cipher suite*.

Definition 2.5 *(Cipher Suite):* A set of cryptographic algorithms and associated implementation requirements on those algorithms used for the generation and processing of cryptographic material in a security ecosystem. While the term *cipher* implies a focus on encipherment/ decipherment, cipher suites in practice include algorithms for key exchange and authenticating data in addition to data encryption. These algorithms may be applied individually or together as part of both external and internal interfaces into the cipher suite. This term can refer to either the algorithms federated in a suite or the software/hardware implementation of those algorithms.

Secure networks must provide a useful context around the generation of these cryptographic materials. This includes the way in which networks provide configurations and data into cipher suites, and how the outputs of those suites are carried in the network itself.

2.3.2 Configurations

The term *configuration* can have several meanings when applied to cryptographic algorithms and the cipher suites that define and implement them in a network. Configuration, in this context, refers to the ways in which cryptographic algorithms and cipher suites are the ways in which the parameters of crypto algorithms and cipher suites are selected to produce and consume cryptographic materials.

This configuration can be described in two ways: configuration inherent in the cipher suite definition, and configuration provided by applications making use of the cipher suite.

Definition 2.6 *(Cipher Suite Configuration)*: The parameter restrictions or settings for the algorithms defined for a given cryptographic solutions. These elements are part of the *immutable* definition of the cipher suite. Consider the specification of allowable key sizes for an encryption cipher – if a cipher suite were to restrict which sizes may be used, then this would be considered part of the configuration of the cipher suite.

Definition 2.7 *(User Configuration)*: Cipher suites require external information to operate on, to include the key or keys to be used for the security operation and any special context or state associated with the node or network. This might include initialization vectors, salt values, timing information, and other items that help to make individual cryptographic operations unique.

The differences in these configuration approaches are illustrated in Figure 2.6, which shows the concept of multiple cryptographic algorithms (Algorithms 1–3) comprising a cipher suite. Cipher-suite-specific configurations, such as the key sizes allowed to be used with Algorithm 1,

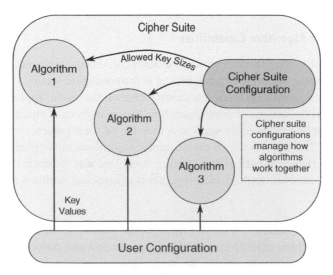

Figure 2.6   Cipher suites support both internal and external configurations.

represent constraints on algorithms that are imposed by the cipher suite itself, regardless of the user of the cipher suite. User configuration, such as key values of an allowed key size, is provided by the user for whom cryptographic materials are being generated (or processed).

The *user* of a user configuration is typically some utilizing service running on a node and tasked with applying security services to a network message under the auspices of a network security protocol or the policy associated with that protocol. In applying these configurations, the utilizing service must ensure that the configurations sent to the cipher suite are appropriate. Appropriate in this context can mean several things, to include the following.

- **Unique**: User configurations often have some uniqueness requirement associated with them. Some algorithms warn that re-using a particular combination of configuration items might lead to a loss of security protection.
- **Fresh**: If there is not a uniqueness requirement, once a user configuration item has been used a certain number of times (or used with a certain amount of user data) that item might no longer be considered appropriate. For example, encryption keys become exhausted once they have been used to encrypt a certain amount of data, or generally used a certain number of times. Networks must know when to stop using exhausted configuration items.
- **Active**: Long-lived configuration items, such as keys, may be compromised such that their use is no longer able to guarantee the validity of the security services they are used to implement. Therefore, nodes must ensure that their configurations are still active in the larger security system and not otherwise revoked or expired.

Focus: AES Key Wrap

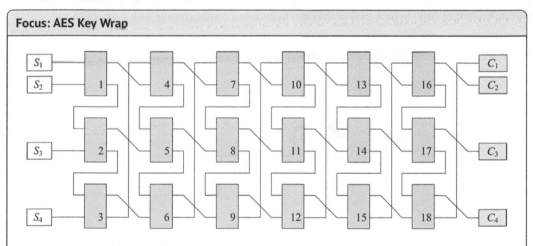

Source: Image Courtesy of NIST/Morris Dworkin [9].

When encrypting information using the Advanced Encryption Standard (AES) [10], one input is an Initialization Vector (IV), which must be different for every encryption performed by the AES algorithm for a given key. Using a unique IV for each algorithm invocation ensures that the same user data, encrypted with the same key, does not produce the same cipher text. When AES is used to encrypt user data, the IV is considered part of *User Configuration*.

Another use of the AES algorithm is to encrypt a session key with a long-term key using a process called AES key wrap [9]. When used for the purpose of key wrapping, the IV input to the AES algorithm is defined to be the fixed value 0xA6A6A6A6A6A6A6A6. For this reason, the programming interfaces to functions within cipher suites that perform key wrap do not accept a user configuration for the IV and, instead, use the fixed value as part of the implementation of the cipher suite itself. Thus, the IV for AES key wrap is considered part of *Cipher Suite Configuration*.

The most common configuration item related to security is the key[3] but these configuration requirements may apply to other data used to govern the use of a cipher suite in a network including policy statements, local state data (such as time), and resource configurations (such as associated with compute, memory, and storage utilization).

2.3.3 Packaging and Transport

The outputs of cipher suite algorithms are placed within a network message and sent with (or in lieu of) original data in the message. Sometimes, in addition to these outputs, user inputs might also be represented in a message to help with security processing at a remote downstream node. In all cases, the set of cryptographic materials that are included in a network message make certain assumptions on the capabilities of the network as they relate to delivery, timeliness, and overall availability.

Network security protocols define both the protocol information that needs to be communicated over a network and the expected behavior of protocol agents at various nodes within the network. As a combination of both data and behavior, network protocols are responsible for discussing how cryptographic materials might be generated, how they are packaged into network messages, and how those messages are processed at various downstream nodes along the message path.

The interactions between user data, a cipher suite, and a network message are illustrated in Figure 2.7. This figure shows user data being prepared for transmission by passing some data through a cipher suite (which may or may not negotiate a configuration with some off-node entity).

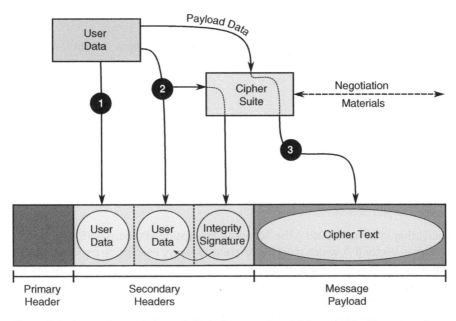

Figure 2.7 Network security protocols package cryptographic materials. The generation of cryptographic materials is separate from the way in which those materials might be represented within a message.

---

3 Keys in this context refer to any one of a number of shared secrets and include concepts such as long-term keys, key-encryption keys, and ephemeral keys.

Network security protocols can choose to process user data in one of three ways: pass-through, annotation, and alteration.

1. **Pass-through**: User data might be copied, directly, into a message without any interaction with a cipher suite. Typically, the determination to pass-through data is a function of security policy.
2. **Annotation**: Cipher suites may produce data meant to accompany user data. This is shown in Figure 2.7 where user data are both directly packed into a message secondary header and, in parallel, given to a cipher suite that generates an integrity signature over the same data and places the integrity signature in a different secondary header.
3. **Alteration**: Cipher suites may produce data that replaces user data. This is shown in Figure 2.7 where user payload data are encrypted by a cipher suite and the resultant cipher text is placed in the message. Unlike an annotation approach, the original user payload data are never represented in the message.

2.4 Summary

To understand the impact on security of network design, this chapter has examined the nature of network design, and the principle of layered architectures that form the basis of most network design activities.

It is easy to overlook the need to perform network stack design – terrestrial networks have settled into a few common architectures and designs, around networking stacks operating in common environments. When a new network is being built, pragmatic intuition is to select from among these existing stacks. If an existing networking stack is unable to operate in a particular environment, there is often a desire to change the environment or accept degraded cost or performance before attempting to invent a new networking stack.

However, as networks are pushed into increasingly remote areas, changing the environment becomes less practical than changing network design. In those cases, a focus on the unique constraints imposed by the environment is important to network design. While constraints differ as a function of the environment, there are some common sources and types of constraints, all of which need to be examined to ensure that no subtle constraints have been missed.

Separate from unique constraints for a particular network design, there are two common design principles that are likely unchanged for any new networking activity: layering and packaging.

Separating networking responsibilities into layers provides multiple benefits for the network designer, implementer, and user. However, there is a balancing point on how many layers are used to produce a given stack. The number and types of layers may be different as a function of the environment and associated constraints. While the presence of a network stack should be presumed, the nature of that stack may be part of the design activity.

Separating cryptographic algorithms, the configuration, and packaging provides a modular way to talk about security at any layer in a network. The mathematical algorithms used to generate cryptographic materials may be usable in a variety of network designs, even though the configuration and packaging of those materials could be quite different for different network security protocols. From this point of view, the strength of a network security protocol lies in its ability to provide appropriate inputs and configurations to strong cryptographic algorithms. Developing new network security protocols does not always imply the need to develop new cryptographic algorithms.

Ultimately, the design of networks for new and challenging deployments might not reuse the same network stacks and implementations that are otherwise popular in resources deployments such as the Internet. To understand why new network protocols, to include security protocols, are needed involves examination of the constraints placed on new types of network deployments.

References

1 Knuth, D. E. [1974]. Structured programming with go to statements, *ACM Computing Surveys (CSUR)* **6**(4): 261–301.

2 Farsi, A., Achir, N. and Boussetta, K. [2015]. WLAN planning: separate and joint optimization of both access point placement and channel assignment, *Annals of Telecommunications-Annales des Télécommunications* **70**(5): 263–274.

3 Birrane, E. and Soloff, J. [2020]. *Designing Delay-tolerant Applications for Store-and-Forward Networks*, Artech House.

4 Rose, S., Larson, M., Massey, D., Austein, R. and Arends, R. [2005]. Protocol Modifications for the DNS Security Extensions, RFC 4035. **URL**: https://rfc-editor.org/rfc/rfc4035.txt.

5 Rescorla, E. [2000]. HTTP Over TLS, RFC 2818. **URL**: https://rfc-editor.org/rfc/rfc2818.txt.

6 Birrane, E. and Heiner, S. [2020]. A novel approach to transport-layer security for spacecraft constellations.

7 Birrane, E., Burleigh, S. and Cerf, V. [2011]. Defining tolerance: impacts of delay and disruption when managing challenged networks, *Infotech Aerospace 2011*, AIAA, p. 1645.

8 Kent, S. [2005]. IP Encapsulating Security Payload (ESP), RFC 4303. **URL**: https://rfc-editor.org/rfc/rfc4303.txt.

9 Dworkin, M. [2012]. Recommendation for block cipher modes of operation: methods for key wrapping.

10 Dworkin, M., Barker, E., Nechvatal, J., Foti, J., Bassham, L., Roback, E. and Dray, J. [2001]. Advanced encryption standard (AES).

3

DTN Security Stressors and Strategies

Any discussion about secured communications within the Delay-Tolerant Networking (DTN) architecture must include analysis of the stressing conditions present in challenged environments. This includes an overview of the strategies employed by the DTN architecture. Since the BPv7 protocol suite is the recommended way to instantiate a DTN architecture, the analyses presented in this chapter capture some of the design constraints levied on BPv7 and BPSec during their development.

After reading this chapter you will be able to:

- Distinguish DTN architectures from other network architectures.
- List reasons why BPv7 requires custom security extensions.
- Identify stressing conditions unique to DTN security.
- Explain common approaches to these unique stressing conditions.

3.1 DTN Constraints

In 2007, the Internet Research Task Force (IRTF) published informational RFC 4838 titled *Delay-Tolerant Network Architecture* [1], which documented the rationale and design constraints inherent in certain types of challenged communications environments. In 2022, the Internet Engineering Task Force (IETF) published standards for BPv7 and BPSec that implement the transport and security portions of the DTN architecture.

> [DTN] is an evolution of the architecture originally designed for the Interplanetary Internet, a communication system envisioned to provide Internet-like services across interplanetary distances in support of deep space exploration.
>
> DTN Architecture, RFC 4838, [1]

The DTN architecture was motivated by the need to extend common networking semantics to communication systems operating across interplanetary distances [1, 2]. This concept of a vastly distributed network has been referred to as the Interplanetary Networking (IPN), the Interplanetary Internet, and the Solar System Internet (SSI). While there are minor differences in the way these concepts are discussed in literature, they all represent a shared aspiration of multi-agency networking across astronomical distances. In the context of security strategies, these concepts can, collectively, be considered equivalent.

Securing Delay-Tolerant Networks with BPSec, First Edition. Edward J. Birrane III, Sarah Heiner, and Ken McKeever.
© 2023 John Wiley & Sons, Inc. Published 2023 by John Wiley & Sons, Inc.
Companion website: www.wiley.com/go/birrane/securingdelay-tolerantnetworks

Both the DTN architecture and the BPv7 protocol suite that implements it have beneficial utility in terrestrial and near-Earth networks. However, they must also operate in the most stressing conditions for which the architecture was created. For that reason, DTN security must be able to function in these most extreme environments.

3.1.1 The Solar System Internet

The SSI concept is illustrated in Figure 3.1. This NASA image represents a mixture of existing and futuristic communications concepts operating using network semantics. In this architecture, a variety of assets from multiple space agencies representing multiple missions cooperate to relay information in both local networks and across long-haul interplanetary distances. Combined with pointing and power limitations, this concept presents a very different infrastructure than that used on the terrestrial Internet. Such a dispersed, sparse, and partitioned network exemplifies the stressing environments that DTN protocols must overcome.

It is, perhaps, easy to dismiss this architecture as fanciful given the complexity, cost, and futuristic nature of the concept. However, there are pragmatic and practical benefits to the SSI. Some of these benefits include the following.

- **Efficient Landers**: Planetary landers transmit data to planetary orbiters over relatively short distances compared with communicating from a planetary surface back to Earth. This reduces the size, weight, and power required by the lander's telecommunications systems.

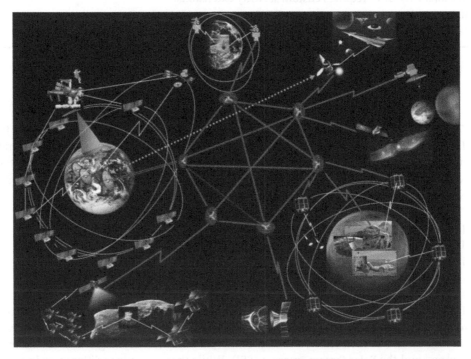

Figure 3.1 The complexities of interplanetary networking exemplify the security challenges inherent to the DTN architecture. Source: NASA.

- **Uninterrupted Communications**: Spacecraft acting as network relays and storage nodes mean that communications to Earth can happen even when orbits prevent line-of-sight communications.
- **Risk Reduction**: Even a sparsely networked solar system can offer significant risk reduction and cost savings by sharing the cost of communications infrastructure across multiple missions operated by multiple space agencies.

3.1.2 Other Challenged Networks

The DTN architecture provides common networking *features*[1] in environments that do not support common networking *infrastructure*. If the DTN architecture could address link disruptions and propagation delays in the SSI concept, it might also be able to address other challenged network scenarios. From its inception, the DTN architecture was built to provide a

> **Common Solutions**
>
> Solutions for the SSI solve other (closer to home) networking problems.

common solution to all challenged networks experiencing these types of delays and disruptions. While interplanetary networks might experience these challenges at larger scales and increased magnitudes, other networks could reliably make use of these protocols for their own deployments.

> [s]ensor-based networks using scheduled intermittent connectivity, terrestrial wireless networks that cannot ordinarily maintain end-to-end connectivity, satellite networks with moderate delays and periodic connectivity, and underwater acoustic networks with moderate delays and frequent interruptions due to environmental factors.
>
> DTN Architecture, RFC 4838 §1, [1]

Each of these networking concepts requires that networks participate in end-to-end data exchange even in cases where there is no instantaneous end-to-end connectivity. Non-DTN architectures view disconnection as an error condition to be waited through. The DTN architecture tolerates these periods of disconnection and continues to make progress along a single link even when an entire end-to-end path is not available. To accomplish this, new algorithms and protocols were needed to operate within these more challenged constraints. See **Focus: Connecting Extreme Earth Environments** on next page.

3.1.3 Tolerant Networking

The DTN architecture must tolerate delays and disruptions as part of the expected operational state of the network. In some extreme networking cases there may never be a time when the network is well connected. Attempting to implement a "recovery mode" wherein traffic is halted waiting for a return to normalcy would cause the network to wait forever. This is a departure from other networking architectures which operationally switch the network *on* and *off* as a function of whether the current environment supports the assumptions required by the architecture.

---

1 RFC 4838 identifies features such as security, addressing, routing, and network management.

Focus: Connecting Extreme Earth Environments

Extreme environments, in this context, refer to places on Earth lacking reliable communications infrastructure. Examples of such environments include oceans, deserts, polar regions, and mountain regions. Most of the Earth's surface lacks a reliable communications infrastructure and the DTN architecture provides one way to bring communications capabilities to these regions.

NASA engineers use DTN to transmit a selfie from the National Science Foundation's McMurdo Station in Antarctica. Source: NASA.

In 2017 NASA engineers transmitted a selfie of themselves holding a photo of Internet pioneer Dr. Vinton Cerf from the McMurdo station in Antarctica through a complex network including NASA near-earth spacecraft, the terrestrial Internet, and the International Space Station using the Bundle Protocol (https://www.nasa.gov/feature/antarctic-selfie-s-journey-to-space-via-disruption-tolerant-networking).

The ability to tolerate intermittent connections provides a new way to think about reaching some of the most remote places on Earth and DTN is being considered for networking mountainous regions, deep-sea networks, and other places where building reliable infrastructure is impractical.

Definition 3.1 *(Delay-Tolerance):* Continuing the function of network services even in the presence of propagation and processing delays and link disruptions that add delays to the delivery of data across the network. This may also refer to the characteristic of hiding this information from applications using the network. In this sense, a delay-tolerant architecture can also be a disruption-tolerant architecture [3]. The DTN architecture relies on protocol capabilities such as store-and-forward transport to achieve delay-tolerance.

Tolerant architectures exhibit fewer changes in externally observable behavior when the network shifts from normal to challenged operations. To do this, the protocols implementing this architecture must understand the specific challenges present in the network and how they may be overcome. An understanding of these challenges is important not only to the definition and construction of protocols, but also to the design and implementation of the applications and extensions made around these protocols.

> **Disruptions cause delays**
>
> A delay-tolerant architecture can adapt to delays caused by frequent disruptions.

3.2 Security-Stressing Conditions

There are many communications environments where the DTN architecture might be usefully deployed. Some of these environments may look similar to other well-resourced networks where delays and disruptions are caused by administrative policy and not physical limitations. Other environments may resemble the most stressing cases already discussed in this chapter and associated with remote locations on Earth and throughout the solar system.

Among this diverse set of potential deployments certain characteristics impose more serious constraints on the design of security services than others. Since the DTN architecture requires a reliable security ecosystem, these security-stressing conditions must be understood and addressed by any security mechanism applied to BPv7. The remainder of this section lists these constraints and discusses their unique impact on network security.

3.2.1 Intermittent Partitioning

A consequence of operating in a disrupted environment is that links between nodes come and go over time. This frequency of this link intermittency may be either periodic or aperiodic.[2] Connectivity loss might be planned (such as when power-cycling radios or changing antenna pointing) or unplanned (such as caused by interference). The impact of link loss might reduce path diversity or completely partition a portion of the network.

> **Sparse Networks**
>
> Sparse networks have less path diversity, so the loss of a link is more likely to partition the network.

When link disruptions partition the networks nodes existing in separated partitions are unable to communicate and, thus, unable to synchronize information across messaging endpoints.[3] To the extent that some security algorithms rely on state synchronization through the network, network partitioning prevents those algorithms from functioning correctly.

Two security-related concepts affected by network partitioning include the establishment of shared secrets and the ability to synchronize state across the network.

3.2.1.1 Secret Establishment
Cryptographic systems require the establishment of some shared, secret information between security endpoints. In this context, secret information refers to either the information used to derive

---

2 Or both, when there are multiple reasons for link disruptions.
3 If the loss of a link does not partition the network nodes can still synchronize information through other paths and might continue to operate with possible extra delays.

security keys used by cryptographic algorithms or the derived keys themselves. The establishment of this information might be performed in advance of any secure data exchange or negotiated in real-time on an as-needed basis.

Understanding how secrets are established across an otherwise untrusted network is important, as certain establishment approaches are more affected by network partitioning than others. There are three ways in which secrets can be established in a network: pre-placement, direct negotiation, and third party negotiation.

Figure 3.2 illustrates three common architectural patterns to secret establishment. While these patterns do not presume a particular algorithm or protocol, examination of these generalized data flows emphasize the degree to which connectivity might be required prior to secured communication.

1. **Pre-placement**: Pre-placed secrets are not impacted by intermittent connectivity because they are presumed to have been shared prior to connectivity issues.[4]
2. **Direct Negotiations**: Direct negotiation involves round-trip message exchanges prior to secure communications to establish a shared secret. Direct negotiation is highly susceptible to connectivity drops because it involves multiple round-trip exchanges over the network.
3. **Trusted Third Party**: Third-party negotiation is a hybrid approach where a message source and destination may reference a third party authority which houses pre-placed keys. In these schemes, there may be more than one third party authority – such as one more "local" to a message sender and another more "local" to a message receiver. Of course, connectivity issues will also impact secret sharing to a third party, but the option to have multiple third parties in a network provides some risk mitigation.

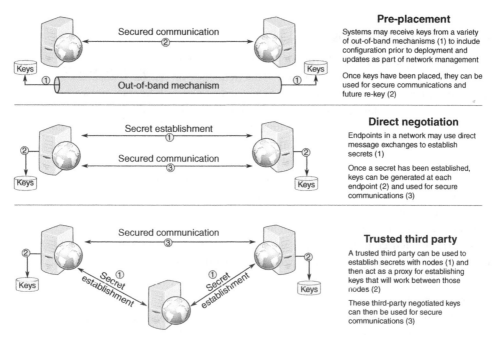

Figure 3.2   Techniques for secret establishment. There is a relationship between some secret establishment techniques and the presumed connectivity of the network.

---

4 Such as the practice of pre-loading keys in a system prior to its deployment.

3.2.1.2 Security State Synchronization

Separate from the initial establishment of a shared secret, certain security mechanisms require that the internal states of senders and receivers change in a synchronized way for the duration of the secured communications session. There are several examples of information that might need to be coordinated across a network to enable secure communications.

One example is a synchronized stream cipher. This type of cipher does not use a single, fixed key for its encryption. Rather, the initial secret establishment is used to initialize some position within a key stream.[5] A particular data item is encrypted using a series of bits from a starting position in that bitstream, and the same series of bits must be used at the message destination to construct the key used to perform the decryption.

Losing synchronization on the key bitstream between the sender and receiver means that the bits used to form the encryption key would not be the same as the bits used to form the decryption key. Similarly, delivering messages out of order would also cause problems in how a destination constructs a decryption key.

Avoid Feedback Mechanisms

Any network algorithm that relies on a feedback mechanism for its proper operation might not function when used in the DTN architecture. The potentially long-term loss of connectivity would delay closing feedback loops for a long enough period of time as to be indistinguishable from network failure. While this is certainly true of security-related algorithms, this also applied to other algorithms, such as those used for monitoring and control.

3.2.2 Time-Variant Topology

The immediate problem of intermittent partitioning is that it prevents two nodes from communicating when they are placed in different partitions. The strategic problem of intermittent partitioning is that it leads to a network topology that changes over time. When the partitioning of the network changes as a function of time, then any given paths through the network might not be constructed from contemporaneous links.

Definition 3.2 *(Non-contemporaneous Path):* Any path whose links are not active during the same time period. Such paths may require messages to be stored at nodes waiting for a future link to be established. Since such a path does not require overlapping individual links, these paths may also be changed to track with topology changes.

To understand how security might be affected, some analysis is needed on how the individual links of a non-contemporaneous path are discovered within the DTN architecture. For any node in the network, link connectivity changed can be asserted, predicted, or discovered.

- **Asserted connectivity** happens when the cause of the connectivity change stems from a known, planned action as communicated from an authoritative entity. If a radio is scheduled to be turned off to save power, then any link requiring access to that radio will lose connectivity. The fact that the radio would be turned off can be asserted to other nodes on the network well in advance of the event. Similarly asserted events include antenna pointing and/or link occultation as a function of commanded motion.

---

5 A pseudo-random bitstream that is sampled to create key information.

- **Predicted connectivity** happens when patterns of link behavior can be inferred from some period of observation. For example, a solar-powered device might learn that its link is lost due to low power during regular parts of the day when the device's solar panels are in shade and at night. Alternatively, nodes might learn that certain times of day see peak data usage and predict link congestion. In addition to predicting the loss of a link, new link opportunities can also be predicted. Consider the case where a node learns the routes of local busses that pass by with open access points. In this case, the route of a bus is regular enough that a node might come to rely on such a link for exfiltration.
- **Discovered connectivity** happens when a device can measure, in real time, the state of a link. Link discovery is the primary mechanism of topological discovery in low-latency, high-reliability networks because the duration of the link is very large relative to the time taken to discover and use the link. This behavior is similar in a DTN, with two additional observations. First, a discovered link must also be assigned an expected "time to live" for the purposes of understanding how the discovered link contributes to the future state of the topology. Second, discovery is also used to verify asserted and predicted connectivity. If a connectivity prediction is incorrect, that error can be determined when the predicted link connectivity (or predicted loss of connectivity) is not confirmed through the discovery process.

Figure 3.3 illustrates different types of connectivity that may be encountered in a network. Consider a data mule node (N) moving through a network at four distinct points in time (t_0, t_1, t_2, and t_3). At time t_0 node N can only see stationary node S1. However, presuming that node N understands its planned motion, it would also understand that at time t_1 stationary nodes S4 and S5 would be visible and that at time t_2 node S6 would be visible. These future connections can

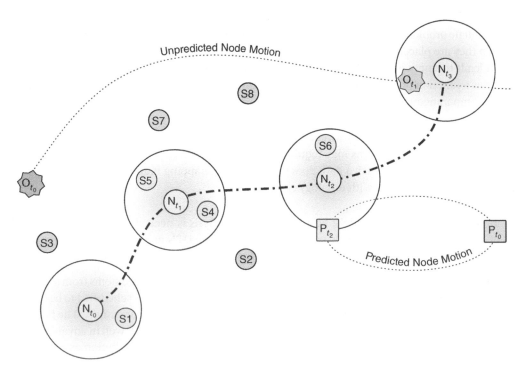

Figure 3.3   A network topology evolves over time. A given mobile node (N) may pass stationary (S), periodic (P), and opportunistic (O) contacts that change the topology of the network.

be *asserted* based on knowledge of the network topology and the motion of the node N. Separately, at time t_2 is can be *predicted* that periodic node P should also be visible so long as P's motion is not perturbed. Finally, other nodes, such as opportunistic node O may represent network nodes whose behavior cannot be asserted or predicted. At time t_3, node N can *discover* that node O is within communications range, whereas that may not have been predictable earlier.

Every node in a DTN must make storage, routing, and security decisions as a function of both the believed current topology of the network *and* the presumed future topology of the network. To the extent that topological changes have security significance, the future state of the network must be considered when deploying security services. Ultimately, topology variation imposes unique challenges for the establishment of security tunnels, key management and selection, and security policy enforcement.

3.2.2.1 Secure Tunnels

A tunneling protocol is used to pass a message from a tunnel source to a tunnel destination, typically through some other network using different protocols or administrative controls. Tunneling is used as a means to create a logical network over an existing network using encapsulation. Secure tunnels refer to the concept of applying security to network tunneling protocols.

A common example of a secure tunnel is a Virtual Private Network (VPN) wherein a remote corporate computer (such as a corporate laptop located at an employee's home) connects to the corporate intranetwork over the Internet. In this case, the remote computer generates corporate network traffic which is encapsulated for transmission over the Internet and unpacked at some gateway between the Internet and the corporate intranet. Applications on the remote computer have no knowledge that they are not directly connected to the corporate intranetwork.

The use of encapsulation to create *secure tunnels* within a network relies on the predictable availability of the endpoints of the tunnel. This means that a secure, encapsulating message from a *tunnel source* must be addressed to a *tunnel destination* that can extract the encapsulated message. When the tunnel endpoints correspond with the source and destination of the message, changes to topology are largely inconsequential; the encapsulated message and the encapsulating message need the same fixed network points. However, when the tunnel endpoints differ from the encapsulated message endpoints, the tunnel is coupled with the topology of the network. Changes to the network topology might break tunnel semantics.

Figure 3.4 illustrates three different ways to terminate a secure tunnel using either messaging endpoints, fixed boundaries, or free boundaries. In the three samples provided in this figure, the topology of the network may change such that tunnel traffic at time t_1 takes a different path (represented by the dashed line) than tunnel traffic at time t_0.

Tunnels established between *messaging endpoints* are immune to non-partitioning topology changes because secured traffic is only evaluated at the destination and topological change as discussed in this section does not change the destination. Tunnels can also be established between *fixed boundaries* in a network, such as may exist between two trusted network enclaves.

In this formulation, border nodes (B1, B2) can be established for the secure communication between them over an otherwise untrusted part of a network. So long as B1 and B2 are always the borders to the source and destination, respectively, these security tunnels are similarly unaffected by topological change between the border nodes. Tunnels might be established between *free boundaries* in which one or more border nodes may not be guaranteed as the pathway to a messaging endpoint. For example, when a message destination can be reached by either border nodes B2 or B3, as a function of time, then a tunnel built from B1 to B2 might not accept traffic through border node B3 at a later time.

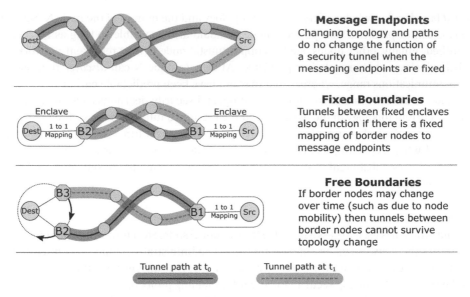

Message Endpoints
Changing topology and paths do no change the function of a security tunnel when the messaging endpoints are fixed

Fixed Boundaries
Tunnels between fixed enclaves also function if there is a fixed mapping of border nodes to message endpoints

Free Boundaries
If border nodes may change over time (such as due to node mobility) then tunnels between border nodes cannot survive topology change

Tunnel path at t_0 Tunnel path at t_1

Figure 3.4 Tunnels should not be tied to topology. Tunnels endpoints must represent fixed access to messaging endpoints.

3.2.2.2 Key Selection

The start of any secure communications session usually begins with the selection of some type of cryptographic key material. One common practice is the selection of a known *long term key* used to protect the establishment or communication process used to derive a *session key* to be used for cryptographic operations on a message.

One important consideration with key selection is that the key selected at one end of a secure tunnel must be appropriately paired to the key at the other end of the secure tunnel. If this is a symmetric key system, the key used at the source must match the key available at the destination. If this is an asymmetric key system, then the used public key must match with the correct private key.

In a low-latency network, key selection is a matter of proper network configuration. If the proper keys are configured secure communications should proceed as expected because message paths are considered contemporaneous and (practically) instantaneous. In a DTN architecture, significant time might pass between the selection of a key at a source and the matching of cryptographic material with a key at a destination. If a message takes very long to reach a destination, keys at the destination might be rotated by the time that message arrives.

Key selection must account for the time-variant topology of the network, to include the expected delivery time of a message. Keys should be chosen as a function of expected valid keys available at a destination node at the time of expected message delivery.

3.2.2.3 Security Policy Configuration

Security policy at a node governs what security services might be applied to what traffic. This policy configuration is often driven by the topology of the network. This topology identifies network elements such as boundaries between different networks and secure tunnel endpoints.

If the topology of the network changes over time such that messages might take alternate paths and, thus, go through alternate gateways then path-based security policies might be invalidated. Security policies that are fixed to the topology of a network may stop working when deployed in a network with topologies that vary over time.

3.2.3 Long-Term Storage

Strategies for handling congestion in networks have been proposed since the inception of networking and, famously, during the congestion collapses encountered on the terrestrial Internet in the mid-1980s [4]. Early, intuitive approaches focused on the use of back-pressure mechanisms to either signal an application to stop producing new data or to signal the sending node of a network to pause and locally buffer data. Other approaches to avoiding congestion define best-effort traffic that can be dropped when the network becomes congested. By the turn of the century, the proliferation of best-effort traffic was itself causing congestion issues making congestion control not only an end-to-end problem but something to be acted upon by intermediate routers [5].

The DTN architecture encounters all of the traditional congestion problems associated with other networks and adds local bulk storage as another resource that can be congested. To wait out period of network partition, DTN transport mechanisms must store-and-forward information until some future contact along a message's non-contemporaneous path. However, store-and-forward can, at times, add to congestion in the network for two reasons.

1. **Queued Backlogs**: Local data storage becomes a second source of traffic in the network. When a new link becomes active on a node both incoming traffic and stored traffic will contend for this bandwidth.
2. **Bottlenecks**: The time-variant topology of the DTN architecture can create bottlenecks in the network where one (or a few) nodes become the only predicted way to cross a network partition. This may cause multiple other nodes to send traffic to this single node well in advance of predicted connectivity which can also exhaust local storage on the node.

The increased likelihood for congestion creates multiple security-related problems, to include those related to securing data at rest and traffic time-to-live.

3.2.3.1 Security-at-rest
Security-at-rest refers to the way in which data is secured while on a node either as part of accumulating that data for transmission, waiting to forward the data, or pending delivery to an application as part of data reassembly at a message destination.

Network security protocols do not typically provide security-at-rest, as this happens after a message is removed from the network. However, in the DTN architecture, network messages might, themselves, be persistently stored. In this case, DTN network security protocols must provide security-at-rest when the protocols themselves are stored locally on a node as part of store-and-forward operations.

Because this secure storage includes network information (to include protocol headers) security-at-rest must be implemented as part of network-layer responsibilities. If a security protocol only secures data while "on the wire" there would otherwise be no security when network data is long-term stored by the network in ways hidden from the application layer.

3.2.3.2 Time-to-live
The time-to-live of a message in a low-latency network is used to limit the total number of networking resources that a message is allowed to use before it is considered to be, essentially, not worth the trouble and deleted. Since network resources are expended in a low-latency network by transmitting and receiving messages, time-to-live is expressed in the number of hops[6] before the message will be deleted if it has not reached its destination.

---

6 A hop is one transmission event from one node on the network to another node.

The DTN architecture allows messages to consume resources even at rest as part of store-and-forward transport. Therefore, simply tracking the number of message hops is not sufficient to prevent messages from not using too many resources; a large message sitting in persistent storage for too long can become a congestion problem.

When messages may be stored long-term in the network, cipher suites, keys, and other cryptographic material must take into consideration that total time that data may spend in the network. Similar to security-at-rest, this is a new security consideration that is not otherwise considered as part of low-latency security protocols.

3.3 Security Strategies

As discussed in Section 3.2, characteristics such as intermittent partitioning, time-variant topologies, and long-term network storage constrain the behavior of a variety of common security activities. These stressing conditions can violate assumptions made by the design or implementation of security protocols and, in doing so, cause these protocols to fail to perform, or perform consistently.

It is certainly possible to reuse existing security protocols and other security-related implementations in a DTN environment when their proper operation is not challenged by that environment. Whether reusing existing security mechanisms or generating new ones, there are certain strategies that can be used to mitigate the challenged of the DTN environment.

The remainder of this section introduces these strategies and discusses how they might be useful in the context of DTN security.

3.3.1 Separate Concerns

> **Cohesive Functions**
>
> Security mechanisms should be highly cohesive and loosely coupled.

The principles of cohesion and coupling are established ways of assessing the modularity of software design [6]. Cohesion measures the relatedness of capabilities within a single software component. A highly cohesive software component comprises internal functions that support a single purpose whereas a poorly cohesive component will attempt to perform many purposes. Coupling measures the relatedness of software components within a system. A tightly coupled system is one whose software components need each other in order to function whereas a loosely coupled system is one where software components can be mixed and matched as needed. Modular software architectures seek to be highly cohesive and loosely coupled. The act of increasing the cohesion within a system is referred to as "separating concerns" and refers to the separation of functions into multiple, smaller functions that each have more singular purpose, or concern.

These principles can be applied to protocol design by making protocols that are of singular purpose and with enough configuration and policy to inputs to allow that single purpose to be applied in multiple networking environments. This approach increases the likelihood that protocol functionality can be re-used across differing network architectures because assumptions on network characteristics can be pushed to policy and configuration.

Figure 3.5 illustrates a conceptual architecture for loosely-coupled, cohesive security processing software. In this figure, individual components are clustered by their single responsibility.[7] Policy components deal with configuration and behavior. Protocol components handle the structural

---

[7] single responsibility is a software design principle which states that highly cohesive software components should have a single purpose or single reason to change.

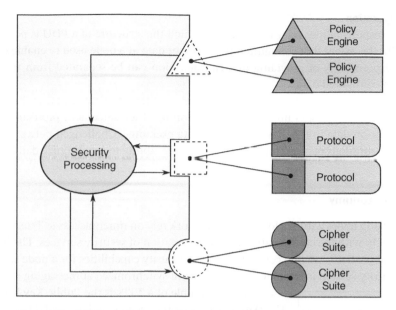

Figure 3.5 Highly cohesive, loosely coupled architectures increase reuse.
Decomposing security protocols into cohesive components allow common
capabilities to be reused where possible and minimize the need to develop
new capabilities for new environments.

encoding and decoding of Protocol Data Units (PDUs). Cipher suite components generate and
interpret cryptographic materials. By keeping these concerns in highly cohesive components, interfaces can be built around them, allowing for security processing software to be built independent
of any particular component implementation.

A network security protocol can be decomposed into at least three separate "single responsibilities": structural, policy, and configuration.

3.3.1.1 Structural
Structural responsibilities for a security protocol refer to defining the data fields of a PDU related to
security information and, thus, the set of information that is available to nodes processing that PDU.
PDU structures that omit critical information for the security protocol under the assumption that
endpoints or other nodes in the system will keep that information synchronized can fail to operate in a DTN architecture. DTN transport protocols can carry information necessary for payload
processing so as to minimize the amount of knowledge required to exist at messaging endpoints.

3.3.1.2 Policy
Policy responsibilities refer to the variety of decisions associated with protocol behavior that must
be specified on a node-by-node basis as part of operating a network. Policy decisions might include
how to deconflict multiple sources of configuration data, how error conditions are handled, when
and what security services are required, and even when a protocol should be used. Protocol specifications often identify areas where policy decisions are necessary and make recommendations on
what those decisions should be in certain circumstances.

The actual expression of policy can be made specific to a network node, or to a network architecture. By separating policy from protocol structure, protocols can be reused in more networking
environments because their behaviors can be customized to the environments in which they are
deployed.

3.3.1.3 Configuration

Configuration responsibilities refer to the way in which the structure of a PDU is populated with information. To the extent that configurations represent data at a node used to enable certain processing, the responsibility of providing that configuration can be separated from the structures populated with configured data.

Different configuration mechanisms can be used in different networking architectures. For example, low-latency networks might use negotiation or other agreement protocols to generate sets of configuration data in real time to support data exchange. Challenged networks might use pre-placed configurations because there might not be any ability to synchronize data in real-time.

3.3.2 Local Autonomy

Security processing nodes within a low-latency network rely on timely access to Trusted Third Parties (TTPs) to help with the configuration and interpretation of security services. The information from these third parties is used to configure general security capabilities for a node separate from the kind of security session information negotiated and synchronized at messaging endpoints.

> **Trusted Third Parties**
>
> Nodes in a network that store information needed by other nodes for establishing secure communications.

A popular example of a TTPs is the Public Key Infrastructure (PKI) used widely on the Internet today. In a PKI, a private key is known only to the entity it represents. Anyone wishing to communicate with that entity must use that entity's public key. This becomes problematic if the advertised public key of an entity is no longer appropriate given an entity's private key. Mismatches can occur when an entity changes its private key, if a malicious actor in a system published incorrect public keys, or if a security policy mandates a change.

Within a PKI, a Certificate Authority (CA) has the responsibility of distributing public keys for entities and ensuring that these keys are kept up to date with changes to entity's private keys. Further, CAs sign the public keys that they distribute so that users can trust the quality and accuracy of the public keys from the CA.[8]

Unlike a low-latency network, DTNs cannot guarantee timely access to TTPs and cannot out-source the verification of cryptographic materials as part of applying security services. This implies that decision points made by a local node can only use the information available at that node. This limitation presents a significant risk for security services, particularly as it relates to frequently-changing information such as the disposition of keys.

Nodes implementing security services may need to make predictions because of an absence of information. The ability to make such predictions, and act on them, requires that DTN nodes implement some level of autonomy. Three examples of how local node autonomy can be applied to security are key appropriateness, local state modeling, and policy application.

3.3.2.1 Key Appropriateness

Nodes must determine the appropriate application of key material (and other configurations) for security functions. Local autonomy is not sufficient to establish and synchronize symmetric keys, but it can determine that a given, existing key be used until an explicit confirmation or revocation is received from a more authoritative source (or sources).

Figure 3.6 illustrates two common reasons why keys might be considered inappropriate over time: key exhaustion and key expiration.

---

8 At least to the level that they trust the CA to hold accurate information.

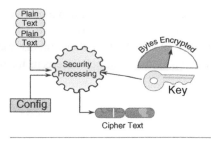

Key Exhaustion

Once cryptographic keys have been used to generate a certain amount of ciphertext, analysis over that set of cipher text could be used to recover the key. When this happens, the key is consiered to be exhausted

Exhausted keys should not be used to generate additional cipher text

Key Expiration

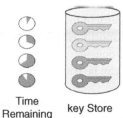

Cryptographic keys may be selected as a function of time within a system. These times may be absolute times representing when a key will be active in a system, or relative based on when a key is first used

Independent of bytes encrypted, if keys are kept in use for too long, they may be compromised

Figure 3.6   Keys lose their effectiveness with use and over time. Keys on nodes must be replenished over time as they become exhausted or are expired.

3.3.2.1.1   Key Exhaustion   Key exhaustion refers to the concept that many cipher suites place limits on the number of bytes of information that can be encrypted by a key before the key is considered not appropriate for further use. Exhausted in this sense refers to that fact that enough material has been encrypted with the key to make certain types of cryptanalysis attacks on the key more practical if that material were collected and analyzed.

3.3.2.1.2   Key Expiration   Key expiration refers to the concept that keys can be considered not appropriate for use solely as a function of time. Time-based expiration may be more pragmatic in DTN security systems when the total amount of information encrypted by a key might not be synchronized across a sender and a receiver communication over a non-contemporaneous path.

Selecting appropriate keys is an important consideration for any key-based security mechanism. This selection becomes especially important in the DTN architecture where keys might need to be selected both as a function of the state of the key at the message sender and the time at which a message might be received at a message destination. Local node autonomy might be needed to make complex determinations of key selection.

3.3.2.2   State Modeling

The determination of whether a node should be the source, verifier, or acceptor of some security service for a message is usually a function of the topology of the network. Nodes serving as boundaries between areas of administrative control or between trusted and untrusted parts of a network are typically given special security roles and responsibilities.

Since the topology of a DTN can vary with time, the security roles of a node can vary with time. In cases where changing security roles are not broadcast by a TTP, local nodes must infer their security roles and responsibilities from information local to the node itself.

3.3.3   Time Awareness

Time awareness in any network can be categorized into absolute and relative time functions.

Absolute time refers to the amount of time that has passed since a fixed and unchanging[9] Epoch. For many terrestrial networks, absolute time is kept using Coordinated Universal Time (UTC) and synchronized through protocols such as the Network Time Protocol (NTP). Other systems, such as spacecraft, use Mission Elapsed Time (MET), which uses a spacecraft's launch time as its fixed and unchanging Epoch.

Relative time refers to the amount of time that has passed since a local event. This differs from absolute time in that by using a local event as an Epoch for a relative timer, the Epoch is reset every time the local event recurs.

> **Monotonic timekeeping**
>
> **Monotonic systems** increment their time by 1 at regular intervals (such as milliseconds).
>
> **Non-monotonic systems** may include negative adjustments, delayed adjustments, or greater than 1 increments to synchronize with external systems.

Low-latency networks use both absolute and relative time measurements to assess the health of the network. For example, if an operation such as a Transport Layer Security (TLS) handshake times out, the network is considered to be in a (temporary) error state and the handshake can be tried again from the beginning.

The DTN architecture differs from low-latency architectures because timeliness is not used to infer network health. DTN data can be annotated with a time-to-live that represents the absolute time at which the data may be removed from the network. However, data may be delivered at any time within its time-to-live and fluctuations in delivery do not imply any more or less capability in the network.

Understanding time and timeliness in a network can provide hints on whether the network is behaving in a regular way. However, the use of non-contemporaneous paths means that DTN security mechanisms must use time differently than in other, low-latency security mechanisms.

> **DTN Timing**
>
> The DTN architecture does not use timing as a way to infer network health.

3.3.3.1 Identification

Absolute times are used by networking and security protocols to help uniquely identify messages, calculate security parameters, and ensure that systems are resilient to replay attacks. Relative time measurements can be used to help uniquely identify items between relative time periods of their initiating events. These uses of time to prevent duplication is predicated on the assumption that time monotonically increases within a system.

However, there is no foolproof notion of timekeeping. Relative time counts can reset due to counter rollover and absolute time can be adjusted forward and backwards as part of synchronizing to some reference clock.[10]

For this reason, within a DTN, time should only be used for message identification when coupled with the identity of the node asserting that time. Any absolute (or relative) time considered independently of the node generating that time makes a flawed assumption that nodes in a DTN architecture have well-synchronized clocks.

---

9 Here, *fixed* means at a specific time and *unchanging* means that this fixed time is never updated.

10 To avoid the potentially negative effects of a backwards time adjustment, NTP prefers to slow down a local clock to synchronize with a reference clock rather than adjusting time backwards.

3.3.3.2 Error Inference

Relative times are used extensively in networks to define performance requirements representing the *normal* operation of the network. If some network operation, such as receiving a positive acknowledgment from a message destination, takes too long, the operation can time-out and be seen as having failed.

This use of time to infer error is based on the assumption that the timeout value is knowable and short enough to allow an error-sensing node time to react to the error in a way that preserves the utility of the data being transmitted. The use of time to infer error conditions must be carefully considered when applied to a DTN architecture.

Some protocols, such as the Licklider Transmission Protocol (LTP) used for point-to-point communications over long distances will suspend time-out timers to account for signal propagation delays so as to not, otherwise, falsely infer error based on delays in receiving acknowledgments.

3.3.3.3 State Prediction

DTNs require that local nodes consider both the time at which a security service is added to a message and the time at which the message might be received at its destination, taking into account the likely time-variant topology.

3.3.4 Atomic Communications

High availability networks can make assumptions related to the notion of message flow between sources and destinations in a network. For example, security protocols are often decomposed into sequential phases such as an initial session establishment phase followed by a data transmission phase and, sometimes, a session shutdown phase.

Unlike high-availability networks, DTNs cannot provide the same guarantees on message order and delivery. Security protocols deployed in challenged environments may need to carry multiple kinds of information in a single message. Security protocols should treat individual messages in a DTN as an atomic, stand-alone communication rather than a part of a larger security primitive.

3.3.5 Threshold Trust

When there is no central authority of security information (such as a TTP) then trust must be derived from consensus; the alternative to central authority is distributed authority. Distributed trust protocols derive information as a function of the number of distributed authorities confirming that information. By requiring some threshold of consensus the likelihood of receiving corrupted or malicious information is reduced.

It is certainly the case that a well-motivated, well-funded, well-resourced malicious actor can take over a majority of distributed nodes in a decentralized trust system. However, if the threshold of trust is focused sufficiently on small information exchanges, and the number of distributed information centers is diverse enough, then the effort of compromising enough nodes to break the trust might not justify the relatively small amount of information affected.[11]

Examples of distributed systems include approaches such a webs of trust, blockchains, and attribute-based encryption (ABE).

---

11 In this sense, all current cryptographic mechanisms are *good enough* in that they can all be compromised given enough resources, but the expenditure of the resources required are not worth the information extracted.

3.3.5.1 Web of Trust

A Web of Trust (WoT) model requires users to provide signed information related to important information, such as security key materials. Nodes in the WoT accumulate key information from multiple other trusted and untrusted nodes in the network. By examining chain of trust relationships and where users agree/disagree on key information, a node may make inferences on the state of keys without otherwise requiring a centralized authority.

The establishment of such a WoT can be useful in the context of a non-contemporaneous paths, especially when there may be additional associations between certain paths and certain users, such as with geographically distributed information [7].

3.3.5.2 Blockchain

> **Blockchain**
>
> A series (chain) of values (blocks) whose Nth value is derived from its $(N - 1)$th value. Changes to a single value in a blockchain can be found by auditing the blockchain.

A blockchain is an ordered sequence (chain) of records (blocks) where the value of any given block is derived, in part, by the value of the previous block in the chain. Blockchains are not dissimilar from synchronized cryptographic streams, except that blockchains do not require endpoint synchronization; the state is embedded in the blockchain itself.

For this reason, blockchains may be useful when applied to intermittent networks, such as DTNs [8]. Blockchains provide two ways of thresholding trust: the information in the blockchain itself and the way in which the blockchain is known to the node.

3.3.5.2.1 Trust from the Blockchain

The contents of the blockchain represent a verifiable history of a series of information as it evolves over time. Were this information to include security information, such as key policies and information, then the contents of the blockchain would represent a intuition of multiple users (all those participating in, and modifying, the blockchain). As such, nodes could examine these chained statements relating to security policy to derive their own policy.

3.3.5.2.2 Trusting the Blockchain

Popular public uses of blockchains involve enlisting multiple distributed nodes broadcasting blockchain information around things such as cryptocurrency transactions. In this sense, malicious editing of a blockchain is made less likely because a local node could receive multiple broadcasts relating to what is considered the most popular[12] perception of the accurate blockchain.

3.3.5.3 Attribute-Based Encryption

ABE evolved from the concept of fuzzy identity-based encryption wherein a single public key could be associated with multiple private keys. In an ABE system, keys can be associated with individual attributes[13] and a node possessing all necessary attributes would be able to process (such as encrypt, decrypt, sign, and verify) information without requiring a single set of public and private keys.

Similar to a standard PKI infrastructure, ABE relies on the concept of a centralized TTP, but there exist distributed versions of ABE that may be appropriate for the DTN environment [9].

---

12 And, thus, most correct.

13 Attributes can include identifiable information associated with a node, such as the country in which it is located.

3.4 Summary

The DTN architecture represents an important new category of networking architecture that emerged as a result of the expansion of telecommunications into increasingly remote environments. These environments encompass sensor networks, underwater networks, highly congested and contested environments, and communications across interplanetary distances.

Attempting to force these environments to support the assumptions and resources needed to maintain familiar Internet networking architectures is rarely practical and often impossible. Rather than force environments to match a network design, designers must adapt the network to match the environment.

The DTN environment is stressful for every part of the networking ecosystem, including transport, routing, management, and security services. These challenges include intermittent connectivity between nodes that happens with such frequency and severity so as to regularly alter the topology of the network. These topological changes (to include network partitioning) happen so regularly so as to be part of the normal operation of the network, and not some special error case. When faced with disjoint network paths, messages will queue at nodes, require persistent storage, and congest bottlenecks.

Not every challenged network encounters every network challenge, but a DTN security solution must be adapted to operate when faced with all of these challenges. While daunting, there are some strategies to help identify potential approaches to this task. Separating out security concerns helps to increase target the reuse of other security approaches The establishment of local autonomy to infer state and apply policy and configuration as a function of that state can reduce reliance on end-to-end synchronization. Incorporating time values in security protocols can help reason about security in cases where instantaneous access to TTPs cannot be guaranteed and trust must, instead, be based on community information, instead. Finally, keeping data and metadata together in single messages creates atomic network primitives that can help nodes reason about how security can be applied to individual data units.

BPv7, as the DTN transport protocol, and its BPSec security extensions support features that allow secure message exchange in these environments. Just as these features enable transport strategies, they also enable certain security strategies. The full implications of BP capabilities, DTN challenges, and security protocols are explored in depth in Chapter 4.

References

1 Cerf, D. V. G., Torgerson, L., Burleigh, S. C., Weiss, H., Hooke, A. J., Fall, K., Scott, K. and Durst, R. C. [2007]. Delay-Tolerant Networking Architecture, RFC 4838. **URL**: https://www.rfc-editor .org/info/rfc4838.

2 Burleigh, S., Hooke, A., Torgerson, L., Fall, K., Cerf, V., Durst, B., Scott, K. and Weiss, H. [2003]. Delay-tolerant networking: an approach to interplanetary internet, *IEEE Communications Magazine* **41**(6): 128–136.

3 Birrane, E., Burleigh, S. and Cerf, V. [2011]. Defining tolerance: impacts of delay and disruption when managing challenged networks, *Infotech@ Aerospace 2011*, AIAA, p. 1645.

4 Jacobson, V. [1988]. *Congestion avoidance and control*, Association for Computing Machinery, New York, NY, USA. **URL**: https://doi.org/10.1145/52324.52356.

5 Floyd, S. and Fall, K. [1999]. Promoting the use of end-to-end congestion control in the internet, *IEEE/ACM Transactions on Networking* **7**(4): 458–472.

6 Allen, E. and Khoshgoftaar, T. [1999]. Measuring coupling and cohesion: an information-theory approach, *Proceedings Sixth International Software Metrics Symposium (Cat. No.PR00403)*, pp. 119–127.

7 Touazi, D., Omar, M., Bendib, A. and Bouabdallah, A. [2017]. A trust-based approach for securing data communication in delay tolerant networks, *International Journal of Information and Computer Security* **9**(4): 324–336. **URL:** https://www.inderscienceonline.com/doi/abs/10 .1504/IJICS.2017.087564

8 Basu, S., Chowdhury, S. and Bit, S. D. [2021]. Using blockchain in intermittently connected network environments, *Blockchain Technology and Innovations in Business Processes* p. 33.

9 Sahana, S. and Apoorva, P. [2017]. Decentralized disruption secure data retrieval in military network using CP-ABE algorithm, *2017 International Conference on Communication and Signal Processing (ICCSP)*, pp. 0519–0523.

4

Delay-Tolerant Security Architecture Elements

The Delay Tolerant Networking (DTN) architecture is implemented, in part, using the BPv7 protocol that has been designed to operate in challenged and contested environments. BPv7 requires that Bundle Protocol Agents (BPA)s understand how to process BPSec security extensions, as the required way to achieve inter-bundle security services. Therefore, the design of BPSec must take into account how to secure information within DTN design constraints.

This chapter describes those architectural elements of DTN security that are derived from analysis of network security practices and the unique properties of DTN transport.

After reading this chapter you will be able to:

- Summarize how protocol capabilities define security architectures.
- List the Bundle Protocol capabilities that make security processing unique.
- Analyze how information is fed to a security mechanism.
- Define how security architectures build resilient security policies.

4.1 Defining Security Architectures

A security architecture is a modular ecosystem, not a single protocol or implementation. These architectures comprise multiple security mechanisms designed to handle multiple security responsibilities. As discussed in Section 3.3.1, a common decomposition of responsibilities is to separate protocol structure, node policies, and cryptographic processing. Modularity in a security architecture is important because of the evolving nature of the cyber security landscape and the increasingly pervasive nature of our communications systems.

4.1.1 Evolving Cyber Threats

Existing cyber threats mature, and new cyber threats emerge, over time. This change leads to security vulnerabilities when cyber threats achieve an effectiveness or a scale beyond that which was originally envisioned when designing a security architecture. As threats evolve, the security architecture must evolve.

> **Opportunism**
>
> Threat actors watch for changes to security posture to better target the timing and types of attack.

One example of maturing existing cyber threats can be seen by examining network security attacks associated with the COVID-19 pandemic. During this time,

Securing Delay-Tolerant Networks with BPSec, First Edition. Edward J. Birrane III, Sarah Heiner, and Ken McKeever.
© 2023 John Wiley & Sons, Inc. Published 2023 by John Wiley & Sons, Inc.
Companion website: www.wiley.com/go/birrane/securingdelay-tolerantnetworks

companies switched rapidly to on-line work accommodations in ways which were not designed as part of most corporate security architectures. Consequently, there were sharp increases in cyber attacks during this time. In March 2020 security researchers identified a 600% increase in fishing attacks [1]. The World Economic Forum identified 30 000 COVID-19 related cyber attacks in the first three months of 2020 and other security firms identified a 30 000% increase in related attacks by September, 2020 [2]. During the pandemic, Google reportedly was blocking 18 million scam e-mails daily [3]. The well documented spikes in cyber threat activities around the pandemic are not unprecedented, with other spikes happening around significant events such as natural disasters and political unrest.

All of these cases represent times when some security mechanism in a security architecture might be overcome by events. Policies may be relaxed when integrating new technologies. Configurations might be changed to include less security[1] (or outsource security functions) to increase network performance.

Many of the attacks seen during the COVID-19 pandemic involved scaling existing attacks. Of more concern is the rise of new cyber threats based on emerging technologies. For example, the progress seen in developing artificial intelligence has been adapted for increasing the effectiveness of cyber attack [4]. Combinations of traditional and new attack methods make it difficult for existing security mechanisms to adapt and new methods for preventing, detecting, and minimizing these emerging attacks remains an open research area.

4.1.2 Novel Capabilities

Another benefit of modular security architectures is the ability to adapt them for new networking deployments. Where possible, existing mechanisms should be reused as they represent known behavior and capability. In the event that a new security mechanism is developed, best practices of other mechanisms should be evaluated and reused as much as possible.

The design constraints present in the Delay Tolerant Internet (DTN) architecture present unique challenges to the way in which DTN transport mechanisms can be secured. Just as a new transport protocol is needed to implement this architecture, new security mechanisms must be developed to work with this transport.

However, other existing mechanisms within the architecture not affected by delay and disruption, such as cipher suites, policies, and configurations can be reused or adapted. When new security mechanisms are developed, the requirements, design, and operational experience of other, existing mechanisms can serve as valuable guidance.

4.2 IP Security Mechanisms

As the largest, most successful internetwork in the history of humankind, it is difficult to overstate the importance of the Internet. The large number of active connections (and the volume of data communicated across those connections) requires both secure and efficient protocols.

Given the diversity of devices, infrastructure, and application data found on the Internet, multiple protocols are involved in any end-to-end communication. However, almost all communications over the Internet involves the Internet Protocol (IP), whose Protocol Data Unit (PDU) is termed the packet.

Both IPv4 [5] and IPv6 [6] (collectively referred to as IP) form what is often terms the "thin waist" of the Internet. IP represents a point of convergence between application layer protocols and link layer protocols. Its capabilities are a common foundation on which Internet communications can be based.

---

1 Such as using smaller keys or reducing the types of data encrypted in an intranet.

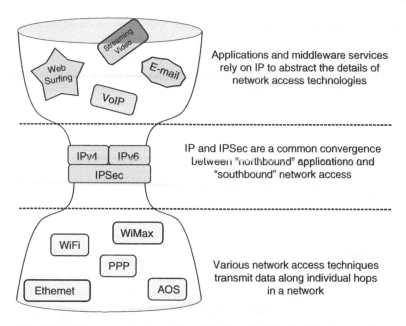

Figure 4.1 IP represents a common element of many networking stacks. The methods used to secure IP packets serve as a good reference point for other security approaches because of the ubiquitous nature of the IP protocol over the Internet.

This concept is illustrated in Figure 4.1. Any single application can communicate over wired and wireless physical networks by having the application generate IP packets (or using a networking stack that will eventually generate IP packets). Similarly, any single data link layer, such as Ethernet, can carry the traffic of multiple applications simply by supporting IP packets.

IP packets are secured using Internet Protocol Security (IPSec) extensions to packets. As the *lingua franca* of the Internet, IPSec must operate be secure and efficient to maintain high data rates and protect against diverse security threats.

The architecture and best practices associated with securing the Internet is both a densely populated and still growing topic on which multiple books can (and have been) written. For the purposes of identifying useful design patterns that can be applied to new security protocol development, the remainder of this section focuses on two important elements of IPSec: the way security services are represented in a packet, and the way in which those services are scoped.

4.2.1 Protocol Structure

One of the key differences between the IPv4 and IPv6 protocols was the expansion of the IPv6 packet structure to include multiple headers and, where necessary, trailers. The original IPv4 design implemented a single header placed in front of an existing piece of data for the purpose of informing how to route that data across networks.

As the Internet grew in complexity, IPv6 designers needed a more expressive way to identify a diversity of operating conditions. Rather than making a single, variable header[2] the IPv6 design allows for multiple, optional secondary headers. Since the purpose of evaluating IP and IPSec is to inform the evolution of

> **IPv6 Headers**
>
> IP handles network complexity by defining optional secondary headers in the packet.

2 Variable length headers make processing much more difficult, especially when considering firmware implementations of networking stacks.

new protocols for DTN security, the examples of IPSec will focus on its application to the IPv6 protocol.

Two of the new headers added to the IPv6 protocol relate to security. The IPSec Authentication Header (AH) [7] is added to a packet prior to the packet payload. The IPSec confidentiality service is applied to a packet by adding both a special Encapsulating Security Payload (ESP) header [8] before the payload and some trailers after the payload.

This design evolution is a useful one for security protocol development for several reasons.

1. Cryptographic materials are easily identified and processed by placing them in headers and trailers.
2. The presence of these headers at as a signal that other contents of the packet have security applied to them.
3. Authentication can be applied without confidentiality.

4.2.2 Security Scoping

Both the AH and ESP headers can be used to protect the entire packet or just the payload of a packet, with the security scope being set as a function of the deployment scenario. Transport mode is used to apply security in cases where the header of the packet (and any secondary headers that appear before the security headers) is understood and safe to transmit in the current network. Alternatively, tunnel mode is used to fully encapsulate a packet so that all of its headers are secured.

Figure 4.2 illustrates the use of ESP to encrypt an IPv6 packet using both tunnel and transport mode. In transport mode, the scope of security services is applied to the payload and some of the

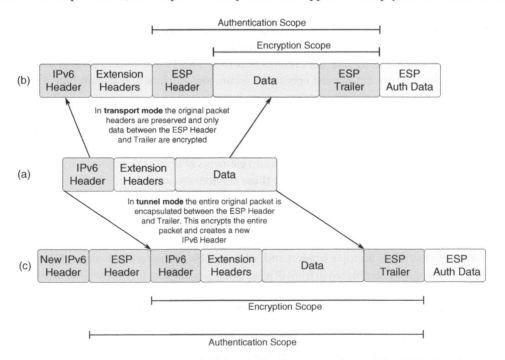

Figure 4.2 IPSec encryption adds special headers to denote security services within the packet. An IPv6 packet (a) can be encrypted in two ways. Transport mode (b) preserves IP headers for regular routing and processing. Tunnel mode (c) is used to encrypt the entire packet and adds a new packet header.

added security headers and trailers. In this mode, the primary header and any other headers prior to the ESP are not secured. Tunnel mode is different, and encapsulates the protected packet as the payload an encapsulating packet. In doing so the encapsulated packet is completely secured and not accessible until it is extracted at some downstream IPSec node.

IPSec transport mode is often used in cases where payload security is required but the header information (including information related to addressing, routing, quality of service) is needed for transport over the end-to-end network. This mode is faster to process because it involves less data processing and does not otherwise increase the size of a packet's payload. While there are multiple advantages to using this mode, the significant *disadvantage* is that the information related to packet delivery is unsecured.

IPSec tunnel mode takes a different approach by encapsulating the entire target packet as the payload of some other packet. In this way, the ESP header and trailers of the encapsulating packet apply to the entire original packet. Not only does this approach allow the entire original packet to be secured, but it also allows for the packet to be sent across different networks that may impose their own addressing, routing, and quality of service information. As such, tunnel mode is used whenever data is transmitted over the Internet or, more generally, across any untrusted network.

Two concerns with the tunnel mode of operation, relating to speed and fragmentation. Since a new packet must be created and cryptographic materials must be generated over the entire original packet, applying tunnel mode in a network takes longer. Another concern when using tunnel mode is that IP packets have a maximum size (IPv6 supports up to 64 KiB.[3]) If a received packet is at its Maximum Transmission Unit (MTU), then any encapsulating packet would, by nature, be larger than the IPv6 MTU.

IPv6 Fragmentation

IPv6 fragmentation exemplifies the difference between IPv6 and BPv7 design choices. IPv6 routers never fragment packets. When a packet is too big the packet is dropped and an error message is sent back to the packet source. This type of feedback mechanism is appropriate for environments such as the Internet, but less so when federating delayed and disrupted networks as part of the DTN architecture.

This focus on headers and payloads makes certain assumptions about the overall security architecture.

- Packet headers only carry information related to network processing.
- Packet sources know the MTU along a path and can size tunnel mode accordingly.
- If one non-security header needs a security service, all such headers need that same service.

While this approach has been very successful for IP packets on the Internet, DTN protocols encode different relationships between and amongst headers and payloads. As such, they may make different assumptions on how security services are represented and how those security services are scoped. However, the scoping of IP security services provides useful lessons learned for the development of DTN security, to include the following observations.

1. Network messages may require different scope based on their use cases.
2. A single security header can carry materials related to multiple parts of a message.
3. Security protocols must consider the impacts of fragmentation when increasing message sizes.

---

3 Networks may allow support for *jumbograms* which increase the MTU for a packet to approximately 4 GiB.

4.3 DTN Transport

The most motivating capabilities of the DTN architecture involve intelligent queuing of information to better tolerate disruptions and delays. This store-and-forward capability is applied to bundles being transported through the network. Therefore, these benefits are considered to be implemented by those mechanisms supplying DTN transport.

Focus: LunaNet

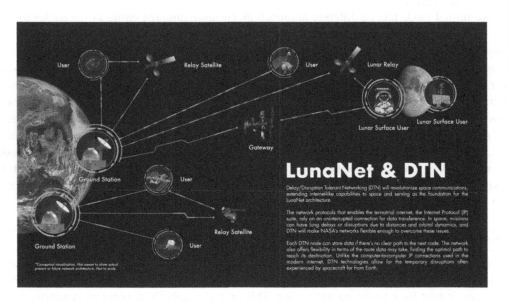

Source: NASA.

LunaNet refers to NASA's concept for a cislunar networked architecture built, in part, using DTN transport. This network is designed around a service-based architecture that would allow for human and robotic exploration of the Moon by NASA and its partners [9].

The extensible nature of this network allows multiple government and commercial organizations to incrementally build and deploy increasing complex sensor capabilities in ways that distribute the economic cost and risk of lunar spacecraft.

This diverse mixture of assets may experience the types of network partitions and delays for which the DTN architecture has been built and LunaNet is seen as one example beneficial use of DTN technologies. By proving these network concepts in cislunar space, they can be baselined for networks being envisioned around Mars and elsewhere in the solar system.

DTN transport is implemented using the BPv7 transport protocol as secured using BPSec. Just as the design of IPSec was influenced by the new capabilities of the IPv6 protocol, the design of BPSec is influenced by the unique capabilities of the BPv7 protocol. The remainder of this section provides an overview of those unique characteristics of BPv7 that influence the way in which it is secured.

4.3.1 The Bundle Protocol

Whenever possible, the responsibility for handling delays and disruptions in a network should be abstracted from the applications using that network. Special behavior required to operate in challenging networking conditions should be implemented within network layers. Just as user applications are unaware of the intricacies of Ethernet framing, they should be unaware of signal propagation delays.

When multiple applications might require delay-tolerant behavior, implementing that behavior in one place (such as in a library or in a networking driver) provides several benefits.

- Individual application designs can be simplified because they rely on the networking layer to handle delays and disruptions.
- Relying on the network to handle these conditions provides a uniform response – the system will behave predictably regardless of the applications running on the networked platform.
- Network implementations can be optimized to the platform they run on and the data links being used. Efficiency in the network is not tied to the efficiency of any one application development.

Any network application developer who has opened a Transmission Control Protocol (TCP) socket to achieve reliable, in-order data delivery can appreciate how protocols and central implementations simplify application development. Adapting to delay-tolerant behavior is not different; a tolerant network protocol such as BPv7 provides a common solution for all delay-tolerant behavior for the similar developer experience of opening a TCP socket.

> **A Common Approach**
>
> BPv7 standardizes the use of storage to tolerate networking delay and disruption.

4.3.2 Format

The BPv7 [10] PDU is the *bundle*. Each bundle is a collection of internal structures called *blocks* which represent either header or payload information.

The structure of the bundle and the description of the major fields of various bundle blocks are illustrated in Figure 4.3. Every bundle is comprised of a primary block and a payload block that represent the traditional primary header and user payload of most networking protocols. Between these two required blocks exist zero or more *extension blocks* representing additional annotative information related to the handling of the bundle, the user data within the bundle, or the network itself.

The single primary block of a bundle contains information necessary to identify and process the bundle in the network. BPv7 Bundle Protocol Agents (BPAs) use this information to make routing and quality-of-service decisions. Once created, the primary block is immutable. Optional extension blocks provide a way to annotate the bundle with additional information and handling instructions. Finally, the single payload block shares the same format as the extension block and is used to hold data being transported in the network.

For a more complete description of the BPv7 bundle format, see Appendix C.

4.3.3 BP Capabilities

BPv7 has been designed to support the design constraints inherent in the DTN architecture. In doing so, the capabilities available in BPv7 both provide new options for tolerant data transport

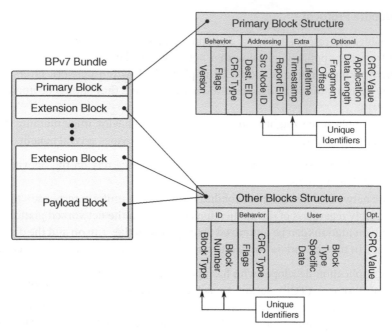

Figure 4.3 The BPv7 bundle format. Bundles are modeled as a collection of blocks. Each bundle starts with a primary block and ends with a payload block and supports zero or more extension blocks in between. BPv7 does not define any trailer blocks.

and new mechanisms that can be used for securing that transport. Similar to the way in which the design of IPv6 informed the design of IPSec, the design of BPv7 informs the design of BPSec.

The four capabilities that have the most significant impact on the design and operation of BPSec are the extension block mechanism, store-and-forward behavior, the use of convergence layers, and late binding. The remainder of this section describes each of these capabilities.

4.3.3.1 Extension Blocks

> **Extension Blocks**
>
> Extension blocks are similar to IPv6 extension headers, but more capable.

Extension blocks insert annotative information into a bundle that can be used by downstream BPAs to assist in bundle processing. These blocks are conceptually similar to IPv6 extension headers, but with some notable differences.

- Multiple extension blocks of the same type may appear in a bundle.
- Unknown extension block types may passed through a BPA unprocessed or removed from the bundle.
- Extension blocks may appear in any order in a bundle.
- Extension blocks are not restricted to carrying network information.
- Extension blocks may be secured independently of one another.
- Extension blocks may be added or removed from a bundle by intermediate nodes in the network.

The BPv7 specification documents three extension blocks that must be understood by every BPA:

1. **Previous Node**: The previous node block documents the identifier of the most recent BPA that forwarded the bundle.
2. **Bundle Age**: The bundle age block documents that the amount of time that the bundle has existed in the network.

3. **Hop Count**: The hop count block documents the number of times that the bundle has been forwarded.

In addition to these three block types, other extension blocks can be defined through standards organizations or by convention in the context of a particular network deployment.

4.3.3.2 Store and Forward

One of the most challenging design constraints associated with the DTN architecture is the need to operate in the presence of non-contemporaneous data paths. A bundle, traveling along such a path, will reach a node such that the link representing the next hop does not yet exist. At this point, the bundle must wait until the next expected next hop[4] exists at the node.

Because bundles might be waiting for their next hop for an extended period of time, and because primary storage[5] is small relative to secondary storage,[6] the bundle is likely to be stored at the BPA while it is waiting to be forwarded.

In cases where a network is partitioned and there is no known future connectivity, bundles might be stored until an appropriate transmission opportunity arises, until the bundle expires and is removed from the network, or until the local BPA makes a policy decision to remove the bundle to preserve storage for higher priority traffic.

4.3.3.3 Convergence Layer Adapters

BPv7 is added to a networking stack to implement the DTN architecture for nodes operating in challenged environments. As part of a stack, it is expected that other protocols might exist "above" and "below" the BPv7 layer. The protocol existing immediately below the BPv7 layer serves as a convergence layer for bundles

Definition 4.1 *(Convergence Layer)* A protocol that abstracts the details of lower-level processing from a user of that protocol. Convergence layers define common interfaces for network applications. For example, IP serves as a convergence layer for the Internet.

BPv7 is not designed to be the "lowest" protocol in a network stack; it expects some other protocol to handle the lower level transport concerns.[7] However, BPv7 differs from other network protocols in that the use of a particular convergence layer might change as a function of time in support of time-variant routing.

When a BPA receives a bundle for which its next link is immediately available, the bundle will be encoded for transmission over the convergence layer most appropriate for that link. However, if the same bundle were to be stored for a future transmission instead, then the convergence layer necessary for that future link would be used instead. The BPv7 specification defines the concept of multiple adapters for convergence layers so that a BPA can select amongst them when it is time to forward a bundle.

> A Convergence-Layer Adapter (CLA) is a node component that sends and receives bundles on behalf of the BPA, utilizing the services of some "integrated" protocol stack that is supported in one of the networks within which the node is functionally located.
>
> BPv7, RFC 9171 §3.1, [10]

---

4 Here, next hop refers to not only the next hop but also some alternate future hop.
5 Primary storage refers to RAM and other fast, volatile memory technologies.
6 Secondary storage refers to slower, persistent storage such as a solid-state recorder or hard disk.
7 It is even possible to have BPv7 act as its own transport protocol when encapsulating bundles within other bundles.

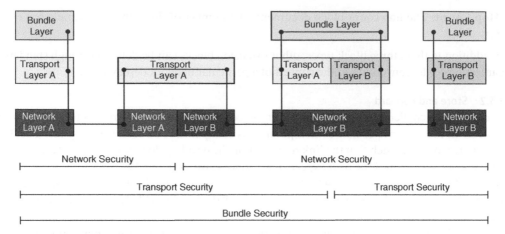

Figure 4.4 Bundles flow over multiple transport and network layers. Transport and network security coexist with bundle security.

Figure 4.4 illustrates a common network stack showing a bundle overlay that crosses two different transport domains (transport layer A and transport layer B) and two different network layers (network layer A and network layer B). Note that transport layers and networking layers may have different boundaries, resulting in different security scopes. These different scopes are outlined at the bottom of this figure, showing that network layer security, transport layer security, and bundle layer security have different endpoints in the network.

4.3.3.4 Late Binding Endpoints

Every layer of a protocol stack includes some mechanism of addressing information communicated through that stack, and BPv7 is no exception. The addressing element for bundles is termed the Endpoint Identifier (EID) in recognition of the fact that BPAs act as the endpoints of message exchanges.[8] EIDs can be unique and used to make routing decisions across a bundle network.

Defining EIDs as their own namespace differs from how transport addressing schemes work on the Internet, particularly concerning the User Datagram Protocol (UDP) and the TCP. With UDP and TCP, addressing consists of identifying a port number that is relative to the underlying IP address of the encapsulating packet. TCP and UDP messages cannot be routed solely as a function of their port numbers as these numbers are not guaranteed to be unique within the network – for example, HTTPS traffic commonly uses port 443 for many users each at unique IP addresses.

BPv7 requires EIDs to be routable addresses because a bundle may be sent over any one of several different CLAs and these CLAs may have different naming and addressing schemes. The underlying naming and addressing mechanism might not be known until the time of bundle transmission, because changes to the topology of the network might cause bundles to take a path different than the one calculated at bundle creation. This just-in-time mapping of a constant EID address to a convergence layer address is considered *late binding* because the EID is not bound to an underlying address at the time of bundle creation.

4.4 A BPv7 Model for DTN Security

A DTN security architecture must address the security challenges present in challenged networks using the capabilities of the transport protocol carrying security information. This section explains

---

8 Even waypoint BPAs use EIDs for addressing, as they may generate and consume their own bundles.

Table 4.1 BP features enable security strategies.

	Separate concerns	Local autonomy	Time awareness	Threshold trust	Atomic communications
Extension blocks	●	◑	○	○	●
Store and forward	○	◑	●	●	○
Convergence layer adapters	●	◑	○	○	●
Late binding endpoints	●	◑	○	○	●

The magnitude of the relationships between BP features and security strategies can be categorized as small (○), medium (◑), and large (●). This magnitude implies the level to which a certain BPv7 feature is required for the successful implementation of a given security strategy.

how BPv7 capabilities (discussed in Section 4.3.3) can be used with the DTN security strategies (discussed in Section 3.3).

Table 4.1 summarizes the ways in which various BPv7 capabilities can be used to implement various security strategies. The remainder of this section examines each of these relationships in more depth.

4.4.1 Extension Blocks Implications

Using extension blocks to carry information outside of a bundle's payload provides a useful mechanism for separating concerns, providing inputs to local autonomy, and allowing for atomic communications. Extension blocks allow security implementations to *separate concerns* by encapsulating multiple types of information within separate blocks in the bundle. Blocks can represent diverse types of information including annotative information relating to the payload, the bundle, the network, the user, and the link. This granularity can be used to better compartmentalize information relating to security protocol configuration, policies, and protected user data. The information carried in extension blocks can serve as an important source of information for *local autonomy* engines running at the node. This is particularly beneficial for blocks carrying network, link, configuration, or other nonuser engineering data. At scale, the set of information across all extension blocks across all received bundles can be used to infer both the state of the node and the local state of the network – even after the bundle containing these extension blocks has been forwarded, delivered, or otherwise removed. Finally, extension blocks allow bundles to support more *atomic communication*. Placing diverse data in an extension block, rather than in multiple individual bundles, reduces the overall number of bundles in the network. More importantly, it ensures that all data needed to process a bundle is kept within the bundle. In this way, extension blocks reduce dependencies on in-border bundle delivery. Securing the single bundle secures all the communication associated with that bundle.

4.4.2 Store and Forward Implications

Storing information at a node provides some resiliency to transient disruptions in a network. From a security perspective, allowing for the accumulation of bundles at a waypoint nodes has several security implications as well.

If stored bundles can be inspected by a local node then *local autonomy* engines have an additional, evolving data set. A secured bundle store can be used to inform state estimation of both the local node and those upstream nodes that previously processed the bundle. BPAs storing bundles must have a sense of *time awareness* as it relates to both network time and relative time. Time-variant routing over non-contemporaneous paths requires an understanding of both when new links are expected to arrive, and when stored bundles have expired. The bundle store also provides a mechanism for evaluating *threshold trust*. Such trust mechanisms involve information collected from multiple sources within the network over time. This information can be collected by dedicated threshold protocols, statistical analysis of metrics associated with the stored data, or by deep inspection of the data stored on a node. Persistent storage is particularly enabling for this purpose because it implies that data may be resident on the node long enough to be analyzed, versus very high-rate data passthrough that may not support deep bundle inspection and associated metric collection.

4.4.3 Overlay Implications

The BPv7 use of EIDs creates an overlay network where bundles can be transmitted over a variety of lower level constituent networks. Securing the bundle overlay follows several security practices, to include separating concerns, enabling local autonomy, and allowing for atomic communications.

Specifying routing, management, and security information using EIDs allows implementations to *separate concerns* by focusing on the operation of the bundle layer, separate from its underling protocols. Bundles can be secured regardless of whether they will, later, be secured by IPSec or some lower-level data link security mechanism. Further, any concerns associated with changes to the physical networks are abstracted by binding the same EID to different physical addresses as-needed. While a bundle might use a different CLA to be sent, it can keep the same EID and base security operations off of its EID.

The stability provided by a consistent EID allows for *local autonomy* processing based on bundle contents, independent of the underlying status of the physical network. Therefore, local autonomy can operate on the logical relationships between bundles as a function of their EIDs, independent of the current underlying network topology. Further, this separation allows local autonomy to play a larger role in the selection of how bundles are provided to convergence layers and, thus, the binding of logical EIDs to physical networks and network addresses.

Finally, this separation allows for bundles to maintain *atomic communications*. When paired with an underlying protocol, bundles may be of arbitrary sizes and not otherwise constrained to the MTU of a particular network. Allowing bundles to stay larger because underling protocols will perform services such as fragmentation and delivery enables BPAs to keep as much information as possible together in a single bundle.

4.5 Scoping Bundle Security

Bundle Security
BPv7 is the DTN transport mechanism, so DTN security is the securing of BPv7 bundles.

The design of security extensions for bundles is, as discussed earlier, influenced by the capabilities of the bundle structure and the normative behavior of a BPA. Extension blocks provide the ability to accurately identify information in a bundle. Behavioral elements of BPv7, such as store and forward, provide the ability to incorporate network-layer security at rest. The design of bundle security extensions must accommodate all of these capabilities, and more.

Definition 4.2 (*Security Scope*) The manner in which information within a bundle is packaged and provided to a security mechanism. This is separate from the manner in which a bundle is constructed, or how information is otherwise communicated in a network. Security scope focuses only on the interface between a BPA and some security mechanisms resident at that BPA.

Separate from the mechanisms available for crafting a bundle security, another important design element is the scoping of that security within a bundle. As discussed in Section 4.2.2, IPSec supports two scopes: tunnel mode and transport mode. Since there are similarities between IPv6 packet security and BPv7 bundle security, it would follow that bundle security might use similar scopes, for the following reasons.

- As with IPSec in transport mode, DTN security must provide security to a bundle payload,
- As with IPSec in tunnel mode, DTN security must provide security across extension blocks and the payload block.
- Additionally, DTN security must be applicable to some blocks (primary, extension, or payload) and not others.
- Additionally, DTN security must provide different services to different blocks.

Discussing DTN security scope is more complex than whether all PDU headers are included in security operations or not. BPv7 extension blocks carry a more diverse set of information and support different rules for how those blocks are added and removed while in the network. Bundle security must be customized to the extension blocks that are present in a bundle.

Therefore, the scopes of bundle security have more fidelity than present in IPSec tunnel and transport modes. Within BPSec, these scopes are referred to as encapsulation and augmentation.

4.5.1 Security by Encapsulation

Encapsulation refers to the practice of wrapping multiple independent items into a single encapsulating item. The technique is popular in networking (and computer science in general) for both information hiding and reducing processing complexity. Processing a single, large data item is often less computationally intensive than processing multiple small data items because processing setup and completion steps only need to be performed once.

When applied to bundle security, encapsulation refers to the practice of collecting all items in a message and presenting them as a single data item for processing by a security mechanism. The results generated from the security mechanism would relate to the encapsulated data item and not any of the individually encapsulated data items.

Definition 4.3 *(Security Encapsulation)* The application of security services to an opaque data set that is built from the union of otherwise discrete data elements. Encrypting a payload when the payload represents a structured set of data, such as the protocol data unit of some other protocol, is an example of applying security using encapsulation.

This approach to security result generation has both benefits and challenges when considered as a practice for security BPv7 bundles.

4.5.1.1 Benefits
Encapsulating the input into a security mechanism has two compelling benefits: increased processing speed and fewer special cases.

4.5.1.1.1 *Processing Speed* While the processing power of network devices continues to increase, even in constrained environments, so does the number and size of messages communicated through those networks. Maintaining these increasing data rates, even when applying security,

requires that any security solution be able to be performed quickly and increasingly with some type of hardware assistance.

Encapsulation allows for increased processing speed by reducing the time needed to compute the inputs upon which security software must act. Consider the case of a payload that consists of a PDU of some higher-layer protocol. While the payload contains structured data that *can* be differentiated, there is no need to perform this differentiation, and the entire payload can be input to security software without needing to expend compute cycles parsing it.

Similarly, consider a case where a message contains multiple secondary headers which also need some security applied to them. Since all secondary headers in a message appear between the primary header and the payload, the range of headers can be calculated rapidly and the contents of individual headers do not need to be processed. The starting position and total length of headers can be rapidly calculated and the resulting block of information copied as a single input into security software for processing.

4.5.1.1.2 *Algorithm Simplicity* Inherent to the concept of encapsulation is that multiple types of information can be simplified for security processing. This not only increases the speed at which these inputs can be conditioned, but also reduces the number of special cases that need to be considered by security processing software.

When merging all extension blocks into a single encapsulating data object, node software would not check for whether certain blocks are present or have any specific data in them. Encapsulated data is typically considered opaque for any security mechanisms operating over that data.

This simplifies the logic associated with security processing, which reduces the likelihood of processing errors, misconfiguration, or introduction of a security vulnerability.

4.5.1.2 Challenges

The notable challenge to encapsulation is that it destroys the ability to differentiate data. If distinct blocks are treated as a single opaque element, any security applied over that element might be over-scoped. This potential error is of particular concern in the DTN architecture where extension blocks carry diverse information and need diverse security services applied to them.

Overscoping security involves including information in a security operation which can decrease the overall use of the message in the network. To illustrate the challenges of over-scoping, consider the case of overscoping an authentication service and then the case of overscoping a confidentiality service.

4.5.1.2.1 *Overscoped Authentication* Overscoping an authentication service refers to one of two conditions:

1. Calculating a signature over an encapsulated item which includes both items that should not change and items that are allowed to change.
2. Calculating a signature over multiple items that might be checked by different nodes in a network with different security keys or authentication algorithms.

Consider a scenario where all extension blocks in a bundle are encapsulated into a single element over which an authentication signature is calculated. This process implies that every extension block in the bundle is now cryptographically bound to every other extension block. Any extension block change, such as adding/removing a block, changing the order of blocks, or updating a block value will cause the authentication mechanism to fail.

4.5.1.2.2 Overscoped Confidentiality Overscoping a confidentiality service also cryptographically binds potentially dissimilar information together. However, unlike overscoped authentication, overscoped confidentiality has problems even when the contents of the encapsulated data should not change.

Consider the case of a BPv7 bundle that includes different types of information in different extension blocks. Some extension blocks might carry quality of service information while others might carry semantic information about the state of the node that created the bundle to aid the processing of the bundle at its destination.

In this case, the extension block holding the quality of service information might need to be referenced throughout the network to assure that the bundle is processed correctly. However, an extension block carrying the state of the source node in the bundle might need to be encrypted to prevent other parts of the network from looking at potentially sensitive operational data.

Were all extension blocks encapsulated together and secured as one, then all header information would be encrypted. In this state, no node would be able to look at the quality of service information without decrypting all blocks at every node, which might subvert the benefit of encryption.

4.5.2 Security by Augmentation

This refers to the practice of augmenting a bundle by adding additional blocks.[9] This contrasts with the encapsulation approach because, using augmentation, all of the original blocks in a bundle are identifiable even after the bundle has been secured.

Definition 4.4 (Security Augmentation) The application of security services to one or more blocks such that the cryptographic result of those services is added to the bundle as a new block. Augmentation preserves the original bundle blocks.

Calculating two authentication signatures over two extension blocks, and the placing the resultant signature in another extension block, is an example of the security by augmentation approach. In this case, the total number of blocks in the bundle increases, and the original blocks in the bundle remain individually identifiable in the bundle.

As shown in Figure 4.5, augmentation also allows for the collection of independently-secured information in a bundle. In this illustration, the contents of multiple packets can be aggregated into a single bundle, with the security of each packet's data preserved. This type of aggregation would be impossible using the IPSec mechanisms that secure IPv4 or IPv6.

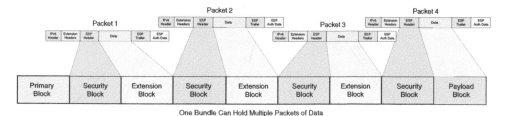

Figure 4.5 A single bundle can carry multiple sets of secured information. When multiple packets of information carry information related to a data exchange (packets 1–4) they can also be carried as extensions and payload of a bundle. The association of cryptographic materials to protected data remains the same in either approach.

---

9 In this case, blocks to carry security information.

Similar to the encapsulation approach, augmentation is a mixture of both benefits and challenges.

4.5.2.1 Benefits

Using information augmentation to scope DTN security operations has two compelling benefits: resilience and diversity.

4.5.2.1.1 Security Resilience
The overscoping of security services that can occur with encapsulation leads to a more fragile security system. Secure operations can be made more resilient by focusing them only on the blocks within a bundle that need to be secured.

This does not mean that, overall, less security is available to a bundle. Instead, it means that multiple invocations of a security mechanism run over individual extension blocks provide a granularity that allows security to be targeted for the type of block being secured.

4.5.2.1.2 Security Diversity
Diversity here means that different security operations can be applied to different blocks within the bundle. Representing a diverse set of security inputs allows different security mechanisms to receive different security mechanisms.

Figure 4.6 illustrates a case where there exist multiple extension blocks in a bundle, where some blocks require authentication only and others require confidentiality. Different security services might use different cipher suites, or similar cipher suites and different keys. Overall, this figure shows two main types of diversity for security in a bundle:

1. **Different Security Algorithms**: Different extension blocks may have different security services applied to them.
2. **Multiple Security Results**: Even amongst the set of extension blocks receiving the same security service, capturing individual security results allows different BPAs in the network to process these services as needed.

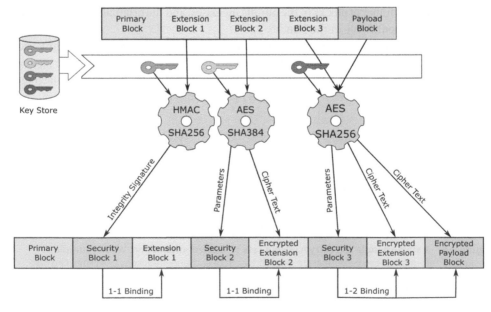

Figure 4.6 BPSec enables security diversity within a bundle. Different security algorithms can be applied within a single bundle.

4.5.2.2 Challenges

Allowing this level of flexibility in security processing brings its own challenges that are almost exactly the inverse of the benefits of the encapsulation approach discussion in Chapter 4.5.1.1. Specifically, this nuanced approach requires more processing to complete and adds complexity in the security processing at each node.

4.5.2.2.1 *Processing Impacts* Treating each block as an individual input to potentially different security mechanisms requires the additional processing of the contents of a block. This includes both syntactic and semantic processing.

1. **Syntactic processing** involves identifying the start and end point of an individual block so that the appropriate bytes can be extracted. While individual blocks within a bundle are likely to be relatively small, there may be several such blocks requiring multiple size calculations per bundle.
2. **Semantic processing** involves determining what security services are needed in what circumstances for what blocks. At a minimum, this could require parsing elements of blocks such as their type.

The computation required for this type of security application is likely higher than that needed for encapsulation schemes. The extent to which this becomes a processing burden is driven by the overall computational ability of the network node, the data throughput on the node, and the amount of time that a bundle might be resident on the node.

Very high rate devices with low computational abilities may see processing delays applying complex security. Alternatively, very low rate systems, or systems where data is stored on a node for extended periods of time, may operate on stored bundles without concern for any processing delays, because processing delays may be short compared to bundle wait times.

4.5.2.2.2 *Algorithm Complexity* Separate from the increase in processing complexity, an augmentation approach requires the setting of policy and other configuration information at nodes throughout the network. The application of per-block policies and maintaining potentially different configurations for different cipher suites adds complexity to the algorithms that must process this security information.

4.6 Policy Considerations

Security policies determine the way in which security services are applied within a network. This includes which security services are offered at different nodes, what traffic should be secured in what way, and the myriad of events that must be handled as part of that processing.

While some networks might negotiate policy in real-time, DTN transport is rarely about to guarantee that kind of closed-loop exchange. From a security perspective, real-time negotiation of the policy between endpoints cannot be made as a precondition for the application of security. Even policy must be tolerant to delays and disruptions.

This has a few implications for the design of a security architecture, related to the way in which security policy statements can be communicated in a BPv7 network.

4.6.1 Configuration

If the security policy for a particular message exchange cannot be negotiated on an as-needed basis, then that policy must exist either as a result of some preconfiguration of information prior to the need for that exchange, or the policy must be encapsulated in some message which accompanies secured messages in the network.

Figure 4.7 illustrates some ways in which policy can be communicated to a node without relying on real-time data exchange.

In-band mechanisms Figure 4.7a refer to configuration of bundle security as codified in received bundles. Policy configuration from bundles could be represented either as the payload of a bundle (e.g. configuration is the reason the bundle exists in the network) or in extension blocks that are transmitted with a payload not otherwise dedicated to configuration.

Additional mechanisms in this illustration show alternatives that are more mission and network dependent. For network devices that implement a local administrator access or console Figure 4.7b configurations may be directly updated on the node through manual administration. This type of activity is more common on data servers but can also be found in deployed network devices that are periodically collected/updated at regular intervals.

Certain nodes may support local autonomy Figure 4.7c that can be used to auto-generate configurations. For example, on-board orbit propagators on a spacecraft may generate configurations relating to spacecraft contact opportunities over time without any other input from an administrative console.

Finally, as the bundle layer often sits over some convergence layer there may exist "lower level" ways to communicate configurations to a node over a network Figure 4.7d. One example of this capability is *emergency commanding* whereby a special radio command might trigger the reset of a device whose networking layer has stopped functioning.[10]

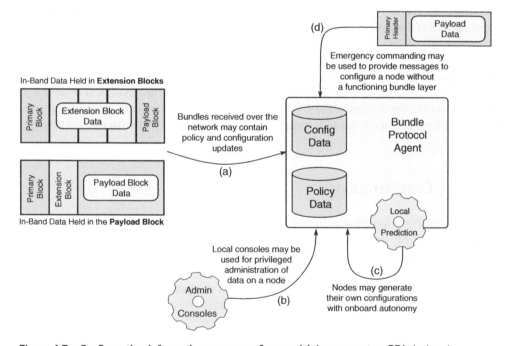

Figure 4.7 Configuration information may come from multiple sources to a BPA. In-band mechanisms for retrieving configurations to a BPA come from the bundle network itself (a). However, local console access to a BPA (b), autonomy running on the local BPA (c), and even emergency commanding to the node (d) provide out-of-band mechanisms for configuration.

---

10 A closer-to-home example of such special commanding is the Wake-on-LAN setting of many desktop computers whereby a sleeping computer can be woken upon receiving a special packet over a network.

4.6.2 Late Binding

Some security policies are driven by the characteristics of the network itself, such as the association of security roles and responsibilities derived from the topology of the network. When network characteristics change, these security policy elements must also change. While topological change can occur in any network, the likelihood is increased in the DTN architecture due to the long-lived nature of bundles and the store and forward behavior of BPAs.

Any attempt to set a static policy at the time a bundle was inserted into the network could be out-of-date at a later point, were network changes to occur over the lifetime of the bundle. For that reason, DTN security architectures need to consider a late-binding approach to security policy.

Definition 4.5 *(Policy Late-Binding)* The process by which a receiving node determines the policy to be applied to a message as a function of the configuration of the node and information in the message itself. This allows the policy to be described at the time of message receipt versus the time at which the message is created. The term late-binding is used here to indicate that the policy of a message is applied at the time when the policy is needed at a node, and not at the time the message is first created.

4.7 Summary

Modular security architectures allow for the most reuse of security mechanisms as new networks are built, and as existing networks encounter evolving security threats. Just as security architectures should not be built from scratch with each new deployment, existing security architectures should not applied to new networks without careful analysis.

One important consideration when examining security architectures is the set of capabilities present in the protocols used to transport security information across the network. As the most successful internetwork on the planet, the Internet's use of IP as secured by IPSec is a motivating exemplar to follow in this regard. The evolution of the IPv6 packet structure from the IPv4 packet structure changes the ways in which the IPSec protocol worked, and the information that it secures.

Similarly, the unique capabilities of BPv7 impact the way in which bundles can be secured. Features such as extension blocks, store and forward operation, CLAs, and the late binding of EIDs all represent an evolution of transport capabilities beyond what is typically seen on the Internet. Building a security protocol for bundles can use these capabilities to implement good DTN security practices.

An equally important consideration for developing a DTN security ecosystem is how security within a bundle can be scoped. Unlike IP, the diversity of information that can exist in BPv7 extension blocks makes whole-bundle encapsulation a poor choice for networking, Similarly, scoping akin to IPSec transport mode would treat multiple extension blocks the same, which leads to overscoped integrity and authentication. Bundle security by augmentation provides a way to avoid overscoped security, although it may increase processing times in certain networks.

Bundles also have the property that they may be stored for long periods of time at an individual node, which potentially changes the networking interfaces over which the bundle might be transmitted and the overall topology and state of the network at the time of that transmission. To support resiliency to these changes, security policies must be focused on the state of the network at the time a bundle is received, not the state of the network at the time the bundle is created. This may involve multiple ways to configure policy on bundle nodes.

References

1 Shi, F. [2020]. Threat spotlight: coronavirus-related phishing, *Barracuda Networks*, March 26, 2020. URL: https://blog.barracuda.com/2020/03/26/threat-spotlight-coronavirus-related-phishing.

2 Lallie, H. S., Shepherd, L. A., Nurse, J. R., Erola, A., Epiphaniou, G., Maple, C. and Bellekens, X. [2021]. Cyber security in the age of COVID-19: a timeline and analysis of cyber-crime and cyber-attacks during the pandemic, *Computers & Security* **105**: 102248. URL: https://www.sciencedirect.com/science/article/pii/S0167404821000729.

3 Tidy, J. [2020]. Google blocking 18m coronavirus scam emails every day. URL: https://www.bbc.com/news/technology-52319093.

4 Kaloudi, N. and Li, J. [2020]. The AI-based cyber threat landscape: a survey, *ACM Computing Surveys* **53**(1). URL: https://doi.org/10.1145/3372823.

5 *Internet Protocol* [1981]. RFC 791. URL: https://www.rfc-editor.org/info/rfc791.

6 Hinden, B. and Deering, D. S. E. [1998]. Internet Protocol, Version 6 (IPv6) Specification, RFC 2460. URL: https://www.rfc-editor.org/info/rfc2460.

7 Kent, S. [2005]. IP Authentication Header, RFC 4302. URL: https://www.rfc-editor.org/info/rfc4302.

8 Kent, S. [2005]. IP Encapsulating Security Payload (ESP), RFC 4303. URL: https://www.rfc-editor.org/info/rfc4303.

9 Israel, D. J., Mauldin, K. D., Roberts, C. J., Mitchell, J. W., Pulkkinen, A. A., La Vida, D. C., Johnson, M. A., Christe, S. D. and Gramling, C. J. [2020]. LunaNet: a flexible and extensible lunar exploration communications and navigation infrastructure, *2020 IEEE Aerospace Conference*, IEEE, pp. 1–14.

10 Burleigh, S., Fall, K. and Birrane, E. J. [2022]. Bundle Protocol Version 7, RFC 9171. URL: https://www.rfc-editor.org/info/rfc9171.

5

The Design of the Bundle Protocol Security Extensions

Prior chapters presented the challenges inherent in the Delay Tolerant Internet (DTN) architecture, the strategies for building modular security ecosystems, and the capabilities of the Bundle Protocol version 7 (BPv7) transport layer. Together, this information bounds the design of security extensions for BPv7. Bundle Protocol Security (BPSec) must operate in the most challenged environments where BPv7 might be used, be compatible with as many security mechanisms as possible, and utilize the unique features of how BPv7 information is transported in the network.

This chapter discusses the design of BPSec from the experimental protocols that came before it to the design principles that allow it to satisfy its required functions.

After reading this chapter you will be able to:

- Trace the history of the BPSec protocol from its research origins.
- Interpret lessons learned from prior experimental bundle security protocols.
- Explain the security services provided by BPSec.
- Analyze how BPSec design principles fulfill DTN security needs.

5.1   A Brief History of Bundle Security

BPSec [1] has been designed to operate using the capabilities of bundles as they are exchanged within networks conforming to the DTN architecture. The sense of Delay Tolerant Internet (DTN) security requirements and bundle capabilities, themselves, evolved through research and testing of experimental protocols. In this way, both BPv7 [2] and BPSec represent the application of lessons learned from this prior experimentation.

The Internet Research Task Force (IRTF) conducts research related to the long-term evolution of the Internet, including the specification of experimental protocol specifications. Experimental protocols are:

> …published for the general information of the Internet technical community and as an archival record of the work, subject only to editorial considerations and to verification that there has been adequate coordination with the standards process…
> The Internet Standards Process – Revision 3, RFC 2026 §4.2.1, [3]

Two important experimental Request for Comments (RFCs) were released by the IRTF relating to experimentation for the bundle protocol suite: the Bundle Protocol version 6 (BPv6), and the Bundle Protocol Security Protocol. These research works served as the precursor for BPv7 and BPSec, respectively.

Securing Delay-Tolerant Networks with BPSec, First Edition. Edward J. Birrane III, Sarah Heiner, and Ken McKeever.
© 2023 John Wiley & Sons, Inc. Published 2023 by John Wiley & Sons, Inc.
Companion website: www.wiley.com/go/birrane/securingdelay-tolerantnetworks

5.1.1 Bundle Protocol Version 6

The Bundle Protocol version 6 (BPv6) was published by the IRTF in 2007 as experimental RFC 5050 [4]. BPv6 supported many features, some of which were included in BPv7 and some which were either removed altogether or migrated out of the core specification into extension blocks.

An analysis of the evolution of bundle structure and behavior is important to understand how the concepts of delay-tolerant information exchange evolved over 15 years separating the publication of BPv6 in 2007 and the publication of BPv7 in 2022. Those protocol elements that were not carried into BPv7 must be understood to prevent including similar elements in the design of BPSec.

5.1.1.1 Changes from BPv6 to BPv7

This section describes those differences between BPv6 and BPv7 that inform the design of bundle security services. The differences discussed next include both elements of BPv6 not included in BPv7 and new design elements added to BPv7.

5.1.1.1.1 Add Node Identifiers The BPv6 specification did not have a formalized definition of a *node* separate from an *endpoint*. A BPv6 Bundle Protocol Agent (BPA) could lose its identity if it were assigned new endpoint(s). The concept that a BPv6 BPA could change its identity presented significant challenges for source-based security schemes. BPv7 introduced the concept of a *Node Id* separate from any endpoints on a BPA. BPv7 security sources can be associated with the *node* providing the security service and, thus, less impacted by changes to the endpoints resident on a given BPA.

5.1.1.1.2 Remove the EID Dictionary The BPv6 primary block included an optional dictionary data structure mapping EID values to compressed integer offsets. The dictionary concept allowed blocks within the bundle (to include the primary block itself) to reference EIDs more compactly using dictionary offsets. Replacing multiple instances of a verbose EID with smaller dictionary references resulted in smaller bundle sizes. However, this also meant that the primary block could change hop-by-hop as new EIDs were added to the bundle.

Dictionaries presented several security issues. End-to-end integrity could not be calculated over a changing dictionary, and changing the dictionary affected other blocks in the bundle. An encrypted block that included a dictionary reference would have no way of knowing if the Endpoint Identifiers EIDs were changed along the bundle path. Upon decryption, the EID pulled from the dictionary reference might be different than the EID that was placed in the dictionary at the time the block was encrypted. This would be a hard-to-detect, significant security vulnerability. Generally, avoiding use of the dictionary concept was an early lesson learned with BPv6.

5.1.1.1.3 Immutable Primary Block The dictionary structure was not the only element of a BPv6 primary block that could be modified. Other elements, such as the current custodian EID of the bundle were also removed. The BPv7 primary block is, intentionally, immutable to support end-to-end integrity both over the primary block itself, and any other bundle blocks that should be cryptographically bound to that primary block.

5.1.1.1.4 Added Optional CRCs Not every integrity service requires cryptographic materials. In certain network environments, integrity is only used to detect non-malicious data corruption. BPv7 extension blocks support an optional Cyclic Redundancy Check (CRC) to identify block corruption. These CRCs are not signed, but also do not require the data overhead of a BPSec security block or the processing overhead of a cipher suite implementation.

5.1.1.1.5 Allowed Block Multiplicity BPv6 only allowed a single instance of a given block type to exist in a bundle. If a waypoint BPv6 BPA needed to add an extension block to a bundle that already had an extension block of that type, the BPA could only append its information to the existing extension block. This becomes an issue if a BPv6 BPA signs or encrypts a block of a particular type and another downstream node needs to add information to that block.

BPv7 avoids this issue by defining a *Block Number* separate from the block type. No two blocks in a BPv7 bundle may have the same block number, but two blocks may have the same block type. This allows BPAs to add their own instances of a block type to a bundle.

Learning from BPv6

BPv7 applied lessons learned from over a decade of experience using the BPv6 experimental protocol. BPv6 pioneered several useful features, such as the extension block and store-and-forward semantics. It also included certain features that made processing difficult, like EID dictionaries. BPv7 added missing features such as node identifiers, CRCs, and allowed for multiple instances of a block type in a bundle.

5.1.2 Bundle Protocol Security Protocol (BSP)

In 2011, four years after the publication of RFC 5050, the Bundle Security Protocol (BSP) was also published by the IRTF as experimental RFC 6257 [5].

The BSP addressed perceived integrity and confidentiality needs for a BPv6 bundle operating within the DTN architecture. As an experimental protocol, BSP was implemented by a variety of researchers and potential users, to include those looking to deploy BPv6 on embedded devices [6].

Similar to BPv6, a series of lessons learned were collected regarding the BSP specification and the general nature of DTN security.

5.1.2.1 BSP Benefits

The experimental BSP protocol included multiple innovations that were included in the design and use of the BPSec, including the reliance on extension blocks to carry cryptographic material and support for integrity and confidentiality services.

5.1.2.1.1 Block Granularity The extension block concept represents an important way to communicate information for both BPv6 and BPv7. The BSP used extension blocks to both represent security services and the targets of those security services. Defining security blocks allowed security to be added to a bundle even after the bundle was created.

Cryptographic material in the experimental BSP was carried in dedicated extension blocks, with some security blocks dedicated to cryptographic operations over the payload, and other security blocks dedicated to such operations over other extension blocks. By separating cryptographic material for different target blocks, the experimental BSP enabled the use of different cipher suites to secure different blocks in the same bundle.

5.1.2.1.2 Unique Services The experimental BSP defined different security blocks for integrity and confidentiality over the bundle payload. By supporting different block types, these services could be differentiated to expedite bundle processing.

5.1.2.1.3 Encapsulated Parameters and Results Many cipher suites include parameters that are applied at a security source and must be understood at a security acceptor. Some of these parameters can be communicated in plaintext and others must be encrypted for communication (or are presumed to be pre-shared over the network). Similarly, the results generated by cipher suites must be carried in an efficient and attributable way.

The experimental BSP used the concept of parameters and results within individual security blocks to minimize the changes to the bundle payload. The results of integrity mechanisms, such as signatures, could be stored in a security block and, therefore, not complicate the structure of the bundle payload. Cipher suite parameters such as session keys, nonces, and other values could similarly be kept in these blocks.

5.1.2.2 BSP Lessons Learned

As an experimental protocol, BSP also supported design choices that complicated implementation and testing efforts. This section describes the lessons learned that directly impacts the design of BPSec.

5.1.2.2.1 Decouple Security and Routing Security blocks in the experimental BSP supported an asserted "security destination." Supporting such a field implies that the security destination is different from the bundle destination. Populating such a field requires that the security destination be known in advance. If the security destination is not the bundle destination, then it must be a node along the path that a bundle will take through the network. Therefore using this field couples security and routing.

Given the time-variant topology of the DTN architecture, tying security to a specific route can lead to several problems. One common problem with this approach is illustrated in Figure 5.1. This figure illustrated encrypting a message until it reaches one of two border node (Nodes D_1 and D_2) into a secure enclave. Routing at the security source calculates a path through D_1, so D_1 is set as the security destination (Figure 5.1a). However, topology changes break the original path (Figure 5.1b) and the bundle is received at the alternate border note, D_2 instead (Figure 5.1c). The bundle passes through D_2 without decryption and reaches the destination still encrypted (Figure 5.1d).

The BPSec intentionally omits the concept of a *security destination* relying, instead, on downstream nodes to determine whether they should or should not be accepting security services in a bundle. In cases where a bundle source (or security source) needs to ensure that a particular BPA is the network is passed-through, a tunneling approach can be used. Tunneling practically decouples security and routing in that security blocks have no path encoding in them. Instead, bundles are encapsulated by other bundles and the source and destination of the encapsulating bundle becomes the endpoints of the tunnel.

5.1.2.2.2 Avoid Special Payload Processing The experimental BSP defined two security block types specific to the payload – the Payload Integrity Block (PIB) and the Payload Confidentiality block (PCB) – as well as a single block type for extension blocks – the Extension Security Block (ESB). This implied that security services over a payload would be different than security services over an extension block.

However, information in an extension block could have the exact same security needs as information in a payload block. That realization, coupled with BPv7 defining the payload block as having the same structure as an extension block, removed the need for special payload-specific security blocks.

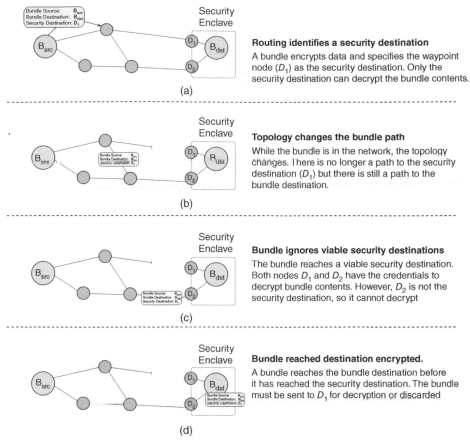

Figure 5.1 Bundle security cannot be tied to routing. The time-variant topology of the DTN architecture will change the routes taken by long-lived bundles. Associated security processing with a bundle path (a) will cause problems when the path changes (b) and a different candidate security node receives the bundle instead (c). In that case, a bundle may arrive at its bundle destination before its security destination (d).

5.1.2.2.3 Security Scoping The experimental BSP placed very few restrictions on the dependencies between security blocks and their targets. Multiple unrelated PIBs could be added to a bundle containing different signatures over the payload. Multiple unrelated PCBs could be added to the bundle representing nested encryption of the payload. In general, an integrity security block could be used to hold a signature over a confidentiality security block which, itself, could be encrypting some other integrity security block. Limits were needed on how much nested security scoping should be allowed within a bundle.

Limit Integrity Scope Allowing overlapping integrity mechanisms introduces verification issues when multiple mechanisms do not agree. This situation is illustrated in Figure 5.2 where two integrity blocks carry two different signatures over a payload. One signature added by Node 1 and a second added by Node 2. A receiver at Node 3 would need to process nine different possible outcomes, shown in Table 5.1. When both integrity mechanisms yield the same result (verify or reject) the behavior of Node 3 is clear. In cases where the mechanisms either disagree or one or both expected mechanisms are missing from the bundle, determining the correct course of action is much harder.

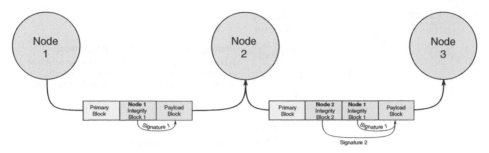

Figure 5.2   Redundant signatures over a block lead to processing ambiguity. In this simple example, policy at Node 3 would have to handle nine distinct use cases associated with whether the integrity blocks existed or not, and whether their signatures verified or not.

Table 5.1   Handling dual integrity signatures.

Node 1 Block	Node 1 Signature	Node 2 Block	Node 2 Signature	Node 3 Action
✓	✓	✓	✓	Verify
✓	✓	✓	∅	Unknown
✓	✓	–	–	Unknown
✓	∅	✓	✓	Unknown
✓	∅	✓	∅	Reject
✓	∅	–	–	Unknown
–	–	✓	✓	Unknown
–	–	✓	∅	Unknown
–	–	–	–	Unknown

Security policy at Node 3 would need to handle cases of present (✓) or missing (–) blocks and valid (✓), invalid (∅), or missing (–) signatures. This complexity grows exponentially with each allowed overlapping security operation.

Limit Confidentiality Scope   A similar situation exists if multiple confidentiality blocks were applied to the same target, such as a bundle payload. This situation is illustrated in Figure 5.3. In this illustration, were Node 1 to encrypt the payload and were Node 2 to separately encrypt the payload then Node 3 would need to decrypt in a particular order, perhaps by examining state in the two confidentiality blocks. However, if Node 2 *also* encrypted the Node 1 confidentiality block, then Node 3 would not have the information necessary to decrypt in the proper order. While this example is shown for the payload, it would apply to the multiple encryption of any bundle block.

5.1.2.2.4   Rethink Hop-by-Hop Authentication   The BSP defined a Bundle Authentication Block (BAB) that could be used to hold a per-hop integrity mechanism over a bundle as a whole. The intuitive benefit of the BAB presumes that the bundle is unmodified from the time at which the bundle is transmitted to the time at which the BAB is processed. However, there are some cases where BAB processing is made difficult or impossible in a network.

Topology Changes   If a BAB's integrity mechanism hard-codes who the receiver of the BAB will be, then the BAB has coupled routing and security. If a bundle is stored for a period of time at a

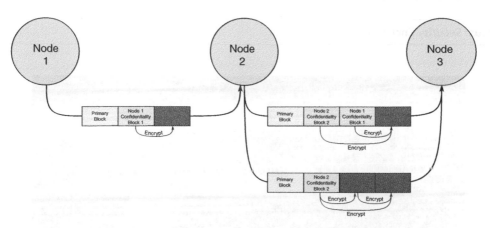

Figure 5.3 Multiple encryption requires ordered decryption. When nodes encrypt the same information, a destination must carefully decrypt in the same order as the contents were encrypted, but the encryption process may hide the information needed to preserve this ordering.

sender such that it is transmitted to a different next hop than anticipated, either the entire BAB must be recalculated just-in-time or the next hop might not have the proper configuration to verify the bundle contents.

Fragmentation If a bundle is fragmented after the BAB is added, then the bundle might not be reassembled until it reaches the bundle destination. At that point, the bundle might be several hops away from the BPA that added the BAB. This may cause the BAB to fail its own processing of the bundle, similar to the topology change described earlier.

5.1.2.2.5 *Preserve Secured Blocks* The BSP encrypted extension blocks in a bundle by replacing the encrypted blocks with the ESB. In this way, the existence of extension blocks would be hidden from a BPA. This type of complete encryption might be useful in certain circumstances, but results in multiple ESBs. Multiple ESBs increases the difficulty of determining decryption order and whether important extension block types are missing in a bundle. With BSP it is impossible to signal that an extension block exists but is otherwise encrypted.

5.1.2.2.6 *Support Multiple Targets* The design of BPv6 did not include a way to differentiate multiple extension blocks of the same type in a bundle. Therefore, only one instance of each block type is allowed in a bundle. In cases where applications might define an extension block type such that there might be more than one instance of the block in a bundle, the experimental BSP would be unable to specify how security would be applied to one such block independent of another.

5.1.2.2.7 *Tolerate Changes to Block Ordering* Because of the nesting associated with certain security block interactions, BSP security blocks were required to be included in a specific order after the primary block in a bundle – thus forming a *stack* of chronologically applied security operations. However, both BPv6 and BPv7 do not guarantee block ordering between a bundle source and destination and, thus, security block ordering created an artificial requirement not otherwise present for a bundle protocol agent.

Focus: Security Enclaves

Source: *Christina Morillo/Pexels.*

The term security enclave refers to the conceptual establishment of a portion of a computer network with a different security posture than the network at large. These enclaves attempt to have firm boundaries such that traffic within the enclave is treated separately than traffic outside of the enclave.

Corporate intranets and data centers are examples of entities that establish security enclaves. Security within an enclave may rely on local resources such as business credentials and may involve closer tracking of data into and out of the enclave. For performance reasons, data within an enclave may be unencrypted.

Home wireless networks are considered by their users to be "secure" for internal traffic, such as between a home laptop and a home file server. However, the tendency to intermingle external and internal traffic on home networks makes them poor examples of enclaves.

This concept of a security enclave can also exist in the DTN architecture. This is a primary reason why security acceptors in a network might differ from bundle destinations. The nodes in a DTN architecture that process security mechanisms may be acting as border nodes between trusted and untrusted portions of the network or internetwork.

5.2 Design Principles

Analysis of both the BSP benefits and lessons learned were foundational to the design of BPSec. This analysis identified five design principles that needed to be incorporated into the design of the BPSec. These design principles are as follows.

1) Block-Level Granularity
2) Multiple Security Sources
3) Mixed Security Policy
4) User-Selected Security Contexts
5) Deterministic Processing

The mapping of BSP benefits (from Section 5.1.2.1) and lessons learned (from Section 5.1.2.2) to these five design principles is provided in Table 5.2 and each of these principles is discussed in detail in the remainder of this section.

5.2.1 Block-Level Granularity

Security extensions for BPv7 must acknowledge the varied nature of the blocks comprising a bundle. Extension blocks might carry different data (with different consumers of that data) than the required primary and payload blocks. These different types of information might persist in a bundle for different periods of time and types of information co-located in a bundle may need to be secured separately.

> **BP Block Receivers**
>
> Bundle destinations are associated with bundle payloads. Individual extension blocks may be used by BPAs other than the bundle destination.

Examples of the types of information that can be found in an extension block are illustrated in Figure 5.4 and include the following.

- **User data** includes both the user-supplied payload block and any extension blocks carrying user-supplied information. Protecting user data is the fundamental purpose of a secured network, and networks exist for the user data they transmit.
- **Bundle metadata** refers to information needed to properly handle the bundle during its time in the network. This information is not directly associated with user data but may be associated

Table 5.2 BPSec design flows from lessons learned.

	Multiple security sources	Mixed security policy	User-defined security contexts	Deterministic processing
Block granularity	●	◐	○	○
Distinguished services	○	◐	●	●
Encapsulated parms/results	●	◐	○	○
Decouple security and routing	●	○	●	○
Avoid special payload processing	●	○	●	○
Tolerate block ordering changes	○	●	○	○
Rethink hop-by-hop authentication	●	○	●	○
Preserve secured blocks	●	○	●	○
Support multiple targets	●	○	●	○

The magnitude of the impact of lessons learned on BPSec design constraints can be categorized as small (○), medium (◐), and large (●). This magnitude implies the level to which a certain BPSec design principle was developed because of experience with the BSP.

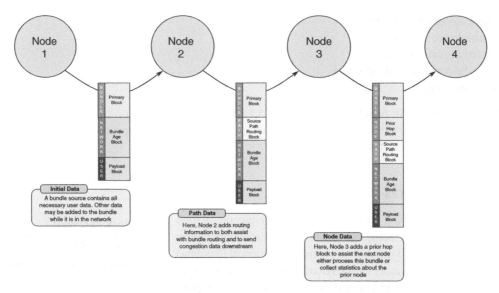

Figure 5.4   Bundles can carry a variety of data. BPAs may add node, path, and other network information to the user data associated with a bundle. Annotating existing bundles allows networks to communicate more information with less traffic overhead.

with how a particular set of user data is allowed to consume networking resources. Securing bundle metadata protects against the improper utilization of network resources.

- **Node information** informs other BPAs about the local state known to a node. Conceptually similar to the low-latency network practice of synchronized session information, these extension blocks can propagate node information downstream in the BPv7 network. Securing node information protects against the misuse or misattribution of information in the network.
- **Path information** includes information collected about the paths between messaging endpoints. This information allows nodes to collect information about the path between endpoints for the purpose of inferring timing information, routing information known to traversed nodes, and congestion predictions. This information is sometimes encapsulated as payloads for specific measurement protocols, but can also be carried in extension blocks for other traffic. Securing this information protects the way in which network resources are scheduled and utilized as a function of their predicted availability.
- **Network information** includes the union of node and path information for multiple nodes, to include information that was predicted, asserted, or otherwise discovered. Similar to path information, these data can be carried either as the payload of specific network measurement and configuration protocols or as blocks within a bundle.

5.2.2   Multiple Security Sources

Any BPA may add, or remove, extension blocks from a bundle. BPAs add extension blocks to assist with communicating information across the network and other BPAs remove extension blocks when they determine they are the appropriate receiver of that information.

Definition 5.1 *(Security Source)*   A BPA that adds a security service to a bundle. In particular, the security source of a given security operation is the Node containing the BPA that added that security operation to the bundle. Similar to a bundle source, a security source must be associated with a network node.

Blocks in a bundle should always have some type of protection whenever the bundle may be communicated through an untrusted portion of the network.[1] However, this does not imply that all security sources in a bundle are equivalent to the bundle source.

> **Security Sources**
>
> A security source for a security operation in a bundle might be different from the bundle source.

There are three occasions where a BPA might add a security block to a bundle (and thus be a security source of a block in that bundle): when a block is added, when an unsecured block is received, and when an existing block is processed and updated. Examples of each of these operations at a single BPA are illustrated in Figure 5.5.

1) **New Blocks**: The most straightforward case of a BPA becoming the security source is when it adds some extension block to a bundle and determines that this added block needs BPSec protection. This is the case for any security block added at the bundle source.

2) **Existing Blocks**: A BPA may become the security source of security on an existing block if some prior BPA (such as the bundle source) failed to protect a block when it was originally added to the bundle. Unsecured blocks may be present if there is a configuration issue at the originating node or if the node receiving the bundle represents some boundary between less-secured and more-secured portions of a network. Consider a case where a network might only apply integrity protections for local data and choose to encrypt that data only at some boundary node. In this case, the bundle source might only add an integrity service to some target block whereas the border BPA may add a confidentiality service over the same target block.

3) **Updating Blocks**: In rare cases, BPAs may alter the existing contents of a target block, requiring that any existing security service for that block be processed/removed and a new security service applied to the modified target block instead. This operation is problematic because it destroys any preexisting security on the block. More often, security blocks may choose to add new blocks of the same type to a bundle rather than updating existing blocks. For example, the Previous

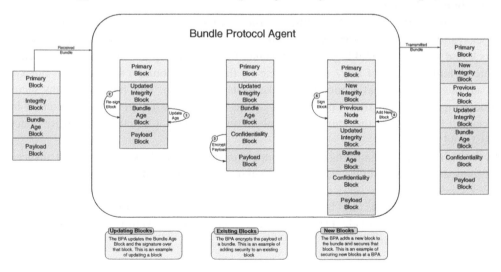

Figure 5.5 Multiple security sources coexist within a single bundle. A BPA receiving an existing bundle might add security blocks when updating and re-signing an existing block (steps 1 and 2), when protecting an otherwise unprotected block (step 3), or when adding a new block (step 4) that needs protection (step 5).

---

1 With rare exception, any network should be considered an untrusted network.

Node block is not updated on each hop – the received Previous Node block is processed/removed as part of receiving a bundle and a new Previous Node block is added prior to forwarding that bundle.

5.2.3 Mixed Security Policy

Every security block in a bundle, regardless of its source, represents an expression of security policy. If multiple security blocks exist within a bundle, they do so because multiple security policies have been applied to that bundle. Therefore, BPSec must accommodate for the expression and co-existence of multiple security policies.

Definition 5.2 *(Security Policy)* The determination that a given BPA must add, check, or process security associated with one or more blocks in a bundle. By default, the bundle destination is the acceptor of any bundle security block. All other cases of security block processing occurs in accordance with policy as configured at network nodes.

While any individual node can support a set of security policy statements, it is the entire ecosystem of such statements across a network that provide security blocks with the decentralized semantic information they need to function properly. Any individual BPSec security policy statement must address three areas related to the processing of security blocks: security block manipulation, role identification, and event handling.

1) **Security Block Manipulation**: Security policy is used to determine what security services are required to exist in a bundle. Because security blocks are the manifestation of a security service in a bundle, policy statements must exist that determine when a security block should be added to a bundle and when a security block should be removed from a bundle. When adding a security block to a bundle, policy must specify how the security block is configured and what other blocks in the bundle they target.

2) **Role Identification**: BPAs have different roles as they relate to the processing of security blocks (and, thus, security services) for a bundle. A BPA's security role is specific to the security block and the bundle, making the establishment of a security policy a multi-BPA effort. The time-variant topology of BPv7 networks means that a given bundle source might not know in advance the path that a bundle will take to its destination. This means that any security processing other than that which occurs at the bundle destination must be asserted by a receiving BPA and not the source BPA. Role identification as a function of security policy is illustrated in Figure 5.6.

3) **Event Handling**: The handling of security operations events (such as successful and unsuccessful processing) may be different as a function of node capabilities, bundle characteristics, target block type, and security service offered. BPv7 defines a flag-based mechanism for specifying how a bundle as a whole (and blocks individually) should be treated in certain circumstances. Security policy must provide a similar enumeration of related event types and configuration reactions to those events.

5.2.4 User-Defined Security Contexts

Generally, cipher suite selection occurs within some larger *security context*. Over the terrestrial Internet, the Transmission Control Protocol (TCP)/Transport Layer Security (TLS) example presume a context of low-latency, high-availability, security-session-based information exchange.

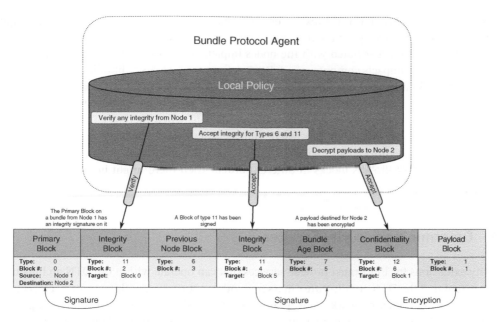

Figure 5.6 Security processing is driven by local security policy. A BPA might act as the security source, verifier, or acceptor of blocks in a bundle as determined by information known locally to the BPA.

These same assumptions might not hold for other networks in which BPv7 might be deployed. BPv7 networks might find themselves operating within some other security context.

Multiple cipher suites can be used to generate cryptographic material to be carried in bundles. Generally, support for multiple cipher suites is considered to be a beneficial feature of security protocols used in multiple domains that, themselves require different cryptographic approaches.

In low-latency networks, cipher suite selection typically happens as part of some round-trip negotiation occurring during the establishment of a security session. For example, cipher suite negotiations that happen as part of a TLS handshake happen within the context of establishing a (TCP) session with a remote node.

BPSec may be deployed in scenarios where, like TLS, low latency data exchange is possible and some end-to-end information can be synchronized. It is also possible to deploy BPSec in scenarios where, unlike TLS, there is never a guaranteed round-trip exchange between endpoints prior to securing bundles. Because of the overall diversity of BPSec deployments, it is not sufficient to simply select which cipher suites are to be used for the creation and consumption of cryptographic material. Some statement on the context within which the cipher suite is being used must also be part of the information carried with the bundle.

The design of BPSec must include some mechanism by which users can signal the context in which security services (and cipher suites) are being applied. In cases where BPv7 networks support low-latency negotiations, BPSec should enable this behavior. However, BPSec must also capture cases where BPv7 networks cannot support this type of low-latency exchange.

5.2.5 Deterministic Processing

Block-level security provides BPv7 bundles with increased fidelity and flexibility, but this benefit comes at the cost of additional complexity. Where possible, BPSec must both preserve block granularity and reduce the processing complexity.

Processing complexity, in this context, refers to both avoiding ambiguity and bounding computation speed. Computation speed is a function of the compute capabilities, architecture, and other implementation details associated with the nodes implementing BPv7 and BPSec. To the extent that protocol design impacts compute speed, BPv7 defines the formats of bundles and the formats of extension blocks and, thus, the format of BPSec security blocks.

Unlike compute speed, the design of BPSec can, and must, prevent ambiguities in the application of security services. Ambiguity in BPSec processing happens when there is a question about the order in which security blocks are processed, and the extent to which block dependencies exist within a bundle. If security blocks within a bundle have dependencies among themselves, and if nodes in a network process blocks in different orders, then there might be inconsistencies in security processing.

5.3 Determining Security Services

Security extensions for BPv7 must provide services that can operate within the DTN architecture. To the extent that the DTN architecture differs from other networking architectures, DTN security services differ from other network security services. This section provides an overview of general security capabilities and how they are influenced by the DTN architecture, followed by a discussion of the two security services that must be implemented by BPSec.

5.3.1 General Security Capabilities

Security extensions, such as those provided by BPSec, exist to enable network security capabilities. Applications using a network presume that certain security capabilities exist and the mechanisms implementing the network's security ecosystem provide those capabilities. In any modular security ecosystem there is rarely a one-to-one mapping between a security mechanism (such as BPSec) and a general security capability. These capabilities are more often achieved by multiple security mechanisms working together.

There are a few conventional listings of security capabilities that need to be examined in the context of securing BPv7 networks. Perhaps the most well-known is the *CIA Triad* of Confidentiality, Integrity, and Availability, but others include Non-repudiation, Authentication, and Authorization. Together, these capabilities describe the desired behavior of a network security ecosystem.

Security Capabilities

- **Confidentiality** hides message contents from unauthorized nodes.
- **Integrity** ensures messages were not modified since creation.
- **Availability** provides networking resources when needed.
- **Non-repudiation** provides proof of message delivery.
- **Authentication** indicates data was sent from the claimed sender.
- **Authorization** restricts access to nodes with proper permissions.

5.3.2 Out of Scope Capabilities

The extent to which security capabilities can be implemented (and how they are defined) is driven by the nature of the networks they secure. The unique characteristics of the BPv7 transport mechanism impact how these capabilities are valued within the DTN architecture. In particular, concepts such as Availability, Non-repudiation, Authentication, and Authorization are considered differently due to the potentially delayed and disrupted nature of BPv7 networks.

5.3.2.1 Availability

One defining characteristic of the DTN architecture is that the network may be constantly partitioned; network paths are non-contemporaneous and messages must use store-and-forward techniques. This concept of store-and-forward changes the definition of network availability.

In the most generic terms, network availability means protecting the network so that networking resources are able to be provisioned by approved users. For non-delayed networks, the resources being made available are processing and bandwidth. The network is considered available if a node has the compute resources to process a message and the bandwidth to send that

> **Staying Available**
>
> Network availability means preserving resources for approved network users.

message. Attacks such as denial-of-service impact availability by overwhelming the compute and bandwidth resources of a node.

Availability in the DTN architecture is interpreted differently – bandwidth resources may not be present because the network is naturally partitioned. When using store-and-forward techniques, the network is considered available if there is enough local storage to hold a message until a future transmission opportunity. Denial-of-service attacks against BPv7 nodes would need to exhaust storage on the node, not just temporarily remove bandwidth from the node.

5.3.2.2 Whole Bundle Authentication

Applying security capabilities such as authentication to an entire message assumes that all of the information in the message is required for the processing of that message. While this is a reasonable assumption in non-BPv7 networks, bundles might carry helpful but nonessential information in extension blocks. The corruption of such blocks may or may not be seen as an issue with forwarding, delivery, or other processing of the bundle.

For this reason, applying an authentication mechanism[2] over an entire bundle is both unnecessary and possibly detrimental when bundles contain either optional blocks or changing blocks.

5.3.2.2.1 Optional Blocks
Optional blocks represent cases where extension blocks carry helpful but optional information, such as information related to network link state. If information in the block was corrupted, other blocks in a bundle can still be verified and processed without issue. The concept of *optional* blocks is, in part, the rationale for the BPv7 *Discard block if it can't be processed block processing control flag*. Blocks with this processing flag set can be removed from the bundle without requiring that the entire bundle be removed from the network.

5.3.2.2.2 Changing Blocks
Changing blocks represent cases where the contents of an extension block are expected to change while in the network. While the Primary and payload block of a bundle should not change between a bundle source and destination, other blocks may. Examples of changing blocks include the Previous Node and Bundle Age blocks, both of which are expected to change at every bundle hop in a network.

Were a network to apply whole-bundle integrity or authentication in the presence of optional or changing blocks, then the bundle would fail to verify after the first hop in a network (due to

---

2 To include the use of a signed integrity mechanism for authentication.

block changes) or at any hop where an optional block was corrupted. Attempting to calculate a whole-bundle mechanism at each hop is computationally intensive and, in the case of authentication, would prevent carrying anything originally signed by the bundle source.

Whole-bundle authentication is not included as a BPSec service for the following reasons.

1) The concept of authenticating transmission between adjacent nodes implies that a transmitter understands its next hop enough to use the appropriate authentication mechanism. In a low-latency network where Protocol Data Units (PDU)s are built just-in-time for transmission this is a safe and common assumption. However, in a BPv7 network, a bundle may be constructed and secured before it is transmitted. While a bundle is stored, a next hop may change as a function of the dynamic topology of the network.

2) The definition of "next hop" is ambiguous in an overlay network as nodes that are adjacent to each other in an overlay may not be adjacent in the physical network. To fully protect against per-transmission corruption, whole-PDU authentication mechanisms should be handled by the network layers performing the transmission.

3) The corruption or loss of individual blocks may be inconsequential to the processing of a bundle because blocks may include different types of information. However, whole-bundle authentication would fail in the event that an unnecessary block is corrupted during bundle transmission.

4) The generation of security results for individual bundles can be cryptographically bound to other blocks in a bundle. For example, when encrypting a payload block, other blocks in a bundle can be included as additional authenticated data. In this way, users can select which blocks must be authenticated to properly interpret a specific secured block.

5.3.2.3 Whole Bundle Non-repudiation

Even in the challenged, time-variant environments in which a BPv7 bundle might be deployed there may need to be some indication that information was received. However, and similar to the analysis of whole-bundle authentication, the question remains: what information needs to be acknowledged? When a bundle contains optional or changing blocks, then bundle non-repudiation is very difficult to define.

For this reason, non-repudiation of an entire bundle, from a security perspective, is not a dedicated security service supported by BPSec. This does not mean that BPv7 networks cannot support acknowledgements or other proofs of delivery. The BPv7 specification includes status reports and other mechanisms to enable signaling delivery of a bundle at a destination. However, there is not a required definition of "bundle" in this context – a destination may signal a bundle is received even when the bundle contains fewer, more, or different blocks than present at the bundle source.

5.3.2.4 Resource Authorization

Similar to the concept of DTN architecture availability, the scope of DTN authorization extends to include permissions associated with the use of on-board resources like storage and compute cycles. These authorizations are important not only for the proper use of security mechanisms, but also for the regular application of any bundle processing.

For this reason, resource authorization is considered part of the configuration, policy, and local autonomy used to implement BPAs in a network. While security extensions such as the BPSec play an important role in the communication of this information, authorization, itself, is not a BPSec security service.

5.3.3 BPSec Capabilities

BPSec provides two fundamental security capabilities relating situations where blocks must have integrity or *both* integrity and confidentiality. The BPSec does not provide a mechanism for confidentiality without an integrity mechanism.

The general requirement for integrity in all situations is, in part, adhering to norms for best security practices. But, also, recognizes that storage resources may be a carefully managed resource in a store-and-forward network. Adding integrity mechanisms to all BPSec blocks enables, in some circumstances, the use of integrity verification by BPAs with the intent of removing corrupted traffic early before it is allowed to consume network storage,

5.3.3.1 Plaintext Integrity

Certain blocks in a bundle may need to be inspected by other BPAs in a network. One important example of such a block is the bundle's primary block, which BPAs must review to understand how to route and prioritize the bundle in the network. While such blocks may need to be viewed they must also be protected from change.

The BPSec plaintext integrity service can be applied in situations where a block must be protected from modification but otherwise be processed by random BPAs in a network. This protection is not a problem unique to bundles and integrity checks are built into multiple layers of a networking stack. To understand the value of an integrity mechanism at the bundle layer, consider some of the several sources of block corruption in a bundle, such as from processing errors, environmental conditions, and intentional manipulation.

- **Processing errors** occur when software or hardware unintentionally changes data values. Data corruption due to these kinds of errors is commonplace, particularly when working with extremely high-rate data transfers and sensitive electronics. This type of processing error cannot be detected by link-layer integrity mechanisms because the error can be made by the software processing the bundle *before* the bundle is sent to the link layer for transmission/forwarding.
- **Environmental conditions** can alter data as a function of the conditions of the medium. This can result in corrupted data regardless of the otherwise proper operation of the nodes that transmit and receive the data.
- **Intentional manipulation** changes data through misconfiguration or manipulation. Most often, manipulation can be considered as malicious attempts to change data as it traverses a network.

Figure 5.7 illustrates the situation where a bundle has a single block corrupted as it is transmitted from *Node*1 to *Node*2. If multiple blocks in a bundle are groups behind a single integrity mechanism, such as provided by an underlying link layer (Figure 5.7 – "No Bundle Integrity") or by a single operation over multiple blocks (Figure 5.7 – "Multi-Block Bundle Integrity"), then the entire bundle might be discarded if any part of the bundle is removed. This situation can be avoided by calculating integrity over individual blocks or collections of related blocks (Figure 5.7 – "Per-Block Bundle Integrity").

The use of the term plaintext in this context indicates data that is input to some integrity mechanism. This plaintext might be encrypted by some other process, or it might represent cleartext that has never been encrypted at all. In other case, this input data is not encrypted by the BPSec plaintext integrity service.

> **Plaintext and Ciphertext**
>
> **Plaintext** refers to any input to a cipher suite, regardless of whether it was previously encrypted. **Ciphertext** refers to the output of a cipher suite.

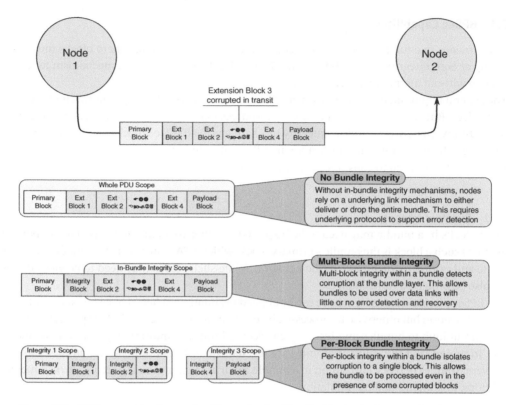

Figure 5.7   BPSec provides fine-grained integrity checking for blocks. By finely tuning integrity protections to individual blocks in a bundle, BPAs can detect exactly where a bundle has been corrupted and continue processing data when the corrupted block is not essential for overall bundle processing.

5.3.3.2   Authenticated Confidentiality

When blocks in a bundle must be hidden from unapproved BPAs in a network, the contents of those blocks must be encrypted. BPSec authenticated confidentiality operates by replacing the plaintext contents of a block with calculated ciphertext and, additionally, applying some integrity mechanism over that ciphertext (and possibly other associated data). Similar to the cases shown in Figure 5.1, authenticated confidentiality should be scoped similarly to plaintext integrity, and for similar reasons.

When applied by BPSec to a target block in a bundle, confidentiality is achieved by replacing portions of the original contents of the target block (plaintext) with the encrypted output of a cipher suite (ciphertext).

The BPSec requires the use of *authenticated* confidentiality, which means applying an authentication mechanism over information related to the confidentiality service. This authentication includes, at a minimum, the ciphertext produced by a cipher suite, but may also include other, non-encrypted data that must be considered unmodified before ciphertext can be allowed to be decrypted. Cipher suites with these properties are referred to supporting Authenticated Encryption with Associated Data (AEAD).

Table 5.3 BPSec services implement some, but not all, security capabilities.

	Plaintext integrity	Authenticated confidentiality
Confidentiality	○	●
Integrity	●	○
Availability	◐	◐
Non-repudiation	◐	◐
Authentication	◐	◐
Authorization	○	○

BPSec capabilities are a part of the DTN security ecosystem. BPSec directly (●) implements certain security capabilities while indirectly supporting (◐) others. Some security capabilities are outside of the BPSec scope (○) and must be handled by other networking mechanisms.

5.3.3.3 BPSec Services and Capabilities Mapping

The capabilities of plaintext integrity and authenticated confidentiality support some, but not all of the general security capabilities discussed in Section 5.3.1.

The integrity and confidentiality capabilities of BPSec implement the DTN security needs for integrity and confidentiality. However, both of these BPSec capabilities also indirectly support other DTN security needs. Non-repudiation and authentication are both enabled by signed integrity mechanisms over either plaintext or ciphertext – both with the option to include additional authenticated data. Availability is indirectly supported by detecting corrupted data early and removing corrupted blocks and bundles from the network.

> **AEAD Cipher Suites**
>
> Cipher suites that verify the integrity of both ciphertext and other plaintext as part of a successful decryption process.

BPSec does not implement an authorization mechanism, but authorization mechanisms deployed in a BPv7 network may be securely communicated through the network using BPSec (Table 5.3).

5.4 Protocol Comparisons

The design of BPSec has been influenced by the constraints imposed from the DTN architecture, the unique capabilities of BPv7, and lessons learned from the BSP. This design was compared with the design of existing, operational protocols to see whether specific BPv7 features and BPSec capabilities have beneficial precedent in other domains. The protocols examined in the design of BPSec include three transport protocols (IPv6 [7], TCP [8], and QUIC[3] [9]) and three security extensions (IPSec, TLS [10], and Datagram Transport Layer Security [DTLS][11]).

The BPSec design features that were compared include placing security information in headers, allowing nodes to add headers to messages, bundling different types of data in a single message, securing information over multiple segments, and allowing the network to select different cipher suites. The review of this analysis is provided next.

---

3 As standardized in the IETF, QUIC is not an acronym but the full name of this sessionless transport protocol.

Table 5.4 BPSec borrowed features.

	IPv6	TCP	QUIC	IPSec	TLS	DTLS
Security in headers	○	○	○	●	●	●
Nodes add headers	●	○	○	○	○	○
Data bundled	○	○	●	○	○	●
Segmented security	○	◑	○	◑	◑	◑
Selectable ciphersuites	○	○	○	●	●	●

The level of reuse of protocol features can be categorized as no reuse (○), partial reuse (◑), and significant reuse (●). Reuse in this context refers to design reuse, and does not imply that features reused for BPSec are syntactically compatible with other protocol features.
Source: Adapted from Birrane and Heiner [12].

- **Security in Headers**: Placing security information in special headers is a common construct among terrestrial Internet protocols. For example, IPSec defines both an Authentication Header (AH) and an Encapsulating Security Payload (ESP) header. Similarly, both TLS and DTLS include security information in headers around the application data being secured.
- **Nodes Add Headers**: The concept of expressive secondary headers is a significant design difference between the IPv4 and IPv6. The ability to add extension headers to IPv6 increases both the flexibility and capability of the protocol. While this increases the complexity of IPv6 processing, expressive secondary headers are beneficial and well used features.
- **Data Bundling**: Several protocols operating over the terrestrial Internet require a round-trip workflow where information must be synchronized at endpoints. A significant speedup offered by protocols such as QUIC (and the DTLS protocol which can be used to secure it) is the bundling of information to reduce the number of round-trip messages in a network.
- **Segmented Security**: The ability to treat different portions of user data differently is explored in other protocols. As a streaming protocol, TCP segments are, by definition, segments of user data which can be secured using TLS and a similar observation can be made with DTLS. While IPSec headers can apply to whole packet payloads, by their position they do not cover other IPv6 headers. However, only BPSec provides this level of individual security within a single message.
- **Selectable Cipher Suites**: Most security protocols provide a mechanism for selecting the most appropriate cipher suite. This allows security to be configured to the compute capabilities and security features of a specific node, but also limit the set of cipher suites to allow for interoperability.

Table 5.4 summarizes areas where important aspects of BPSec either follow or diverge from the designs of other protocols. From this table, most important elements of BPSec have some precedent in other protocols, though none of these pre-existing protocols provide all of the BPsec features.

5.5 Summary

The design of BPSec was influenced by a combination of lessons learned from experimental security protocols, expressed needs for operating in DTN environments, and observations of useful features from existing protocols.

The experimental BSP proposed for BPv6 introduced several valuable features that persisted into BPSec, to include the application of distinguished security services applied to individual bundle blocks. Some experimental features of the BSP were avoided in the BPSec based on implementer experience. These included the coupling of security and routing and several issues related to dependencies between and among blocks in a bundle. Ultimately, the experimental BSP was an important historical precursor to BPSec as it proved many of the techniques needed to secure a BPv7 network.

The BPSec design is built around five guiding principles relating to how bundles must be secured in a network. Security must be able to target individual blocks and BPAs other than the bundle source may both add blocks to a bundle and add security to blocks in a bundle. Since different BPAs may augment the bundle, different policies may be present in the network at once, representing different behaviors and using different cipher suites. A significant new concept provided by BPSec is the reliance on security contexts – a topic covered in greater detail in other chapters. Finally, this overall set of security processing must result in deterministic application of security; the BPSec cannot be so flexible as to result in non-uniform processing behavior in the network.

In line with these principles, another aspect of BPSec design is its provided security services. While there are multiple services present in an overall security ecosystem – to include confidentiality, integrity, availability, authentication, and authorization – BPSec only needs to provide a small subset as part of a bundle. The two security services of plaintext integrity and authenticated confidentiality provide the basis for implementing other security capabilities in a network.

Many of the features applied to the design of BPSec are inspired by similar features from other protocols. Common protocols deployed over the terrestrial Internet, such as IPv6 and IPSec, TCP and TLS, and QUIC and DTLS, allow for security in headers, nodes to add headers, bundling data together, and applying security to portions of user data with selectable cipher suites. While none of these protocols performs the same function as BPSec, their use of BPSec design elements provides some endorsement of BPSec's unique combination of abilities.

References

1 Birrane, E. J. and McKeever, K. [2022]. Bundle Protocol Security (BPSec), RFC 9172. **URL:** https://www.rfc-editor.org/info/rfc9172.

2 Burleigh, S., Fall, K. and Birrane, E. J. [2022]. Bundle Protocol Version 7, RFC 9171. **URL:** https://www.rfc-editor.org/info/rfc9171.

3 Bradner, S. O. [1996]. The Internet Standards Process – Revision 3, RFC 2026. **URL:** https://www.rfc-editor.org/info/rfc2026.

4 Scott, K. and Burleigh, S. C. [2007]. Bundle Protocol Specification, RFC 5050. **URL:** https://www.rfc-editor.org/info/rfc5050.

5 Farrell, S., Weiss, H., Symington, S. and Lovell, P. [2011]. Bundle Security Protocol Specification, RFC 6257. **URL:** https://www.rfc-editor.org/info/rfc6257.

6 Tselikis, C., Poulakidas, A., Aggelis, A. and Ladis, E. [2013]. An efficient implementation of the bundle security protocol for DTN-enabled embedded devices, *Journal of Applied Mathematics and Bioinformatics* **3**(1): 163.

7 Hinden, B. and Deering, D.S. E. [1998]. Internet Protocol, Version 6 (IPv6) Specification, RFC 2460. **URL:** https://www.rfc-editor.org/info/rfc2460.

8 Transmission Control Protocol [1981]. RFC 793. **URL:** https://www.rfc-editor.org/info/rfc793.

9 Iyengar, J. and Thomson, M. [2021]. QUIC: A UDP-Based Multiplexed and Secure Transport, RFC 9000. **URL:** https://www.rfc-editor.org/info/rfc9000.

10 Rescorla, E. [2018]. The Transport Layer Security (TLS) Protocol Version 1.3, RFC 8446. **URL**: https://www.rfc-editor.org/info/rfc8446.

11 Rescorla, E. and Modadugu, N. [2012]. Datagram Transport Layer Security Version 1.2, RFC 6347. **URL**: https://www.rfc-editor.org/info/rfc6347.

12 Birrane, E. and Heiner, S. [2020]. A novel approach to transport-layer security for spacecraft constellations.

6

The BPSec Security Mechanism

The BPSec security mechanism refers to components existing within individual BPv7 bundles and at Bundle Protocol Agents (BPAs) in a BPv7 network. This mechanism implements the design of security extensions for BPv7 networks and is an important part of the overall DTN security ecosystem.

This chapter explores each of the components that comprise the BPSec mechanism, how this mechanism might be expanded upon in the future, and how this implementation satisfies the desirable properties of a bundle security solution.

After reading this chapter you will be able to:

- Identify the key components of the BPSec mechanism.
- Trace the lifecycle of a security operation.
- Explain how security contexts are represented in security blocks.
- Describe the differences between integrity and confidentiality blocks.
- List the concerns associated with adding other security block types.

6.1   The BPSec Mechanism

A mechanism is a system of individual components working together to accomplish a common purpose. This concept well describes the security extensions for BPv7 – the BPSec *mechanism* is implemented as a series of individual components that work together to secure information in bundles.

The components comprising this mechanism are Security Operations (SOps), security contexts, and security blocks. SOps describe the relationship between blocks in a bundle and the security services that act on those blocks. Security contexts generate and process the cryptographic materials associated with those SOps. Security block carry information about security contexts and the results of SOps in bundles.

Part of the proper operation of these components is their ability to be configured as a function of the local data and policies of the Bundle Protocol Agents (BPAs) in the network. While policy and configuration are an important part of the overall DTN security ecosystem, they are not themselves part of the BPSec mechanism. As such, policy and configuration are covered in greater detail in other chapters.

Securing Delay-Tolerant Networks with BPSec, First Edition. Edward J. Birrane III, Sarah Heiner, and Ken McKeever.
© 2023 John Wiley & Sons, Inc. Published 2023 by John Wiley & Sons, Inc.
Companion website: www.wiley.com/go/birrane/securingdelay-tolerantnetworks

6.2 Security Operations

An SOp represents the application of a security service to a target block within a bundle. As such, these operations represent the atomic unit of security within the BPSec. BPAs implement security by adding, verifying, or accepting operations within the network.[1]

6.2.1 Notation

To assist with the description of policies as they relate to SOps, BPSec defines an abstract notation to compactly describe the relationship between security services and the target blocks they operate on. Shown as Eq (6.1), the SOp notation uses a function-based syntax to envision an SOp as a function of a security service and the target of that service.

$$OP_{sec}(Service_{sec}, Target_{sec}) \tag{6.1}$$

This notation is useful for describing relationships between and among SOps in a bundle, but is otherwise too limited to represent the processing of SOps at a BPA. For example, this notation does not include parameters associated with security services or whether a BPA would be adding, verifying, or accepting the operation. Therefore, the OP_{sec} notation is only used as a shorthand for operations existing in a BPv7 bundle and otherwise abstracted from how BPAs process those operations.

While limited, the utility of having a notation can be shown by describing security applied to a bundle with a primary block and a payload block. A likely common security application would be to calculate an integrity mechanism over the primary block[2] and encrypting the contents of the payload block. These two distinct SOps would coexist in a single bundle and could be compactly notated as shown in Eqs. (6.2) and (6.3).

$$OP(Integrity, Block_{primary}) \tag{6.2}$$

$$OP(Confidentiality, Block_{payload}) \tag{6.3}$$

6.2.2 Security Operation States

SOps have a simple lifecycle describing the duration that their security service exists in the bundle. This simple lifecycle can be interpreted as a series of states that the operation can find itself in, relating to how it has been processed by a BPA.

These states include the following.

- **Inserted**: Any SOp present in a bundle is considered to be in the *inserted* state for as long as it stays resident in the bundle.
- **Rejected**: An SOp that has been removed from a bundle.
- **Accepted**: An SOp is accepted when it is verified by a BPA and its target block has its security service removed.

As discussed later in Section 6.2.4, there is a potential one-to-many mapping of SOps to security blocks within a bundle. As such, the *lifecycle* of an SOp might differ from the lifecycle of any individual security block carrying that operation. Similarly, the lifecycle of a security block might differ from the lifecycle of a bundle.

---

1 The determination of the role a BPA plays in any given SOp is given as a matter of BPA security policy and configuration.
2 Encrypting the primary block would make the bundle difficult or impossible to route. Generally, unless tunneling, security services over a primary header are restricted to authenticating and not hiding.

6.2.2.1 Inserting Security Operations

Whenever a BPA adds a security service to an existing bundle, it creates an SOp associated with the application of that service to a target block in the bundle. From that point onward, the SOp is considered to remain in the *inserted* state as long as it remains represented in the bundle.

An SOp in this state has the following properties.

- The security operation is represented in the bundle. In particular for confidentiality, this means that the target block of this security operation remains encrypted.
- The security operation has persevered up to the current point in the network. This means that the BPA inserting the security operation was able to produce appropriate cryptographic material and that this material has survived[3] all evaluation by any prior inspecting BPAs in the network.
- The SOp is still required for the bundle. A future BPA might expect (or require) this SOp for proper processing.

Timing Security Operations

A BPA might create a bundle (or add blocks to an existing bundle) long before the bundle is transmitted. BPAs must decide whether security is applied when the bundle/block is created or when the bundle is ready for transmission. It is **strongly recommended** (but not required) that SOps be applied on bundle block creation to enable security at rest.

6.2.2.2 Rejecting Security Operations

BPAs in a network might choose to evaluate SOps in a bundle as a matter of their local security policy configuration. This verification might happen in an attempt to remove corrupt information from the network early. This verification might happen as part of final acceptance of an SOp before removing it from the bundle.

If the verification fails, then the SOp may be rejected. Rejecting an SOp involves processing the SOp, security block, target block, and possibly the bundle itself in accordance with policy. At a minimum, this processing requires that the SOp be removed from the bundle.

Rejection can happen for a variety of reasons, to include syntactic manipulation of the bundle blocks associated with the SOp, or semantic issues arising from security policy misconfiguration. Rejection based on configuration can occur when a BPA determines it is required to verify an SOp, but does not support the configuration necessary to interpret the cryptographic material carried by that operation.

6.2.2.3 Accepting Security Operations

While any waypoint BPA in a network may attempt to verify an SOp, only one BPA may accept that operation. Acceptance can be performed by any BPA that represents the point at which an SOp is no longer needed. While this must happen at a bundle destination, it may also happen at other BPAs in the network.

Acceptance of an SOp involves both successful verification and removal of the cryptographic materials from the bundle. For integrity mechanisms, removal of cryptographic information simply involves removing integrity results (such as signed hashes) from the bundle. Removal of cryptographic information for confidentiality mechanisms involves replacing associated ciphertext in the bundle with its original (recovered) plaintext.

Once accepted, the SOp is removed from the bundle. With the exception of handling errors, this is the only way in which an SOp is removed from a bundle.

---

3 It is possible that a waypoint BPA might fail to verify a security operation but choose to allow it to persist in the bundle in the event that a subsequent BPA with different configuration would be able to verify the service.

Table 6.1 BPSec security service and target uniqueness.

	Same service	Different service
Same target	✕	~
Different target	✓	✓

Security operations are uniquely identified by their service and target block, therefore certain combinations of (service, target block) are either always allowed (✓), always disallowed (✕), or sometimes allowed (~) to preserve uniqueness and avoid processing ambiguity.

6.2.3 Uniqueness

> *Security operations in a bundle MUST be unique; the same security service MUST NOT be applied to a security target more than once in a bundle.*
>
> *BPSec, RFC 9172 §3.2,* [1]

An SOp must be uniquely identifiable using only the combination of security service and target block – no two SOps may share these attributes. This uniqueness prevents ambiguities in how security blocks would be processed.

There are four combinations of services and target blocks that represent the different scopes of SOps, as illustrated in Table 6.1.

The various combinations of SOps are both discussed next and illustrated in Figure 6.1.

6.2.3.1 Same Service. Same Target

Applying the same service to the same target block would introduce processing ambiguities. For this reason, it is the only combination of security service and target block that is disallowed by the BPSec.

Consider an SOp that wants to sign the payload block twice, as represented by

$$OP(Integrity_1, Block_{Payload})$$

$$OP(Integrity_2, Block_{Payload})$$

If these two SOps were to coexist, there would be processing ambiguity were one operation to verify the integrity of the payload and another to fail to verify the payload. If a BPA verified $Integrity_1$ but rejected $Integrity_2$, would $Block_{Payload}$ be considered valid or not?

6.2.3.2 Same Service. Different Targets

Applying the same service to multiple target blocks in a bundle is allowed, as doing so does not introduce an ambiguities when processing at a BPA. Consider the case where the same security service, $Integrity_1$ is applied to two different extension blocks resulting in the following SOps in the bundle.

$$OP(Integrity_1, Block_{ext1})$$

$$OP(Integrity_1, Block_{ext2})$$

If a BPA verifies one SOp and rejects another, there is no confusion over what parts of the bundle are affected. For this reason, this formulation is allowed by BPSec.

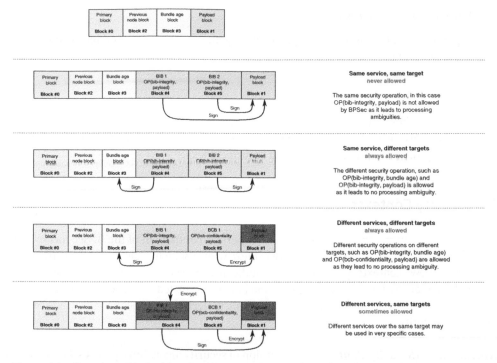

Figure 6.1 Security operations in a bundle must remain unique. Security operations (carried in BPSec security blocks) may target Other Security Blocks (OSBs) in a bundle as long as these operations do not introduce processing ambiguity.

6.2.3.3 Different Services. Same Target

Applying different security services to the same target block is also allowed in certain circumstances that would not otherwise result in ambiguity. The dependencies between and among blocks, to include circumstances where the same block can be the target block of different security services is covered in more detail in Chapter 8.

Consider a situation where one BPA adds an integrity service to the payload block and another BPA wishes to encrypt the payload block. In this case, the bundle would have at least the following two SOps.

$$OP(Integrity, Block_{Payload})$$

$$OP(Confidentiality, Block_{Payload})$$

These operations are unique, but require a deterministic processing order to avoid ambiguity. Namely, the payload block would need to be decrypted prior to attempting to verify the plaintext integrity of the payload block.

6.2.3.4 Different Services. Different Targets

Similar to the case of the same service applied to different target blocks, different services may also be applied to different target blocks without any loss of uniqueness.

For example, choosing to sign the primary block and, separately, encrypt the payload block would result in the following operations being added to the bundle.

$$OP(Integrity, Block_{Primary})$$

$$OP(Confidentiality, Block_{Payload})$$

These operations can be applied and processed together without any ambiguity and can also coexist within a bundle.

6.2.4 Bundle Representation

While BPSec describes security in terms of SOps, there is no such concept in BPv7. SOps, to the extent that they exist in a bundle, must do so in the context of an extension block. The extension blocks that carry BPSec information are called security blocks.

6.3 Security Contexts

> **Security Contexts**
>
> Users may define new contexts describing the behavior, assumptions, and cipher suites used to implement bundle security.

The second important component of the BPSec mechanism is the *security context* within which security services are applied. Context here indicates the set of assumptions, configurations, algorithms, and behaviors related to the generation and processing of cryptographic material.

Definition 6.1 *(Security Contexts)* The union of security algorithms, policies, behaviors, and user-configurable values associated with the population, analysis, and consumption of BPSec security blocks. A security context not only specifies the cipher suites used to generate and process cryptographic materials but also contains specialized parameters and other results that can be used customize the behavior of bundle block processing to match a specific scenario or situation.

6.3.1 Scope

All SOps represented within a bundle occur under the auspices of a security context, and the specifics of this security context must be captured in the security block representing these operations. The security context information captured in the security block identifies the source of the SOp, the context being used, and which parameters (if any) are used to customize the context.

The implementation of security context behavior, and its integration into BPAs, is an important part of the overall BPSec mechanism. This includes associating security contexts with policy statements for SOps sourced by the local BPA and resident in bundles received by the BPA.

6.3.2 Moderation

It is expected that there will be a variety of security contexts that can be used to generate the cryptographic materials carried by the BPSec. However, these contexts cannot be defined in an ad hoc manner, as they are instrumental to both the syntactic and semantic processing of the security block.

Therefore, the first element of security context information captured in the security block is the well-known identifier of the security block itself. BPAs can evaluate this identifier to determine whether it has the appropriate information necessary to process the contents of the security block. If the BPA does not implement the identified security context, it cannot process the security block.[4]

---

4 This condition may or may not be an error, based on associated security policy.

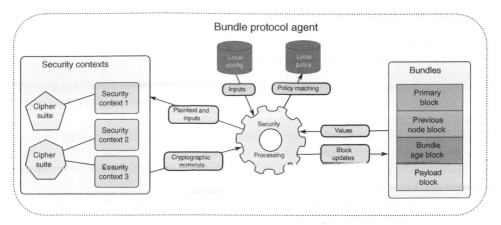

Figure 6.2 Security contexts provide behavior in addition to cryptographic processing. A BPA might implement multiple security contexts. Each of these contexts may use the same or different cipher suites for cryptographic processing. Security processing on the BPA relies on security contexts to implement specialized behaviors.

6.3.3 Application

Figure 6.2 illustrates the interaction between cipher suites, security contexts, and general BPA security processing. In this illustration a BPA must provide security services in a bundle. Security processing relies on local configuration and policy statements to determine what SOps need to be in a bundle. Values from the bundle are passed to appropriate security contexts, which, themselves, use one or more cipher suites for cryptographic processing.

6.4 Security Blocks

The third significant component of the BPSec mechanism is the security extension block. Any annotative information carried in a BPv7 bundle is done so using extension blocks, and security annotations are not different in that respect.

BPSec relies on the capabilities of extension blocks to represent and communicate security information across BPAs in the network. A security block is structured as any other BPv7 extension block, meaning that each security block has a block number and block type, block processing flags, an optional CRC value, and some block-type-specific data.

The presence of a security block indicates that an SOp exists within a bundle. In this way, security blocks signal to a BPA information about the processing and format of other blocks in the bundle. A security block is never processed only on its own – there is always a target block associated with the security block.

Security configuration and security context information must be carried in the bundle to help BPAs process SOps. Security information must be carried in its own dedicated block to help other BPAs apply their own security services. Security results, such as integrity signatures or authentication tags, may also be carried in security blocks to make sure that target blocks are not changed in size[5] as they are secured.

---

5 Keeping bundle blocks to a deterministic size helps to enable the creation of hardware for block processing.

6.4.1 Security Block Features

Security blocks as used by BPSec support certain important behavior which may differentiate them from other blocks in a bundle, as described in the following text.

1. There may be instances of the same block type in a bundle and, thus, block type is not sufficient to uniquely identify a security block. If a bundle has multiple security services it may have multiple security blocks.
2. Security blocks are not meant to be processed alone – they are coupled to their target blocks. This means that a security block cannot be processed if their target block is absent from the bundle. This also means that processing actions of security blocks may alter the contents (or existence) of a target block in a bundle.
3. The failure to process a security block may have consequences for the delivery of the bundle itself. If bundles cannot be secured, they may be removed from the network.
4. The BPSec supports multiple security block types – there is not a single security block type for all SOps. Specifically, the BPSec defines a block type for plaintext integrity and another for authenticated confidentiality.
5. While all BPv7 BPAs can understand BPSec blocks, they are not all required to process security blocks. The decision as to whether a security block is processed or not is a function of security configurations and security contexts.

6.4.2 Security Operation Aggregation

There may exist a one-to-many relationship between SOps and security blocks. A single security block many contain multiple SOps (to reduce block overhead) whereas, to enable more deterministic processing, each SOp can be traced to exactly one security block carrying its information.[6]

If there were to be a single security block for each SOp, then information that is not specific to a single SOp would be repeated in the bundle. This repetition would needlessly increase the size of the bundle.

Examples of information that is not tied to a specific SOp (and thus may be shared by multiple operations in a block) include the following.

- **Extension Block Fields**: The fields of a BPv7 extension block, with the exception of the block-type-specific data, is independent of the SOp(s) within the block-type-specific data. The block type, block number, block processing flags, and any CRCs can be shared by all the operations in a security block.
- **Security Context Information**: Security context information in a security block carries special parameters used to configure the cipher suites that generate the security results either carried within a security block, or used to replace portions of the security target block. If the configuration of a cipher suite is unchanged for multiple SOps – such as using the same key to sign multiple security target blocks – there is no need to repeat this information in every security block.

To reduce the amount of redundant information in a bundle, multiple SOps can be represented by a single security block when *all* of the following special conditions are true.

1. Each of the security operations apply the same security service, which allows the same security block type to represent all of them.

---

6 In certain circumstances a security block might be replicated when fragmenting a bundle, but each replicated security block is a single security block – with a single block number – when the bundle is reassembled.

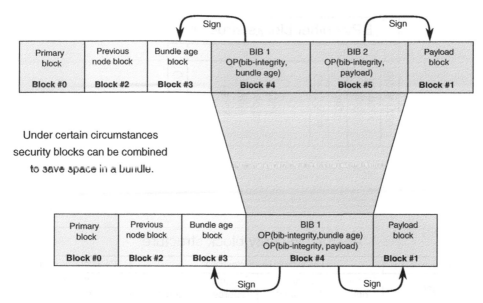

Figure 6.3 Merging Security Blocks. A BPA should merge security blocks whenever possible to avoid carrying redundant information in the bundle.

2. Each of the security operations use the same security context configured with the same context parameters.
3. Each of the security operations have been added by the same security source, meaning that they are all being added by the same protocol agent.
4. None of the security operations share the same security target block. Since the operations all are applying the same service, if they were to apply that same service to the same target block that would violate the uniqueness requirement for these operations.
5. None of the security operations being added match any security operations that already exist in the bundle in a different security block.

The concept of combining multiple SOps is illustrated in Figure 6.3. In this figure, a BPA might have originally created two security blocks (*BIB* 1 and *BIB* 2). If these blocks represent different SOps sharing similar security information, they could be more compactly represented in a single security block.

When each of these conditions is met, a security block can be constructed that preserves, once, all of the shared information and only includes per-security-operation-information for each SOp.

The uniqueness requirement for SOps is unaffected by how these operations might be packaged (or repackaged) into security blocks. While there might be multiple security blocks in a bundle, there must not be any disallowed SOps in the bundle. Similarly, the removal of security blocks from a bundle occurs a part of removing SOps.

6.4.3 The Abstract Security Block

Security blocks share a common block-type-specific data structure termed the *Abstract Security Block* (ASB), illustrated in Figure 6.4. The ASB is considered *abstract* in the sense that the ASB is not, itself, a type of security block, but rather a common structure used by other BPSec security blocks.

The ASB encapsulates the SOp(s) signed by the block. Some of this information is based on information shared by all operations in the block, and other information is specific to each operation.

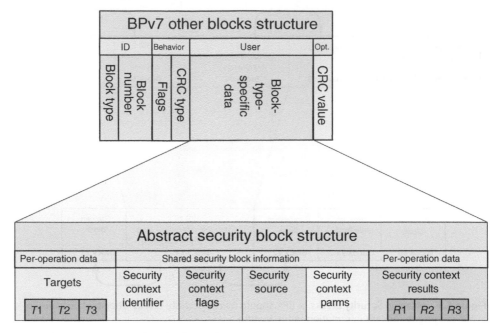

Figure 6.4   The Abstract Security Block. All security blocks share the same block-type-specific data structure.

6.4.3.1   Security Operation Identification

Each security block must contain enough information to identify the SOp being carried in the block. Since an SOp is a combination of a security service (as implemented with a security context) applied to a security target block, this information includes the following security block data fields.

- **Security Block Type**: This identifies the security service being applied. BIB block types apply the bib-integrity service and Block Confidentiality Block (BCB) block types apply the bcb-confidentiality service.
- **Security Target**: Each security operation is applied to a single security target block within the bundle.

SOp identification must, of course, be unique for each SOp. If there were two SOps that shared these data fields, they could not be differentiated and, thus, would be disallowed from existing in the same bundle at the same time.

6.4.3.2   Security Configuration

The security configuration of an SOp provides additional context associated with how cryptographic material was generated or how it should be handled by a security processing BPA in the network. Security configuration information includes the following security block data fields.

- **Security Source Identifier**: Each SOp is added to a bundle by a security source, and that source may affect how the cryptographic materials are selected when verifying or accepting that SOp. In some cases, the security source may be used to determine the verifier and acceptor roles.
- **Security Context Identifier**: This is the identifier of the security context used to generate, verify, and accept the cryptographic material carried in the bundle and associated with a particular security operation.
- **Security Context Parameters**: These parameters are generated at a security source and flowed to the security context implementation at security verifiers and acceptors in the network.

Security configuration might be shared by multiple security operations in a bundle. Operations that are added by the same security source and use the same security context parameterized in the same way can all share this information.

6.4.3.3 Security Results

The security results field of a BPSec security block contains the results generated by a security context as applied to a specific security target block. These results are specific to the SOp; security results cannot be reused for multiple SOps.

6.4.4 Types of Security Information

The data fields of a BPSec security block can be categorized by the type of information they contain. The following categorizations are useful in discussing the rationale for supporting security block multiplicity and providing design guidance on how to implement it within a particular block and bundle.

6.4.4.1 Shared Information

BPSec security blocks are designed to reduce the amount of redundant information carried in a bundle. Where possible, information useful to all of the SOps in the block can be represented just once in the block as shared information.

This set of shared information includes information about the BPA that added block to the bundle and the security context used to generate the materials carried in the block. These common elements are described next.

6.4.4.1.1 Security Context Identifier The well-known, unique identifier of the context. The definition of security contexts cannot be done in an ad hoc manner, as these contexts must be understand by any BPA in a network processing SOps in the security block. Therefore, this identifier must be well known across the network.

6.4.4.1.2 Security Context Flags These flags identify the optional security context fields present in the security block. RFC 9172 defines only one flag, which indicates whether or not security context parameters are present in the security block.

6.4.4.1.3 Security Source

Security Context Flags

Keeping a flag value even with one flag allows future versions of BPSec to accept more fields without breaking backwards compatibility.

The security source identifies the node of the BPA that added the security block to the bundle. The specification of security source is important as it provides an anchor point for policy expression in the BPSec. For example, a BPA could be configured to process SOps from a particular security source.

6.4.4.1.4 Security Context Parameters Each security context might be parameterized with one or more values that assist in the interpretation of security results. Since all of the SOps represented by the block share the same configuration, these parameters (when present) represent shared information. Each security parameter is identified by a parameter identifier and a parameter value.

Parameter Identifier Security context parameters are enumerated as part of the specification defining the security context itself. This identifier field holds the enumeration of the current parameter.

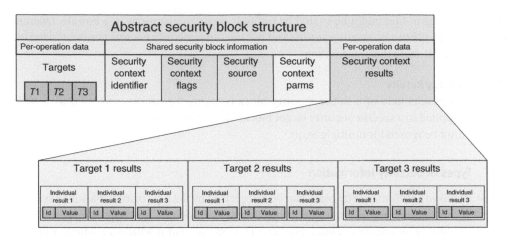

Figure 6.5 Security block results are organized by target block. Each target block in the *security targets* field has a corresponding entry in the security results field. Within each set of results for a target block is a set of individual results for that target block.

Parameter Value This field holds the value for the given parameter, encoded as required by the security context specification.

6.4.4.2 Security Operation Specific Information
There are two types of SOp specific information carried in every security block – information about security target blocks and security results associated with those target blocks.

6.4.4.3 Security Targets
Every ASB starts with a listing of the security target blocks covered by the SOps in the block. As outlined in Table 6.1, the determination of uniqueness is heavily influenced by the security target block of the operation. Placing the set of target blocks covered by operations in an easily accessible field allows BPAs to quickly scan this field to assess whether new SOps can be added to a bundle and whether the block has operations that must be processed. This field contains an ordered set of the block numbers[7] of security target blocks of the SOps in the security block.

6.4.4.4 Security Results
These results represent the set of cryptographic materials associated with the SOps over the target blocks listed in the *security targets* field. This field is a complex set of encapsulated information that is first grouped as a series of target block results that, themselves, encapsulate a series of individual results. This field is illustrated in Figure 6.5.

Target Result Set

> **Target matching**
>
> The number of target results must equal the number of security target blocks in the security block.

An SOp might produce multiple results associated with its target block. For example, it would be possible to define an integrity mechanism that generates multiple signatures over the target block. The target results field is a complex field consisting of all of the individual results generated by an SOp over its target block.

---

7 The BPSec identifies blocks by their block number with the exception of the bundle primary block (block number 0) and the payload block (block number 1).

Individual Result Each individual security result in the target result set is defined as a 2-tuple of information: an identifier of the security result, and the value of that security result. The *Result Id* identifies the type of value being captured in the result field, and is defined by the security context associated with the security block. The *Result Value* holds the value of the result, also as defined by the applicable security context. The Result Id is used only to describe the type of information stored in the Result Value field and does not imply uniqueness. The encompassing result set may have multiple instances of the same Result Id.

6.5 Block Integrity Block

The BPSec Block Integrity Block (BIB) is a security block that holds SOps applying the *bib-integrity* service to one or more target blocks.

Definition 6.2 *(bib-integrity)* This BPSec service generates integrity mechanisms over plaintext information present in a bundle and carried in a BIB block within that bundle. This service is not used to generate integrity over encrypted data, such as data protected by the *bcb-confidentiality* service.[8]

The integrity mechanism used by the BIB is given by the security context associated with the BIB and may represent either signed or unsigned integrity. In this way, BIB can be used to represent authentication (with a signed integrity mechanism) or simply error detection (with an unsigned integrity mechanism).

> **Error Detection**
>
> The use of BIBs with a security context that does not produce signed results has no security value. An unsigned integrity mechanism cannot authenticate the security source. An alternative to unsigned integrity with a BIB is the optional CRC value built into every BPv7 block. Unsigned BIBs should only be used in cases where a CRC value is an insufficient error detection mechanism.

6.5.1 Populating the ASB

The population of the BIB block is illustrated in Figure 6.6. In this illustration, a security target block has its block-type-specific data field input to a security context implementation to calculate a security result ($R1$) held in the BIB block itself. The parameters used by the security context are placed in the shared block information section of the BIB, and the block number of the target block is stored in the security target array ($T1$).

6.5.2 Block Considerations

6.5.2.1 Block Processing Control Flags

A BIB is a BPv7 extension block and its block header information must be set with consideration for how the security-related information of the BIB will be handled. This handling in BPv7 is associated with the setting of Block Processing Control Flags (BPCFs). These flags can be used to determine how the block and the bundle are handled in the event that the security block cannot be processed.

---

8 The bcb-confidentiality service includes an integrity mechanism over calculated ciphertext.

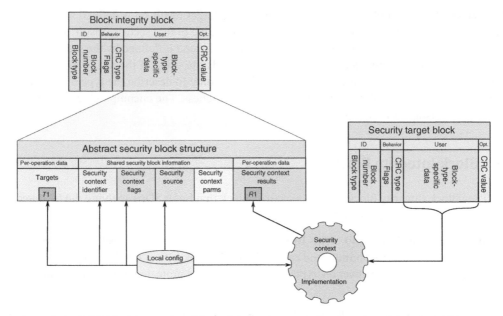

Figure 6.6 A BIB block is populated from the security context and local configurations. BIB blocks contain some of the inputs to, and output of, a security context implementation.

BPCFs were originally intended to handle conditions where a BPA receives a bundle with an unknown extension block. However, BPv7 requires that all BPAs understand BPSec security blocks, so failure to process a security block would only happen when a BPA attempts to process such a block and that processing fails.[9]

If BIB processing flags are set such that a BIB is discarded if it cannot be processed, that means that the target block will be kept in the bundle and, at a subsequent security verifier or security acceptor could reject the target block (or the entire bundle) if a BIB was required. When that happens, the security would be seen as failing because of a *missing* SOp within the bundle. However, if the BIB were never removed from the bundle, then if the BIB failed to verify at a security verifier or a security source, the failure would be caused by a *failed* SOp, rather than a *missing* SOp.

To the extent that these issues have larger impacts relating to security policy, the setting of these flags must be considered carefully at a security source. While Block Processing Control Flags (BPCFs) are helpful in processing some extension blocks in a BPv7 network, they should not be used as a mechanism for enforcing security policy.

Extra Policy

BPCFs only addresses behavior of individual blocks. For example, if these flags require removal of a security block, then the target blocks of operations in that security block might also need to be removed from the bundle. This behavior requires special processing rules beyond what can be expressed by BPCF fields.

The impact of BPCFs on block processing is illustrated in Figure 6.7. In this figure, four BPCFs defined in BPv7 are discussed that perform *No Error Action*, Send Report On Error, *Delete Bundle On Error*, and *Delete Block On Error*.

---

9 Note that correctly determining that a target block has been corrupted is different from failing to process the security block itself.

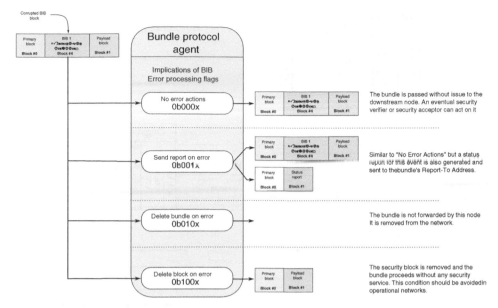

Figure 6.7 Block flags impact processing. Setting block processing control flags impacts how BPAs handle blocks and bundles in various circumstances.

6.5.2.2 Multiple Signatures

SOps are required to be unique within a single bundle and, thus, the operation of $OP(bib-integrity, Block_{Target})$ cannot be repeated for the same target block. However, there are cases where a user may wish to generate multiple integrity results for a single target block such that the verification of the target block against any of the calculated security results is enough to verify the target block.

In examples such as this, it is expected that a multiple-result security context can be developed which would allow a single BIB on a single target block to hold multiple security results.

6.5.2.3 Cryptographic Binding

The plaintext provided to an integrity mechanism includes, at a minimum, some contents of the target block being protected. However, security contexts may choose to use a more expansive definition of plaintext, and calculate that text to include not only portions of the target block, but also portions of other blocks in the bundle.

This would allow the integrity of a target block to be cryptographically bound to some other block in the bundle. For example, binding a target block and a primary block together indicates that the target block could not be extracted from one bundle and placed in another bundle.

6.6 Block Confidentiality Block

The BPSec BCB is a security block that holds SOps applying the *bcb-confidentiality* service to one or more security target blocks.

Definition 6.3 (*bcb-confidentiality*) This BPSec service generates an authenticated confidentiality mechanisms over its target block. The confidentiality service replaces the plaintext of its target block with ciphertext. Separately, this service calculates an integrity mechanism over that ciphertext and any other additional authenticated data to which the target block should be bound.

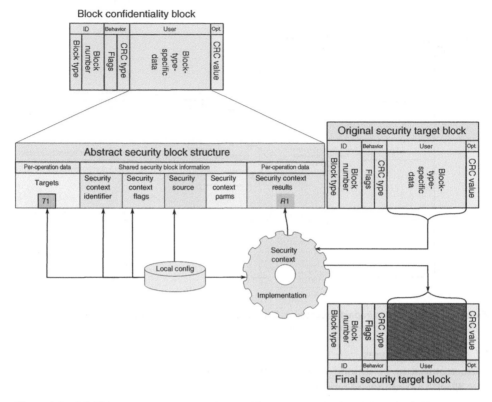

Figure 6.8 A BCB block replaces the plaintext of its target block with ciphertext. BCB blocks contain some of the inputs to, and outputs of, a security context implementation.

The bcb-confidentiality service is complex because it changes its target block. Additionally, because the service is *authenticated* confidentiality, an authentication mechanism is calculated over the ciphertext and the results of this mechanism may be either included in the BCB or coupled with the ciphertext placed in the target block.

6.6.1 Populating the ASB

In this way, the population of the BCB is similar to the population of the BIB, with the exception that the BCB also changes the security target block-type-specific data field.

The population of the BCB block is illustrated in Figure 6.8. In this illustration, a security target block has its block-type-specific data field replaced with ciphertext computed by a security context. Additionally, the BCB holds any additional integrity mechanisms in a security result ($R1$) held in the BCB block itself. The parameters used by the security context are placed in the shared block information section of the BCB, and the block number of the target block is stored in the security target array ($T1$).

6.6.2 Block Considerations

Block processing considerations when constructing a BCB are similar to those when constructing a BIB, with two notable exceptions relating to fragmentation and BCB processing.

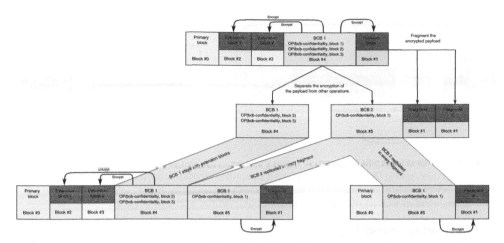

Figure 6.9 BCBs may be replicated in every bundle fragment. BCB blocks should be included in every fragment of a bundle that holds their target blocks. In cases where BCBs hold multiple security operations, a new BCB can be created to hold only the security operation relevant to the fragmented target.

6.6.2.1 Encrypted Payload Fragmentation

When a bundle is fragmented, and the payload block of that bundle has been encrypted, then the BCB indicating the encryption needs to be included in every bundle fragment. This is accomplished by setting the *Block Must Be Replicated In Every Fragment* flag to true for the BCB. Carrying the BCB in every bundle fragment (and every fragment of a bundle fragment) indicates to any BPA that the payload block contains ciphertext and, thus, cannot be processed until the original bundle is reassembled and decrypted.

In cases where a BCB contains SOps, it can be split into one BCB holding the security operating targeting the payload block and the original BCB containing every other SOp, as illustrated in Figure 6.9. In this case, only the smaller BCB targeting the payload block would need to be replicated in every bundle fragment.[10]

6.6.2.2 BCB Processing

A BPA might fail to process a BCB for a variety of circumstances, to include BCBs that use a security context unsupported by the BPA. Whenever any block fails to be processed, one option available to the BPA is the removal of that block from the bundle.

The removal of a BCB from a bundle has significant consequences since the BCB is the sole indication in the bundle that the BCB target block(s) have had their block-type-specific-data field encrypted. Removing a BCB would make it impossible for future BPAs to decrypt the block.

For this reason, BCBs should never be removed from a bundle simply because they cannot process it. This is accomplished by ensuring that the *Block Must Be Removed From Bundle If It Cannot Be Processed* flag is set to 0 when creating a BCB.

Failing to process a BCB at a security acceptor is a different consideration handled by different policy. A BCB may (and should) be removed from a bundle along with its target block if the target block cannot be decrypted at the security acceptor.

---

10 The replication of security parameters across the two BCBs is expected to be less than the replication of multiple security results in every payload fragment.

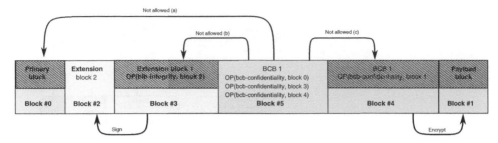

Figure 6.10   Disallowed BCB operations. The three security operations carried in BCB 2 are all disallowed. Encrypting the primary block (a) hides bundle identity. Encrypting a BIB and not the target block of the BIB (b) removes the ability to check the integrity of that target block. Encrypting another BCB (c) removes the ability to decrypt the BCB 1 target block.

6.6.2.3   Appropriate Security Targets

An appropriate target block for a BCB is any block that may have its block-type-specific-data encrypted. There are three types of security blocks that can never be encrypted in this way and, thus, can never be a target block of a BCB.

1. The primary block of a bundle cannot be encrypted, as doing so would prevent the proper processing of the bundle in the network.
2. Other BCBs in a bundle cannot be encrypted, as doing so would hide the fact that other blocks in a bundle have been encrypted.
3. A BIB in a bundle cannot be encrypted unless the BIB and BCB have the same target block and the BIB is being encrypted to handle the special case of protecting the plaintext integrity of the target block when that plaintext had been replaced by ciphertext.

The issues with encrypting these disallowed blocks are illustrated in Figure 6.10.

6.6.2.4   Authenticated Encryption with Associated Data

The bcb-confidentiality service requires authentication of the ciphertext prior to decryption. This authentication is primarily used to ensure that the ciphertext has not changed since it was created at a security source. However, this authentication is also used to ensure that certain attributes of the BCB, target block, and bundle have also not changed during transit.

This more expansive integrity mechanism is accomplished by defining additional *associated data* when calculating an authentication tag for the target block ciphertext. This allows, for example, confirming that ciphertext *and* block processing flags *and* other blocks in the bundle have not been modified. To enable this behavior, any cipher suite defined for use with any security context supporting BCBs needs to implement Authenticated Encryption with Associated Data (AEAD).

Security contexts used by the BCB must specify whether the authentication tag[11] is included in the BCB as a security result or whether it is represented with the ciphertext that replaces the security target block's plaintext.

6.7   Other Security Blocks

> *OSBs may be defined and used in addition to the security blocks identified in this specification.*
>
> *BPSec, RFC 9172* §10, [1]

---

11 Authentication tags are distinct from produced ciphertext.

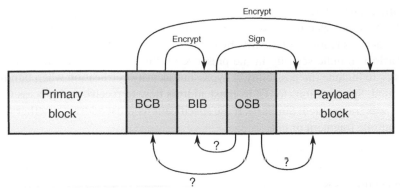

Other Security Blocks (OSBs) must not only consider how they are
applied to their security target blocks but, also, how they interact with
other security blocks present in a bundle.

Figure 6.11 **BPSec allows for the definition of other security blocks.** BPSec
requires that other security block definitions explain the potential complex
relationships between BIB and BCB blocks and any new security block.

The BPSec specification identifies the BIB and BCB security block types, their dependencies, and
associated behaviors. However, the specification make an allowance for OSB types to be defined
over time in the event that the structure of the BIB and BCB is insufficient to carry certain types of
future cryptographic information.

If a need emerges for new structures and/or behaviors beyond that which can be accommodated
by a new security context, new specifications can define OSBs for use in BPv7 bundles. To protect
the processing of BPSec in the possible presence of OSBs, care must be taken to understand the
processing and interactions among and between BPSEc BCB/BIB blocks and any OSBs.

In the event that BIBs and BCBs must coexist with any OSBs, as illustrated in Figure 6.11, it is
incumbent on the specification of these OSBs to ensure that security dependency processing is not
broken and that BIB and BCB processing can continue without requiring modifications to BPSec
protocol constraints, processing rules, or security context implementations.

BPSec places stringent requirements on any other definitions for OSBs, to include the following
behaviors.

- **Clear Separation**: BPSec requires that OSBs not repurpose any of the identifiers defined either
  in BPSec (such as repurposing block types) or in other standards relating to BPSec (such as repurposing the definitions of security context identifiers).
- **Target Behavior**: As a new block in a bundle, OSBs could be the security target blocks of BIBs
  and BCBs. If that presents a problem, the OSB specification would need to clearly identify what
  actions a BPA should take to avoid targeting an OSB. Similarly, OSBs must clearly state whether
  they can target BIB and BCB blocks.
- **Deterministic Processing**: Much of the processing associated with BIB and BCB blocks is
  devoted to ensuring that BPAs process security blocks in a deterministic way. OSB specifications
  must not require any behavior that would affect the existing BIB and BCB processing rules, or
  otherwise lose processing determinism.
- **Canonicalization**: OSB blocks may be canonicalized as any other BPv7 extension block or may
  require custom canonical forms. The BPSec requires that any OSB that could be the target block
  of a BIB or BCB block specify how the OSB should be canonicalized.

- **Conformance**: The behaviors of BPSec are authoritative for processing of BIB and BCB blocks in a bundle. As such, specifications for OSBs cannot require that a BPA perform actions that conflict with actions required of the BPA by BPSec.
- **Fragmentation**: Bundle security in the presence of bundle fragmentation imposes multiple security risks, to include the possibility of bypassing security verification when bundle contents are fragmented. Specifications for OSBs must address how to receive and process OSBs from a bundle that was fragmented. Similarly, these specifications must address whether OSBs can be added to a bundle fragment or not.

Generally, the definition of OSBs is a complex task that is not to be undertaken lightly. The BPSec BIB and BCB blocks, as customized with security contexts, is likely to provide sufficient bundle security for most use cases.

6.8 Mapping

The BPSec mechanism is comprised of three functional components – SOps, security contexts, and security blocks – supported by policy and configuration data at every BPA in the network.

This mechanism must fulfill the design principles of BPSec as defined in Section 5.2. To explain how this is the case, Table 6.2 maps the mechanism components to design principles. This is not a simple 1-1 mapping, as many of the design principles of BPSec must be implemented by various parts of the BPSec mechanism.

Block-level granularity is achieved in the BPSec through the use of SOps and security blocks. SOps target individual blocks by their block number. Security blocks encapsulate SOps so that they, themselves, can be the target of other SOps.

Multiple security sources are achieved by allowing individual security blocks to identify their own security source. In this way, security blocks from various sources can be differentiated within a bundle. Security contexts reinforce this implementation by allowing for different security sources to apply different SOps. Similarly, security policy at BPAs is necessary to understand when a BPA needs to act as a security source.

Mixed security policy differs from mixed security sources in that policy informs the entirety of BPA involvement in security, not just when a BPA acts as a security source. This property is enabled partially by the use of SOps and security blocks, which provide the embodiment of policy in a bundle. However, the security policy and configuration at a node are needed to completely implement

Table 6.2 BPSec mechanism components fulfill design principles.

	Security operations	Security contexts	Security blocks	Policy and configuration
Block-level granularity	●	○	●	○
Multiple security sources	○	◑	●	◑
Mixed security policy	◑	○	◑	●
User-selectable security contexts	○	●	○	◑
Deterministic processing	●	○	●	○

The extent to which a BPSec component supports a design principle can be categorized as not supporting at all (○), partially supporting the principle (◑), and completely supporting the principle (●).

this capability. While policy and configuration and not part of the overall BPSec mechanism, the design of SOps, contexts, and blocks to be configurable through policy statements is what enabled a flexible policy.

The concept of user-selectable security contexts is enable straightforwardly by the definition of moderated security contexts as a first-class element of the BPSec mechanism. Coupled with the ability to select which contexts to be used in which circumstances, the BPSec allows operators to select the tools used to generate and process cryptographic materials across a BPv7 network.

Finally, the rules associated with SOp uniqueness and security block targeting rules ensure that BPSec processing at a BPA remains unambiguous.

6.9   Summary

The BPSec security mechanism is implemented with three unique components: SOps, security contexts, and security blocks. These components have been design to both work together to carry cryptographic information and to support flexibility policy and configuration approaches at BPAs in a BPv7 network. This balance of deterministic behavior and flexible configuration is meant to keep the BPSec security mechanism functioning in the variety of environments where BPv7 might be deployed.

SOps represent the atomic unit of security in BPSec. They are envisioned as an abstract function of a security service applied to a security target block in a bundle. Security operations exist within a bundle according to a simplified lifecycle in which operations may be added, rejected, or accepted while in the network. To keep bundle processing deterministic, these operations must always be unique – no combination of service and target block may be repeated in a bundle.

Security contexts are used to provide a set of environmental assumptions around the use of cipher suites in a BPv7 network. These contexts are used to specify behaviors such as how cipher suite inputs are constructed and how both long-term and ad hoc configurations are generated. In places where BPv7 networks support low-latency operations, low-latency security contexts can be defined. In places, such as with the DTN architecture, that preclude negotiation then delay-tolerant security contexts can also be defined.

The last major component of the BPSec mechanism is the security block, which is the embodiment of SOps and security context information within a bundle. The presence of security blocks in a bundle are used to signal to a BPA that certain security services are present and may need to be processed by a BPA based on local security policy. Security blocks are defined using a common form called the ASB which consists of both shared and operation-specific information. Per-operation information carried information unique to each SOp aggregated in the block. Shared information allows a security block to capture, once, information that is common to all SOps aggregated in the block.

Finally, the BPSec defines two types of security block. The BIB block is used to implement a plaintext integrity service and the BCB block is used to implement authenticated confidentiality. These blocks carry SOps applying *bib-integrity* and *bcb-confidentiality* services, respectively. The BPSec also allows for the specification of OSBs that can be used to implement new security services. Using any OSB in a bundle with BPSec blocks requires that the specifications defining these OSBs provide detailed information on how they should (or should not) interact with BPSec blocks.

The three components of the BPSec mechanism fulfill the design principles of a DTN security mechanism by allowing block-level granularity for security processing applied by multiple BPA across the BPv7 network. BPAs may implement a mixture of security policies and apply multiple, moderated security contexts. Despite this significant flexibility, these components have been designed to allow for deterministic processing.

Reference

1 Birrane, E. J. and McKeever, K. [2022]. Bundle Protocol Security (BPSec), RFC 9172. URL: https://www.rfc-editor.org/info/rfc9172.

7

Security Block Processing

Bundle Protocol Agents (BPAs) in a BPv7 network apply certain levels of block processing to every block in a bundle. This processing occurs when blocks are added to the bundle, when they are processed at waypoint BPAs, and when they are removed from the bundle.

As extension blocks, BPSec security blocks follow this same set of events at every BPA in the BPv7 network. However, BPSec security blocks are, themselves, aggregations of the various security operations resident in the bundle. Each of these operations implement their own security operation lifecycle. Security operations may be added, processed, and removed from security blocks in a method similar to how security blocks are added, processed, and removed from bundles.

This chapter discusses the processing of BPSec security blocks, both as containers of individual security operations and as generic BPv7 extension blocks.

After reading this chapter you will be able to:

- Identify how BPAs add, process, and remove extension blocks from a bundle.
- Describe how security blocks specialize these processes to address security operations.
- Explain the mechanisms by which multiple security operations can be represented in a single security block.
- Analyze the requirements for representing multiple operations in a single block.

7.1   General Block Processing

There is a complex relationship between extension blocks and the bundles in which they exist. In some cases, extension blocks represent critical information without which the bundle may have no meaning. In other cases, extension blocks represent best-effort data whose delivery is inconsequential relative to its bundle.

The processing of any block can be decomposed into two activities: the processing of the generic extension block by any BPv7 agent and the processing of the block-type-specific information contained within that block. Both types of processing, which are illustrated in Figure 7.1 impact how bundles are secured using BPSec.

This general processing must be considered because it will be applied to all BPSec security blocks prior to any BPSec-specific processing. Similarly, the target blocks of security operations are also blocks in the bundle which will also first undergo some generic processing. Therefore, the way in which a Bundle Protocol Agent (BPA) creates, updates, and removed blocks from a bundle will impact the security operations present in that bundle. Only after a block has passed through generic

Securing Delay-Tolerant Networks with BPSec, First Edition. Edward J. Birrane III, Sarah Heiner, and Ken McKeever.
© 2023 John Wiley & Sons, Inc. Published 2023 by John Wiley & Sons, Inc.
Companion website: www.wiley.com/go/birrane/securingdelay-tolerantnetworks

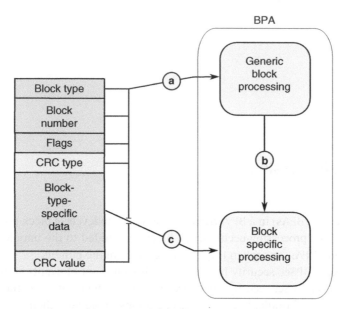

Figure 7.1 BPAs process extension blocks in two ways. Every BPA first examines the common portions of an incoming extension block (a) to ensure that the block type is recognized and has a correct CRC value (if present). Once an extension block is matched with a block-specific processing function (b) then the block-type-specific data associated with the block can be processed (c).

BPv7 processing can its block-type-specific data fields be accessed and security-specific processing be performed.

7.2 The Extension Block Lifecycle

The general relationship between a bundle and an extension block can be illustrated using a generic block lifecycle. This lifecycle represents the progression of an extension block from its creation to its destruction, and through its serialization to, and deserialization from, on-the-wire encodings.

Unlike the bundle which carries them, blocks may be added and removed from the bundle at any point along a bundle path. Relative to a specific block, a BPA can be identified as being either a source, processor, or acceptor of a block.

Definition 7.1 *(Block Source)* The BPA that adds a block to a bundle. This is not necessarily the same as the bundle source, as blocks may be added to bundles while they are in the network.

For required blocks (e.g. the primary block and the payload block), the block source is always the bundle source, since those blocks must always be present in a bundle. Extension blocks may be added by waypoint BPAs in the network. When a waypoint BPA adds an extension block to a bundle, it is considered the block source.

Definition 7.2 *(Block Processor)* Any BPA that processes the block-type-specific data of a block. Block processors must be able to parse these contents as a function of the block type.

For example, a BPA that routes information based on the contents of the primary block is a block processor for that block. If that same BPA were to ignore another extension block, such as the

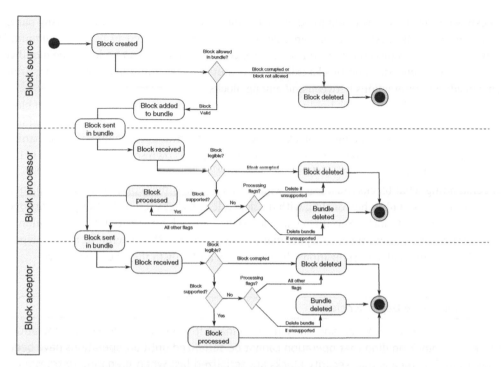

Figure 7.2   A generic extension block processing lifecycle. All extension blocks in a bundle have a generic lifecycle from the time they are created in a bundle, processed one or more times at waypoint BPAs, and finally removed from a bundle by a block acceptor. Notably, a block acceptor does not need to be the bundle destination. For example, the Previous Node block is accepted (and removed) from the bundle at every BPA in the network.

Bundle Age Block, then it would not be a block processor for that block. The designation of block processor for a BPA can occur as a function of block type, and sometimes as a function of block number. Under normal circumstances, block processors do not remove blocks from a bundle.

Definition 7.3   (Block Acceptor)   A BPA that removes a block from a bundle as a function of the processing of the block-type-specific data field of that block.

In this block lifecycle stage, a block processor is distinguished from a block acceptor by the result of successful processing. The bundle destination is, by default, the block acceptor of every block present in a received bundle.

The actions taken through this lifecycle are illustrated in Figure 7.2. This illustration covers the lifecycle of an extension block, starting with its creation at a block source. If the contents of a block are successfully created and added to a bundle, then the block will be in the bundle when it is sent into the network until it encounters a block processor. If the block processor encounters errors in processing, the block (or bundle) may be removed as a function of the block processing control flags that have been set for that block. Otherwise, the block continues until it reaches the block acceptor. At this acceptor, the lifecyle is similar to that at the block processor, with the exception that the block is removed from the bundle by the acceptor even in non-error conditions.

7.2.1   Implementation Notes

One interesting observation on the utility of this lifecycle is that it has an impact on the forms that an extension block takes while present within a BPA. When "on the wire," all of the blocks in a bundle exist in a compact, serialized form. When being processed, these same blocks might be deserialized into a data structure for easier reference of block information.

An extension block is considered to be in its *mutable* form when its fields can be easily parsed and updated by a BPA. Alternatively, an extension block is considered to be in its *immutable form* when it has been placed into its on-the-wire encoding and, thus, can no longer be modified. Understanding when extension blocks exist in their mutable and immutable forms is important when considering interactions between and among blocks.

Consider the case of a BIB block holding an integrity mechanism calculated over its target block. In order to calculate the integrity mechanism, the target block would need to be represented in its immutable form. Changing the target block afterwards would invalidate the integrity mechanism. However, the BIB block must be in its mutable form to update its own security results field to hold the integrity mechanism.

Understanding what blocks must exist in what forms at what times complicates the implementation of a BPA, especially when dealing with the dependencies between and among security operations and their target blocks. Serializing and deserializing block information is computationally intensive relative to the amount of data potentially going through a BPA. Therefore, many waypoint BPAs only perform block deserialization when necessary and only to the extent necessary.

Block Forms Create Dependencies

A block must be canonicalized and unchanged to make it the target of a security operation. The security block holding that operation cannot be serialized until all operations have been added to it. This means that security blocks are serialized last when preparing to transmit a bundle.

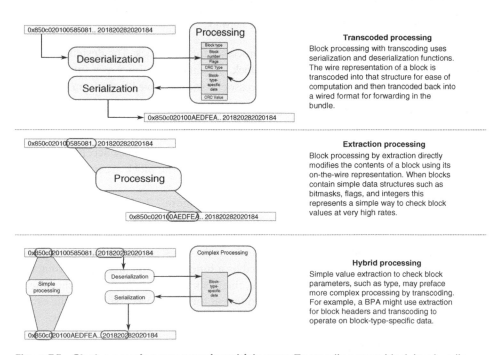

Transcoded processing
Block processing with transcoding uses serialization and deserialization functions. The wire representation of a block is transcoded into that structure for ease of computation and then trancoded back into a wired format for forwarding in the bundle.

Extraction processing
Block processing by extraction directly modifies the contents of a block using its on-the-wire representation. When blocks contain simple data structures such as bitmasks, flags, and integers this represents a simple way to check block values at very high rates.

Hybrid processing
Simple value extraction to check block parameters, such as type, may preface more complex processing by transcoding. For example, a BPA might use extraction for block headers and transcoding to operate on block-type-specific data.

Figure 7.3 Block processing may occur in multiple ways. Transcoding every block in a bundle makes block manipulation easier but at the cost of added computation. Other techniques such as extraction and hybrid approaches attempt to reason about block data while performing little-to-no additional transcoding of the block.

In support of this, BPAs might implement one of three kinds of processing techniques: transcoding, extraction, and hybrid techniques. These three techniques, illustrated in Figure 7.3, highlight ways in which blocks can have all or part of their contents examined in as efficient a way as possible.

7.2.1.1 Transcoding

Extension block transcoding involves alternating between two formats for representing block data. Typically these two formats are a mutable data structure for parsing and manipulation and an immutable byte array for transmission. Transcoding from a data structure to a byte array being the act of *serialization* and the opposite being *deserialization*. In this design, the mutable form of the extension block would be correlated with its existence as a data structure and the immutable form of the extension block would be correlated with its existence as a serialized byte stream.

This approach has the benefit of easier software maintenance over a data structure. However, this approach has the significant downside of requiring the computationally expensive task of converting extension blocks as part of the serialization/deserialization process.

7.2.1.2 Extraction

An alternative to transcoding is to preserve a single on-the-wire format for information in the extension block and to have BPA software operate directly on this format. This approach is popular in firmware and hardware implementations, particularly for extension blocks using fixed-length fields to optimize calculating byte array offsets.

By keeping the representation of the extension block as a byte array implementations do not need to spend the computational time and resources on transcoding to and from a data structure format. As such, implementations may be significantly faster, but may also be more complex.

When using an extraction approach to block processing the concept of mutable and immutable is purely logical, as the extension block never exists in a separate form.

7.2.1.3 Hybrid

A balance between transcoding and extraction can be used in cases where portion of a block can be inspected quickly and then decide on a processing strategy. Information may be *extracted* from extension blocks for certain fields, such as block type and processing flags. This allows for rapid, deterministic identification of the processing required of an extension block. In the event that block-type-specific fields need to be read or modified, the extension block can be deserialized, processed, and then serialized again. This approach preserves the easy of software implementation of transcoding when required and achieves the efficiency of extraction otherwise.

7.2.2 Lifecycle Actions

A brief discussion of the actions taken by a BPA during the major lifecycle stages of an extension block is helpful in understanding the approach taken by BPSec in the processing of security blocks. In this subsection, the major activities of a block source, processor, and acceptor are described. These actions are taken by every BPv7 BPA as extension blocks, including security blocks, are processed.

7.2.2.1 Block Source Actions

An extension block can be added to a bundle by any BPA as part of the processing of a bundle. When adding a block to a bundle, the adding BPA is considered the block source. This source may or may not be codified in the block-type-specific data field of the added block, but is not otherwise represented in the BPv7 extension block format.

The general steps associated with this step are as follows.

1. A block is generated at a BPA in some mutable form and a block type and block number are assigned, block processing flags set, and the block-type-specific field populated.
2. Once added to a bundle, an extension block may be *referenced* by other blocks in the bundle.
3. At some point prior to the transmission of the bundle, the extension block is put into its immutable form with the rest of the bundle. Once the immutable form of the block is generated, a Cyclic Redundancy Check (CRC) value can be calculated over that form of the block.

7.2.2.2 Block Processor Actions

Upon receiving a bundle, blocks within the bundle are examined as part of BPA processing. This can include all, some, or none of the blocks. The general steps associated with this step are as follows.

1. Each block, upon receipt, may be examined by the receiving BPA to determine whether the block should be processed by the BPA.
2. If a required block, such as the primary block or the payload block, is missing from the bundle, then the bundle is treated as malformed and processed in according to policy. This usually means that the bundle is removed from the network.
3. If the BPA understands the extension block type, the block is processed by special software on the BPA. This might include modification of block contents in some cases, which may require recalculation of any integrity fields which cover the block. This might include the removal of the block or bundle in accordance with the block processing control flags is the processing of the block fails.
4. If the BPA does not understand the extension block type, then the BPA will process the block, or the bundle, in accordance with block processing flags. The block might be removed from the bundle, or ignored, or the bundle itself may be removed from the network.

Block processing provides a good use case for the hybrid implementation approach. A BPA might inspect a block's type and processing flags without deserializing the block itself. Then, if the block is to be processed locally by the BPA, it can be deserialized and passed to local block processing software.

7.2.2.3 Block Acceptor Actions

The block acceptor is the last BPA able to process a block in a bundle. The bundle destination is, by necessity, the block acceptor for every block remaining in a bundle. However, other blocks may be *accepted* by other BPAs in the network.

When the block acceptor is not the bundle destination, then the actions at the block acceptor are similar to the actions at a block processor. The only difference being that at the block acceptor, the extension block is always removed from the bundle prior to its retransmission.

The bundle destination is, by necessity, the acceptor of any block remaining in the bundle. Since a bundle will no longer exist after processing at this BPA, all blocks need to be accepted prior to passing the bundle payload to applications resident on the destination BPA.

Unlike block processing at other BPAs, the purpose of processing blocks at the bundle destination is not about protecting networking traffic, but ensuring proper delivery of bundled information at the local BPA.

Focus: Previous Node Block

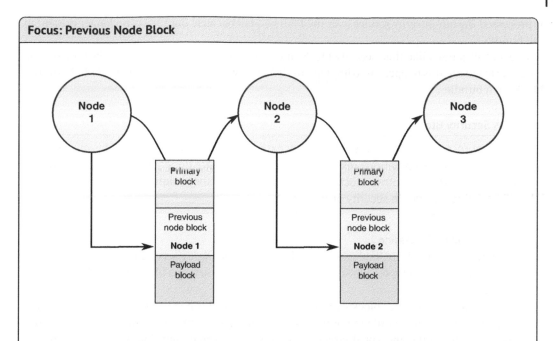

The Previous Node Block (PNB) is an extension block identifying the most recent BPA that forwarded the bundle to its current BPA. The PNB is useful for bundle and block processing in cases where the path a bundle takes is important for its processing, such as when determining whether to accept bundles from a particular BPA in a network.

The PNB is an example of a block which has a very short *lifetime* within a bundle – the BPA which adds the PNB is the block source for the PNB and the very next BPA is, by definition, the block acceptor. At the acceptor, the PNB is removed from the bundle and a new PNB can be added prior to any forwarding of the bundle. In this sense, the block acceptor of one PNB becomes the block source for the next PNB in the bundle.

7.2.3 Security Implications

Security blocks must be correctly processed by generic block processors first so that the security operations resident in their block-type-specific data field can be subsequently processed. In order for this to happen, there are several special considerations related to the behavior of block processors in the present of security blocks and security operation target blocks.

7.2.3.1 Order of Block Evaluation

As collections of security operations, security blocks have an implicit relationship with the target blocks of these operations. Therefore, security blocks, unlike many other extension blocks, have dependencies on other information in a bundle. The success or failure of processing a security block might be based on information external to that extension block.

Consider the case of a BIB carrying a *bib-integrity* operation targeting a PNB. The next hop BPA is, by definition, the block acceptor of the PNB. If the PNB is processed before the BIB, then the PNB will be removed from the bundle as part of block acceptor actions. Then, later, when the BIB is processed the *bib-integrity* service over the PNB will fail because the PNB is no longer present in the bundle.

To avoid situations like this, security blocks must be processed prior to any blocks they secure. One way to ensure this happens is to implement BPAs to always process security blocks before other blocks in a bundle.

Process Security Blocks First

BPAs must process security blocks before other extension blocks in the bundle. This must occur because the security operations resident in security blocks are dependent upon other blocks in a bundle. Were these other blocks to be processed first, they might be removed from the bundle causing an error in security processing.

7.2.3.2 Defer Some Processing

Security operations within BCB blocks apply the *bcb-confidentiality* service to their target blocks. When this occurs, the encrypted block-type-specific data of those target blocks cannot be processed by the block processor.

In this situation, the target block cannot be processed until it has been decrypted. This might not happen for some time while the bundle is in the network. Until that time, the normal block processing action on the encrypted target block must be deferred, and that deferral must not be seen as a processing error.[1]

7.2.3.3 Preserve Security Blocks

Whenever possible, the presence of a security block should be preserved in the bundle whenever the security operations encapsulated by that block remain relevant in the bundle. This means that the failure to process a security operation should not necessarily result in the removal of the security block from the bundle.

Removing a security block (or removing a security operation from a security block) removes the indication that a target block had a security service applied to it. In the case of *bib-integrity* this means the target block may be open to undetected modification. In the case of *bcb-confidentiality* this means that the target block may be seen as corrupted because a BPA would not otherwise know that the target block had been encrypted.

For this reason, the generic block processing control flags of a security block should be set to **not remove** the security block if it fails to process. Keeping a security block even if all of its encapsulated security operations fail both preserves the indication that security exists in the bundle and allows for a more expressive set of actions applied as part of security processing instead of generic block processing.

Examples of security-specific processing for a failed security operation include the following.

1. Special failure reports can be generated by the BPA.
2. The security operation and its target block can be removed from the bundle.
3. The bundle can be removed from the network.
4. The security operation can stay in the bundle, but the security target block can be removed.

---

1 If skipping the processing of an encrypted block is seen as an error, the block could be removed from the bundle or the bundle removed from the network

> **Security Policy Handles Failures**
>
> The Block Processing Control Flags of a security block **should not** be set to remove the block from the bundle if it cannot be processed. Handling security failures is a matter of the local security policy of the BPA. This allows a coordinated response between the security block, its security operation, and the target blocks in the bundle.

7.3 Security Operation Processing

The processing of security blocks happens as part of two distinct lifecycles. As a standard BPv7 extension block, security blocks adhere to the generic block processing lifecycle as discussed in Section 7.2. As a collection of security operations, each security block must also process each operation in accordance with the definition of each security operation – where operations can be inserted and then later rejected or accepted.

To reduce ambiguity when discussing security block processing, security-specific roles are defined to describe processing specific to the security operation lifecycle. Instead of the generic terms *block source*, *block processor*, and *block acceptor*, the terms *security source*, *security verifier*, and *security acceptor* are used instead.

7.3.1 Security Roles

In the context of these security roles, operation processing represents both syntactic and semantic checking of security blocks. Similar to the generic block processing which precedes it, syntactic operation checking ensures that security blocks are populated with valid data within given ranges and representative of data in the bundle and configured at the processing BPA. Semantic checking determines whether desired security operations are present and whether security parameters are used in accordance with local policy. Security block processes cannot succeed if there is a syntax problem and cannot succeed if the security in the block is insufficient or inappropriate as determined by the local BPA's policy and configuration.

> **Security Processing Scope**
>
> Successful security processing indicates that security operations are both syntactically valid and semantically correct.

Just as a BPA, acting as a block processor, may choose to ignore unknown or unsupported extension block types, a BPA acting as a security verifier may choose to ignore the processing of certain security operations. A BPA could choose to process one security operation in a block, accept another operation, and ignore the remainder of the operations in that block.

This interaction between security operations in a block and BPAs acting as security verifiers is illustrated in Figure 7.4 and described in further detail in this section.

These roles, and their accompanying responsibilities, cannot be hard-coded into the design of a network. BPv7 networks, in general, may experience significant topological change over the lifetime of a bundle. Different BPAs might take on different security roles as a reaction to this topological change. A BPA acting as a security acceptor at the time of the creation of a security operation at a security source might have been replaced by a different node acting in the same capacity by the time a bundle with that security operation is received. For this reason, BPSec requires receiver-side policy assertions of security roles and does not hard-code role expectations at the time of creation of a security source.

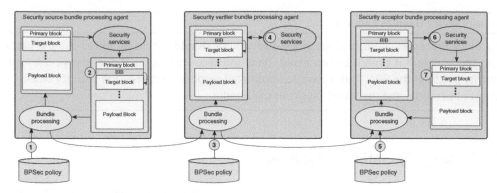

Figure 7.4 Security policy roles. Local policy identifies security source ①, verifier ③, and acceptor ⑤ roles. Sources add security blocks to a bundle ②, verifiers check security blocks ④, and acceptors both check ⑥ and remove ⑦ security blocks.

Separate Security and Routing

A lesson learned (discussed in Section 5.1.2.2) from the experimental, deprecated Bundle Security Protocol (RFC 6257 [1]) was that tightly coupling routing and security fails when network topology changes. BPSec requires receiver-side policy to determine role, which is why the term security acceptor is used instead of security destination. BPSec has **no concept** of destination as that term implies information known in advance.

7.3.2 Security Source Processing

Definition 7.4 *(Security Source)* A BPA that generates one or more security blocks (and associated security operations) for a bundle. This source determines, via policy, when blocks require the application of some security service. Because security operations are required to be unique in a bundle, a given operation will only ever have a single security source associated with it.

As part of populating the block-type-specific data fields, a security source will additionally generate cryptographic material related to the security service as well as potentially modifying the contents of other blocks in the bundle.

The steps taken by a BPA to determine, and fulfill, its role as a security source are provided in the following text. The ordering of these steps is somewhat subjective, and might differ as an implementation matter.

1. **Add or Update Non-security Blocks**: Whenever creating or forwarding a bundle, any block that may require a security service must be present in the bundle. This does not mean that all blocks must be present in a bundle prior to applying security, only that target blocks exist prior to securing them.
2. **Determine Role**: Whenever creating or forwarding a bundle, local security policy must be evaluated to see if the current BPA is configured as the security source of any blocks within the bundle.
3. **Add Security Block**: A security source for a security operation is the block source of the security block carrying that operation. As such, a security block of the proper type, with an appropriate block number and processing flags, must be added to the bundle.

4. **Generate Cryptographic Material**: Part of applying a security service is the generation of material associated with that service. Some of these materials will be stored as security parameters and security results in the block itself. Other sets of cryptographic materials may be carried in the target blocks of the security service.[2]

5. **Populate Security Block**: While still in its mutable form, a security block must have its block-type-specific field populated. This requires updating the Abstract Security Block (ASB) associated with the block. Cryptographic material carried in the security block as parameters and results is added to the ASB at this time.

6. **Modify Target Blocks**: Once cryptographic material, such as ciphertext, has been generated, that material must be serialized and placed in the target block.

7. **Serialization**: At some point prior to bundle transmission, the security block must have any final changes applied to its block header, such as the addition of a CRC value, and serialized into its immutable form for transmission with the rest of the bundle.

7.3.3 Security Verifier Processing

Definition 7.5 *(Security Verifier)* A BPA that ensures that one or more security operations in a security block remains valid. Unlike a security source, a security verifier does not alter the contents of the bundle as part of verification.

The evaluation performed by a security verifier can refer to either a syntactic or a semantic verification of the operation.

Syntactic verification refers to whether the operation is correctly expressed in the bundle. This includes ensuring that associated security blocks and security target blocks are present in the bundle and have the appropriate structure and encoding.

Semantic verification refers to whether the security service represented by the operation is still correct. This includes, for example, verifying integrity signatures to make sure that the contents of a security target block have not been modified since the signature was generated at a security source.

Because a security verifier examines existing information in a bundle, it is a block processor for the associated security block. The purpose of verification is to detect, as early as possible, potential security issues within a bundle to avoid allowing corrupt or otherwise problematic bundles to persist in the network. A security operation may have multiple security verifiers throughout the network. Every waypoint BPA encountered by a security operation might be configured with a security verifier role for that operation.

The decision of whether a given BPA is a verifier for a particular security operation is part of the configuration of the BPA itself. Any BPA may identify as a verifier for a security operation as long as that BPA is neither the operation's source or acceptor. A verifier cannot be the same as a security source as the source authoritatively determines the baseline cryptographic material – there is nothing to verify. A verifier cannot be the same as a security acceptor since an acceptor already includes a verification step.

The general steps taken by a BPA to determine, and fulfill, its role as a security verifier are provided next.

1. **Determine Role**: Similar to a security source, local security policy must be evaluated to see if the BPA should verify a security operation. This determination can be made independently of whether the security operation being verified exists in the bundle.

---

2 For example, when generating a ciphertext version of a target block, that ciphertext is carried in the target block itself.

2. **Verify Existence**: Security verifier must determine whether the received bundle contains the information necessary to process the security operation. For example, if policy requires a *bib-integrity* service over a particular target block, then the bundle would be checked to ensure that both the target block and a BIB with that operation on the target block existed in the bundle.

3. **Verify Configuration**: Verification often requires appropriate configuration at the security verifier. Some configuration for the security operation may be carried in the bundle itself.[3] Other configuration may exist as part of local policy on the BPA, or generated as needed by the verifying security context. The BPA must ensure that all configuration is present, deconflicted, and that configuration values are of the expected size, type, and value range.

4. **Verify Contents**: Once required blocks and configurations have been collected, the security operation itself can be verified. The BPA, using the appropriate security context, target block information, and configuration ensures that any cryptographic information in the security block remains valid. **Note** that when the security operation involved is encryption, a security verify does not perform a decryption.

5. **Process Actions**: The results of verification may generate actions to be taken by the BPA that may alter the contents of the bundle or cause the removal of the bundle from the network.

7.3.4 Security Acceptor Processing

Definition 7.6 *(Security Acceptor)* A BPA that terminates a security operation in a bundle at a particular BPA. Unlike security verifiers there is only a single acceptor because the security operation no longer exists in a bundle after it has been accepted. The processing of accepting a security operation includes first verifying the operation, reversing any changes made at the security source (such as decrypting a target block), and removing the operation from its security block.

Because a security acceptor might remove a security block from a bundle, it is a block acceptor for the bundle. The purpose of acceptance is to confirm that a security operation has fulfilled its purpose and is no longer needed in the bundle for the duration of the bundle in the network. When a security acceptor removes the last security operation from a security block, it also removes the security block from the bundle.

Similar to other BPSec roles, the decision of whether a given BPA is an acceptor for a particular security block is part of the configuration of the BPA itself. Any BPA may identify as an acceptor for a security operation with the caveat that a bundle destination must accept any security operation in any security block in a received bundle.

The steps taken by a BPA to determine, and fulfill, its role as a security verifier are provided next.

1. **Verify**: Prior to accepting a security operation, the acceptor must verify that the operation being accepted is still valid. This involves performing all of the steps regularly taken by a security verifier.

2. **Accept**: If the operation remains valid it can be accepted. For integrity-based operations acceptance involves simply removing the security operation from the bundle. For confidentiality-based operations, acceptance involves decrypting the target block prior to removing the security operation from the bundle. The specific acceptance steps are a function of the security context used in the generation of the security operation at the security source.

3. **Process Actions**: The results of either verification or acceptance may generate actions to be taken by the BPA, similar to processing actions at a security verifier.

----

3 To include the security parameters and results in the containing security block.

7.4   Security Block Manipulation

In simple cases the manipulation of extension blocks by a BPA is constrained to adding, processing, and removing those blocks from the bundle. Security blocks add a layer of complexity to this processing as these blocks represent collections of security operations sharing common parameters. Since security processing is based on security operations, security block manipulation involves the manipulation of security operations.

Security blocks can be manipulated by adding or removing security operations in the block. While inefficient, every security operation could be represented by a single security block. Similarly, if all security operations shared common information they could all be placed in a single security block. The ability of a security block to represent multiple security operations is termed *target multiplicity*.

Definition 7.7   *(Target Multiplicity)*   The property of a BPSec security block that it can contain more than one security operation (security service applies to a security target block). The term *target* multiplicity is used to indicate that the unique information in a security block is that information related to the specific target block in question. Non-target security operation information is shared by all operations in the block.

This property of multiplicity creates the notion of a group of security operations as constructed through a grouping process and, conversely, destructed through an ungrouping process.

Definition 7.8   *(Security Operation Grouping)*   This term is used in two ways to discuss the use of target multiplicity.

1. *(Noun)* The set of security operations resident in a single security block. Operations in a grouping share the common information present in the ASB of the carrying security block.
2. *(Verb)* The act of placing new security operations into a security block, thus creating or expanding the grouping in that block. The opposite of security operation ungrouping.

The management of security operations in a security block is the primary method of security block manipulation. The remainder of this section describes various reasons why this processing might be beneficial, with examples of security operation grouping and ungrouping.

7.4.1   Grouping Security Operations

> *A single security block MAY represent multiple security operations as a way of reducing the overall number of security blocks present in a bundle. In these circumstances, reducing the number of security blocks in the bundle reduces the amount of redundant information in the bundle.*
>
>   BPSec, RFC 9172 §3.3, [2]

Every security operation could be represented in its own security block. Such a one-to-one relationship would introduce multiple inefficiencies related to bundle processing. Potential problems with single-operation security blocks include the following.

1. **Multiple Blocks**: Requiring every security operation to have its own block increases the number of security blocks in a bundle. This complicates the actions taken when generating and receiving bundles at a BPA, as each security block must work through both the generic extension block lifecycle and the security operation lifecycle.

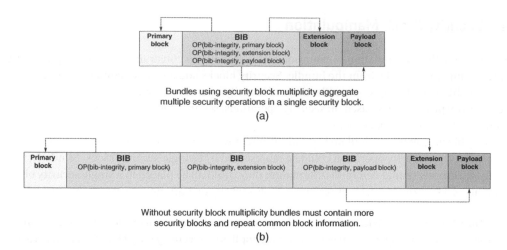

Bundles using security block multiplicity aggregate
multiple security operations in a single security block.

(a)

Without security block multiplicity bundles must contain more
security blocks and repeat common block information.

(b)

Figure 7.5   Target Multiplicity Reduces Bundle Footprint. Target multiplicity results in smaller bundles, less redundant information, and more efficient processing.

2. **Redundant Information**: Security parameters (such as wrapped keys) can represent the bulk of information carried in a security block. When multiple operations share the same security parameters, repeating them per block can significantly increase the size of each block and, consequently, the size of the bundle. Larger bundles require more storage and bandwidth resources for their processing.

3. **Inefficient Processing**: More blocks carrying redundant information increases the amount of computational resources needed for security processing at a node. For example, a BPA might unwrap a wrapped key to process one security block, and then delete that unwrapped key when the block is finished processing. If the next security block uses the same wrapped key (redundant information) then the BPA must unwrap the key again. This type of waste is avoided by grouping security operations behind a single block processing action.

Without target multiplicity, the potential increase in bundle size can become significant. Consider the example of a bundle consisting of three blocks (primary, extension, and payload) all of which need to be signed using an HMAC-SHA256 mechanism. A BPA might choose to use multiplicity or not, as illustrated in Figure 7.5. Without multiplicity, three BIBs would be added to the bundle, one BIB for each security operation: $OP(bib\text{-}integrity, Block_{Primary})$, $OP(bib\text{-}integrity, Block_{Extension})$, $OP(bib\text{-}integrity, Block_{Payload})$. Alternatively, using multiplicity, a single BIB could be added to hold all three operations.

Using some assumptions on the minimal encoding sizes of block information and HMAC-SHA256 keys and results, a simple analysis can be performed to illustrate the benefit of target multiplicity. This analysis is presented in Table 7.1. In this table, the column "Minimal Size" represents the smallest possible security block – one that has no parameters, no CRC value, and no security results and perfect encoding.[4] Such a minimal security block would use 10 bytes in a bundle. In contrast, a BIB holding a single security operation related to an HMAC-SHA256 using a 32 byte key and 32 byte signature would require 79 bytes.

If a single BIB takes approximately 79 bytes, then three single-operation (1-Op) BIBs would require 237 bytes. There is significant repetition in this three-BIB case. The security context parameter holding the 32-byte HMAC-SHA256 key is repeated three times, as are individual array header

---

4 Such a block is theoretically possible if all parameters are presumed cached at BPAs and security results are held in the target block.

Table 7.1 Target multiplicity results in significant size savings.

	Minimal size (Bytes)	1-Op sample BIB (Bytes)		3-Op sample BIB (Bytes)
		1 BIB	3 BIBs	
Block header	4	4	12	4
Security targets	2	2	6	4
Security context Id	1	1	3	1
Security context flags	1	1	3	1
Security source	2	2	6	2
Security context parms	0	34	102	34
Security results	0	35	105	103
Total bytes	10	79	237	149

"Minimal Size" assumes block types and numbers can be encoded in 1 byte, all EIDs can be encoded in 2 bytes, and other identifiers (context ID, parameter ID, and result ID) can be encoded in 1 byte each. "Sample BIB" examples use HMAC-SHA256 accepting a 32-byte key and generating a 32-byte digest.

encodings for arrays carried in the BIB. Alternatively, if a single BIB is created with three security operations in it, using target multiplicity, then the resultant "3-Op" BIB only needs 149 bytes, which is almost a 63% reduction in size. The bulk of this savings comes from not repeating security parameter information, such as the 32-byte key. Ciphersuite-generated materials such as keys and results represent the bulk of information carried in a security block. In the 1-Op BIB case, HMAC-SHA256 information comprises 87% of the block size.

7.4.2 Grouping Requirements

Security operations may only be grouped when the set of shared information in a security block has no conflicts or ambiguities. Specifically, two security operations may be grouped in a security block when the following five conditions have been met.

1. **Same Service**: A security block must apply the same service (e.g. *bib-integrity* or *bcb-confidentiality*).
2. **Same Parameters**: The security context parameters portion of a security block applies to all security operations in the block.[5]
3. **Same Source**: Each security block lists the BPA that added the block to the bundle (the security source). A BPA cannot append an operation to a security block that was created by some other BPA.
4. **Unique Targets**: Every security operation must remain unique, even when collected into a single security block. Target multiplicity is only used to reduce the amount of redundant information carried in a bundle, never to make security operations not be unique.
5. **No Conflicts**: Only security operations that would, otherwise, be allowed in their own security block may be merged into a single security block. If adding a security operation would be prohibited by a constraint or processing rule, the operation cannot be added as part of block multiplicity, either.

---

5 Not every operation need use every parameter, but the same parameters must be available to all operations in the block.

7.4.3 Block Manipulation Algorithms

The grouping (and ungrouping) of operations in a block can happen for a variety of reasons at a BPA. One simple scenario is when a BPA is the security acceptor for one operation in a security block, but not for all of the operations in the block. Once the one operation has been accepted, the security block must be modified to remove the information for just the accepted operation.

There are additional other use cases where a BPA might decide to merge multiple security blocks into a single block (if possible in accordance with the requirements of Section 7.4.2) or even split a single security block into multiple blocks.

This section explains the following three security block processing algorithms associated with the grouping and ungrouping of operations in a block.

1. **Add**: Add a security operation to an existing security block.
2. **Merge**: Merge multiple security blocks into one.
3. **Remove**: Remove one of more operations from a security block.
4. **Split**: Create two or more security blocks from a single block.

7.4.3.1 Add Security Operation

The simplest algorithm related to security block management (other than creating the block in the first place) is the addition of a new security operation into a block. This happens whenever a BPA, through analysis of its security policy, determines that it is the security source of a security operation and that an existing security block can carry that operation.

Adding a security operation to a security block involves the following steps.

1. Verify that the common information for the new security operation matches the common information in the security block.
2. Verify that the security operation is allowed to be in the bundle, considering other operations present in the bundle.
3. Ensure that the security target block is available for processing.[6]
4. Calculate security results using the security context and associated parameters.
5. Apply security results to other blocks (such as the case with encryption)
6. Update the security block by adding the security target block to the block's list of targets.
7. Update the security block by adding the security results to the block's list of results. The ordering of security results must match the ordering of security targets.

7.4.3.2 Merge Security Blocks

In the event that two or more security blocks exist in a bundle such that the operations in those blocks could be grouped together, then the two blocks can be merged into one.

An instance of merging two BIB blocks into a single BIB is illustrated in Figure 7.6. In this figure, BIB 1 and BIB 2 share common information and are only differentiated by elements specific to their security operations (target blocks and security results). BIB 2 can be merged into BIB 1 by adding the target-specific information from BIB 2 into the associated arrays of BIB 1.

Merging security blocks involves incorporating the operations from one block (the source block) into another block (the destination block). The steps for merging a source block into a destination block are as follows.

---

6 This may require splitting an existing security block in accordance with the Split Security Blocks algorithm

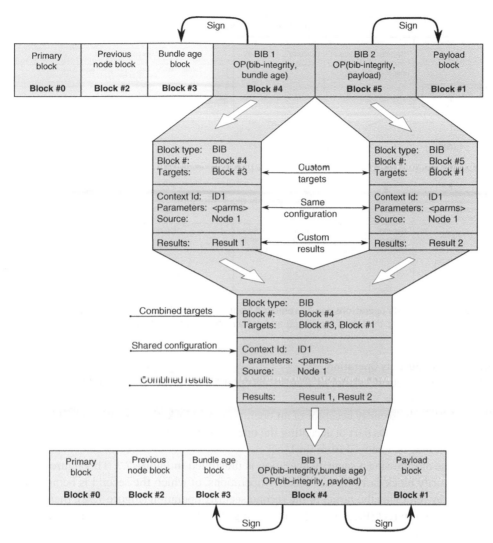

Figure 7.6 Merging security blocks. When the security operations in two blocks have the same common information, they may be merged into a single security block.

1. Verify that the source block is not referenced by other blocks in the bundle. The source block will be removed from the bundle as part of this merge algorithm.
2. Verify that both blocks are of the same block type (e.g. they are both BIB and BCB blocks).
3. Determine the block processing control flags for the destination block. This could be the same as the current flags for the block, a merge of flags with the flags from the source block, or some other mechanism given by the policy of the BPA.
4. Verify that the common security information in both blocks is the same. This includes the security source, security context ID, and security parameters.
5. Append the set of target blocks from the source block set of targets to the destination block set of targets.
6. Append the set of target results from the source block set of target results to the destination block set of targets.
7. Remove the source block from the bundle.

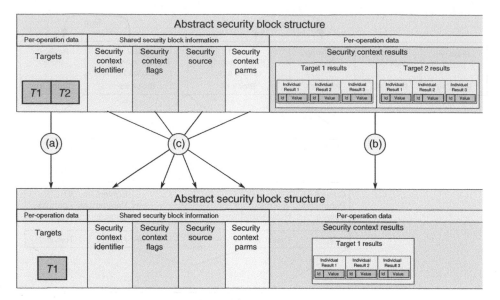

Figure 7.7 **Security operations can be removed from a security block.** Operation-specific information includes the list of targets (a) and associated target results (b), but not shared information (c).

7.4.3.3 Remove Security Operation

A BPA might be a security verifier or security acceptor for some, but not all, of the security operations in a security block.

When a BPA is the security acceptor of an operation in a block, the operation must be removed from the security block as part of accepting the operation. Similarly, if an operation fails to verify, that operation may be removed from the block.

The process of removing a security operation is illustrated in Figure 7.7. This figure shows the ASB of a security block carrying two security operations, of which the second is being removed. This is done by removing the target of the second security operation (Figure 7.7a), then removing the results for that target (Figure 7.7b), and making sure that common information in otherwise unaffected (Figure 7.7c).

The following steps are taken when removing an operation from a security block.

1. Remove the block number of the operation's target block from the security block's list of target identifiers.
2. Remove the security results associated with the security operation from the security block's list of security results.
3. If this was the last security operation in the security block, remove the block from the bundle.

7.4.3.4 Split Security Blocks

There are rare circumstances when some security operations must be ungrouped. This involves moving some security operations out of their security block and into some other security block (and perhaps a new security block created for that purpose).

There are a few potential reasons for why a BPA may wish to split a security block.

7.4.3.4.1 Reduce Block Size Splitting security blocks increases the overall size of a bundle because common information is repeated across blocks. However, splitting blocks also decreases the size of any one such block. This might be important in cases where there is limited memory or other resources that can be applied to the processing of any one block. For example, firmware decoders for security blocks might implement limits on the size and number of operations that can be in a given block.

7.4.3.4.2 Support Bundle Fragmentation When a bundle is fragmented, different extension blocks might be placed in different fragments. In this case, a BPA might split a security block into multiple blocks so that each security block could be kept with the extension blocks it secures.

7.4.3.4.3 Support BIB Encryption BPSec allows for a BIB to be encrypted by a BCB in the special circumstance where the *bcb-confidentiality* service is being applied to the target block of an existing *bib-integrity* service. If the BIB being encrypted includes other security operations, encrypting those security operations would hide integrity information for no reason. In this case, the security operations within a BIB that can be encrypted should be split into a separate BIB for encryption.

When splitting a security operation, the following steps should be taken.

1. The receiving security block is identified (or created if needed).
2. The security operation is added to the receiving security block in accordance with the *Add Security Operation* algorithm.
3. The security operation is removed from the existing security block in accordance with the *Remove Security Operation* algorithm.

7.5 Target Multiplicity Examples

7.5.1 Confidentiality

Target multiplicity for *bcb-confidentiality* services allows a BCB to carry security results for multiple operations. Each of the BCB's security target blocks are encrypted using the same security context and parameters.

The overall process of applying target multiplicity in a BCB is illustrated in Figure 7.8. The enumerated steps in this figure are described in more detail next.

1. A BPA, identified as a security source by local policy, is preparing a bundle to be transmitted. The bundle contains two blocks, an extension block and the payload block, that are required by policy to be encrypted using the same *bcb-confidentiality* service before the bundle is forwarded.
2. The security source generates a single BCB with two security target blocks: the payload block and the extension block. This results in the encryption of both security target blocks block-type-specific data.
3. The bundle is transmitted to its destination.
4. The bundle arrives at its destination, where policy at the receiving BPA determines that it is the security acceptor for both security operations represented by the BCB.
5. Each of the BCB's security operations are processed separately. The block-type-specific data for the extension block is decrypted, then the payload block is decrypted in a similar fashion.
6. After successful processing, the BCB is removed from the bundle.

7.5.2 Integrity

Target multiplicity for *bib-integrity* services allows a single BIB to carry security results, including integrity signatures, for multiple target blocks. Each integrity signature for the BIB's security target blocks is generated using the same security context and parameters.

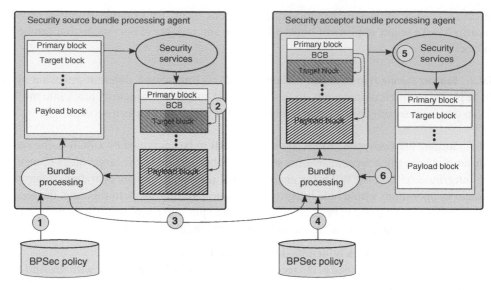

Figure 7.8   Target Multiplicity for Confidentiality.

The overall process of applying target multiplicity in a BIB is illustrated in Figure 7.9. The enumerated steps in this figure are described in more detail next.

1. A BPA is preparing a bundle to be transmitted. The bundle is composed of three blocks: the primary block, payload block, and an extension block. All three of these blocks are required by local security policy to have *bib-integrity* services applied before transmission begins.
2. As the security source, the BPA generates a BIB with three operations that target the primary, payload, and extension blocks, respectively. Security results are generated for each security operation and added to the security block.
3. The bundle is transmitted.
4. The bundle arrives at a BPA configured to serve as a security verifier for the bundle's BIB in order to confirm integrity before transmission continues.
5. The BPA processes the three *bib-integrity* operations represented by the BIB and verifies the integrity of the primary, payload, and extension blocks.

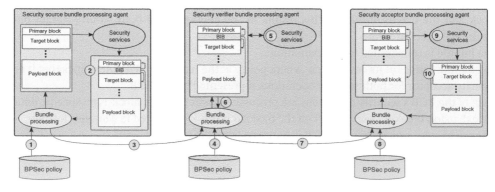

Figure 7.9   Target Multiplicity for Block Integrity.

6. Integrity is confirmed and the BIB is retained in the bundle.
7. Bundle transmission continues.
8. The bundle arrives at its destination which BPSec requires act as the security acceptor for the BIB.
9. Each BIB security operation is processed, allowing for verification of each of the three target blocks' integrity signatures.
10. When the BIB has been successfully processed, the security acceptor removes it from the bundle.

7.6 Common Error Conditions

There are many potential error associated with processing security blocks. This section identifies some common error conditions and how they must be processed in accordance with the BPSec specification. These examples can serve as a guide to explore and define further error conditions that might occur in a particular implementation.

7.6.1 BIB Target Verification Failed at Security Verifier

This error may occur when the security verifier attempting to process an integrity mechanism cannot verify the result. This indicates that the contents of the security target block (or the integrity mechanism itself) have been modified during transmission.

The BPA handles the impacted security target block according to local security policy. This may result in the removal of the security target block from the bundle, removal of the BIB representing the corrupted security operation, or a data generation processing action.

> **Security Operation Event**
>
> This error corresponds to the Security Operation Corrupted at Verifier security operation event.

If the security target block o be removed from the bundle is the primary block or the payload block, the bundle must also be discarded as it is then incomplete.

In all cases, requested status reports may be generated to reflect bundle or block deletion.

7.6.2 Security Block Segmentation Failure at Security Source

This error may occur at a security source that must segment an existing security block in the bundle in order to add the security operation(s) it is required by BPSec policy to represent.

If the existing security block cannot be segmented, the security operation(s) the security source is attempting to add to the bundle is identified as misconfigured.

> **Security Operation Event**
>
> This error corresponds to the Security Operation Misconfigured at Source security operation event.

Local security policy determines how the misconfigured security operations are handled. It is recommended that bundle transmission does not continue beyond the security source node as the bundle does not have sufficient security to continue moving through the network.

7.6.3   Security Block Segmentation Failure at Security Acceptor

> **Security Operation Event**
>
> This error corresponds to the Security Operation Misconfigured at Acceptor security operation event.

A security acceptor that must segment a security block in the bundle in order to process some but not all of the security operations it represents may encounter an error when performing this segmentation. If the security block cannot be split into multiple security blocks for processing, it is identified to be misconfigured by the security acceptor.

The misconfigured security operations are handled according to local security policy.

7.7   Summary

The act of processing a bundle is the act of processing the blocks that comprise the bundle. The processing of blocks that occurs at a BPA can be modeled by defining a lifecycle for individual extension blocks – blocks are added, processed, and removed.

Understanding the generic extension block lifecycle is important because it influences the implementation of block processors and because block process ordering is important to security processing overall. BPA implementations might choose to transcode, extract, or develop a hybrid approach to reviewing the contents of extension blocks.

Simple processing might not require translating an extension block from its on-the-wire representation. More complex processing might require full block deserialization. A hybrid approach where a generic block processor evaluates a block by extraction first and then transcoding block-type-specific data second is a recommended strategy for high-performance applications.

Security blocks differ from other kinds of extension blocks because they are dependent on other blocks in a bundle (their security target blocks). This requires that a BPA impose a certain ordering to block evaluation, that some security processing be deferred, and that security blocks be preserved to enable downstream processing at other BPAs.

Importantly, security block processing is the processing of security operations resident in the block. To understand how a BPA processes a security block, three roles can be defined on an operation-by-operation basis. a BPA may be either a security source, verifier, or acceptor of operations in a security block. A BPA processes a security block by fulfilling its role for each operation in that block.

The ability to represent multiple operations in a single security block, called target multiplicity, can be done whenever security operations share common information. This is beneficial as it reduces the size of a bundle and can make processing more efficient. The grouping and ungrouping of operations in a security block can be accomplished through four block manipulation algorithms that Add or Remove operations, and merge or split security blocks.

References

1 Farrell, S., Weiss, H., Symington, S. and Lovell, P. [2011]. Bundle Security Protocol Specification, RFC 6257. **URL**: https://www.rfc-editor.org/info/rfc6257.

2 Birrane, E. J. and McKeever, K. [2022]. Bundle Protocol Security (BPSec), RFC 9172. **URL**: https://www.rfc-editor.org/info/rfc9172.

8

Security Dependency Management

BPSec offers significant flexibility related to the application of security services to elements of a bundle. Multiple security blocks can be added to a bundle, and each of these blocks represent a collection of security operations.

This flexibility is needed to adapt to the variety of potential deployments of BPv7 in operational networks, to include those conforming to the Disruption Tolerant Internet (DTN) architecture. But, this flexibility leads to dependencies on bundle information, block processing, and overall constraints within the protocol itself.

This chapter discusses the information that security blocks depend on to be present for their successful processing, and the information at a Bundle Protocol Agent (BPA) that is, itself, dependent upon the information in security block.

After reading this chapter you will be able to:

- Describe why blocks need different levels of security.
- Explain the concept of inter- and intra-bundle dependencies.
- Demonstrate how BPSec constraints simplify processing.
- Justify the processing order of security policy.

8.1 Dependency Management

The defining characteristic of the Disruption Tolerant Internet (DTN) architecture is its inability to guarantee timely, end-to-end message delivery. In addition to the regular constraints placed on application design and development, this inability introduces new restrictions on the assumptions made by transport protocols such as BPv7.

If transport protocols cannot presume that common information is pre-shared at messaging endpoints, that information must be carried within the transport mechanism itself. This observation was an important consideration in the development of the BPv7 extension block mechanism. Bundles must allow for their diverse annotation based on the characteristics of the networks in which they are deployed.

Much of the discussion associated with BPSec security extensions center on the design and implementation of capabilities such as block-level granularity for security services. However, the dependencies associated with extension blocks, and the securing of those blocks, has both *structural* and *informational* impacts.

Securing Delay-Tolerant Networks with BPSec, First Edition. Edward J. Birrane III, Sarah Heiner, and Ken McKeever.
© 2023 John Wiley & Sons, Inc. Published 2023 by John Wiley & Sons, Inc.
Companion website: www.wiley.com/go/birrane/securingdelay-tolerantnetworks

Definition 8.1 *(Information Dependency)* A relationship where certain information must be present in order to complete a processing action. This dependency requires the sharing of information between entities in the network. These entities may be copresent in a bundle (such as between two extension blocks) or shared between BPAs on opposite sides of a BPv7 network (such as with security policy settings).

Definition 8.2 *(Structure Dependency)* A relationship where fields or elements of a structure must be copresent to enable the processing of information. This kind of dependency is a specialization of an information dependency, but applied to information incumbent in the data structures used to carry information. For example, a security block has a structural dependency on its target block because the security block both needs the target block information (information dependency) and that information must come from the target block copresent in the bundle (structure dependency).

Figure 8.1 illustrates the concepts of information and structure dependencies in the context of providing a *bcb-confidentiality* service. In this illustration, a Bundle Protocol Agent (BPA) at Node 1 sends a bundle to a BPA at Node 2. This bundle consists of an encrypted payload with the primary block added as additional authenticated data, as indicated by a Block Confidentiality Block (BCB).

The information carried in bundle blocks represents multiple *structure* dependencies. The payload block is dependent upon the BCB to signal to the BPA at Node 2 that it is encrypted. The BCB is dependent upon the primary block and payload block not changing so as to perform authenticated decryption. These types of structural dependencies are the bases for the block processing algorithms outlined in Chapter 7. More significantly, the seemingly simple bundle construction shown in Figure 8.1 represents a complex set of information dependencies.

To achieve the proper security outcome, multiple sets of information must be shared between the BPAs at Nodes 1 and 2, to include the following.

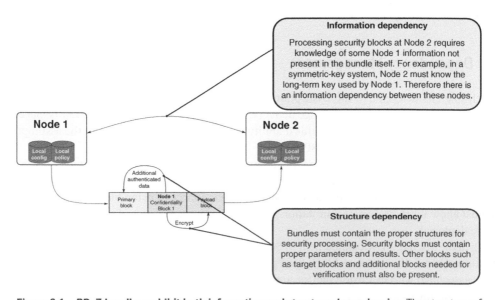

Figure 8.1 BPv7 bundles exhibit both information and structure dependencies. The structure of a BPv7 bundle allows for carrying some data in situ, but other information must also be configured out-of-band of a specific bundle transmission.

1) The BPAs must agree on what security contexts will be supported for this BCB operation. Where the BPA at Node 1 to select a security context unsupported by the BPA at Node 2, no decryption could take place.
2) The keys and other cryptographic materials generated at Node 1 must be shared with Node 2. This may include pre-sharing symmetric, long-term keys as well as carrying information in the BCB to the BPA on Node 2.
3) In determining that it is the security acceptor for certain operations, the BPA at Node 2 must have some sense of traffic coming from the BPA at Node 1.

Understanding the unique characteristics of dependencies in a BPv7 network is fundamental to understanding the design decisions, implementations, and configurations associated with the BPSec.

This chapter discusses dependencies in two ways. The first portion of this chapter explores several types of information and structure dependencies that exist in this security ecosystem. This starts with a discussion of dependencies associated with bundles – both among blocks within bundles, and among bundles within a network – followed by a discussion of security dependencies. The second part of this chapter explains how BPSec constraints and special processing rules help organize the complexity associated with these dependencies. This includes a discussion on handling seemingly conflicting dependencies in BPv7 networks.

8.2 Bundle-Related Dependencies

As the mechanism for securing information within bundles, dependencies associated with bundles are a primary focus of BPSec. The dependencies between and among blocks within a bundle represent a significant portion of the rules and design constraints associated with the construction and processing of security blocks.

> *Note that removal of an extension block will probably disable one or more elements of bundle processing that were intended by the BPA that inserted that block. In particular, note that removal of an extension block that is one of the targets of a BPSec security block may render the bundle unverifiable.*
>
> *BPv7, RFC 9171 §4.4, [1]*

These dependencies are termed *intra-bundle* dependencies as they focus on information within a single bundle.

There are also dependencies among the overall stream of bundles between and among BPAs in a network. When extension blocks carry information not directly related to their application payload, that annotative information is likely to have benefit beyond the individual bundle. These dependencies are termed *inter-bundle* dependencies because they focus on information that is applicable to more than one bundle in the network.

Understanding the nature of both *inter-bundle* and *intra-bundle* information helps clarify BPSec design decisions, configuration needs, and best practices.

8.2.1 Intra-Bundle Dependencies

Because a bundle is a collection of blocks, an intra-bundle dependency refers to the situation where one or more blocks in a bundle are needed to process other blocks in the same bundle. Understanding the relationships between extension blocks is vital to understanding the order in which these blocks should be secured, and how to construct sensible policies to oversee the security process.

The relationships between and among blocks in a bundle can be characterized by the kind of processing they enable, to include payload support, information decoding, processing configuration, and data viability.

Each of these activities can be supported with the addition of extension blocks within a bundle in ways that adjust how security services should be applied to a bundle.

8.2.1.1 Payload Processing

The payload of a bundle represents the reason why the bundle exists in the network. Viewed from this perspective, every other block in a bundle is related in some way to payloads – either in the delivery of the current payload block, or helping to deliver future payload blocks in future bundles.

In the context of a single bundle, a payload dependency refers to the relationship between a payload block and any other block in the bundle that is required for the successful extraction of the payload at the bundle destination. This also includes blocks that might cause a payload (and thus a bundle) to be dropped in the network, as doing so prevents the payload from ever reaching its destination.

Blocks that are required for payload processing should have similar security applied to them as to the payload. If the payload is encrypted, information needed to process the payload may also need to be encrypted. At a minimum, blocks that can affect payload processing should be integrity checked to make sure that malicious block manipulation does not lead to the loss of payload processing.

8.2.1.2 Decoding

When a block carries information needed for the proper decoding of other information in a block, then those blocks share a decoding dependency. Security blocks represent one type of decoding dependency within a bundle.

The encrypted block has a structural dependency with the BCB that encryptes it. This is a structural dependency because, in addition to the information carried in the BCB, the structure of the target block has been altered – its plaintext block-type-specific data field is replaced with ciphertext.[1] Unlike a BCB, Block Integrity Blocks (BIBs) do not represent a decoding dependency on their target blocks, as there is no difficulty decoding the target block.

Decoding dependencies are only used to hide information and, thus, should be used sparingly within a bundle. Too many decoding dependencies may result in confidentiality loops wherein two blocks might rely on each other to be able to be decoded.

BPSec constraints, such as disallowing a BCB to target another BCB, seek to avoid problems related with this kind of dependency.

8.2.1.3 Configuration

Bundles carry information both to the applications resident at bundle destinations and to BPAs processing information from extension blocks. In both cases, bundle information can be used to inform the local configuration of the bundle destination, of any block acceptor, or any other BPA in the network.

In cases where BPAs rely on information carried within a bundle for configuration, there is an information dependency between these blocks and the BPAs they help configure. For example, if an extension block were to carry information about the state of the network topology, as known by the block source, every BPA in the path traversed by that bundle could inspect that block and use it to update its local configuration.

---

1 And that ciphertext may have a different size than the plaintext.

Sometimes, configuration information needs to be protected from other BPAs in the network – such as when the configuration includes privileged items such as security keys. Other times, configuration items should keep their integrity, but otherwise be viewable by all BPAs in a network to assist in information dissemination throughout the network.

8.2.1.4 Assessment

An assessment dependency exists whenever information within a bundle is required to process other information in the same bundle. In this dependency, a BPA cannot fully process a block without the contents of some other block. Assessment is differentiated from configuration in this case as assessment information is focused on the processing of blocks in the same bundle, whereas configuration refers to the changing of data on a BPA independent of the processing a given the bundle.

Integrity mechanisms are a common example of this type of assessment dependency. The security operation $OP(bib\text{-}integrity, Block_{target})$. In this case, the integrity mechanism carried in the BIB for this security operation must be considered as part of assessing the viability of the target block. The target block cannot be considered valid if its integrity signature has not been verified.

Security is not the only time when there exists an assessment dependency between information in a bundle. One non-security example of such a dependency is the Bundle Age Block used to determine how long the payload of a bundle should persist in the network. The timing information carried in this block helps to determine whether the bundle has exceeded its lifetime. If so, the bundle can be removed from the network. In this way, the information in the Bundle Age Block is used to assess the validity of the bundle payload.

In cases where block types encode assessment dependencies, security policies should not prevent this assessment. For example, if a Bundle Age Block were encrypted, then it could not be used to assess bundle age without being decrypted at a security acceptor. By the time a bundle reached such as acceptor, the bundle could have been in the network for significantly longer than its expected lifetime.

8.2.2 Inter-Bundle Dependencies

The information present in blocks can have an impact and utility beyond the bundles in which they are transmitted. When the information in one bundle impacts the processing of other bundles, then there exists an inter-bundle dependency. Since the structure of one bundle does not impact the structure of another bundle, inter-bundle dependencies are always examples of an information dependency.

Figure 8.2 illustrates one concept of inter-bundle dependencies. In this figure, bundles generated from a BPA at Node 1 and routed through BPAs at Node 2 and Node 3 on their way to some other bundle destination. These bundles include multiple types of information, such as:

- Updated path information (bundle B1),
- Policy updates (bundle B2),
- Link state measurements (bundle B3), and
- User data (bundle B4).

This example presumes that the information held in bundles B1, B2, and B3 have some impact on the ability to deliver the information for bundle B4. For this reason, we can say that B4 has inter-bundle information dependencies on bundles B1, B2, and B3. If the BPA at Node 2 were to reorder the transmission of bundles, perhaps due to retransmission, then this four-tuple of bundles

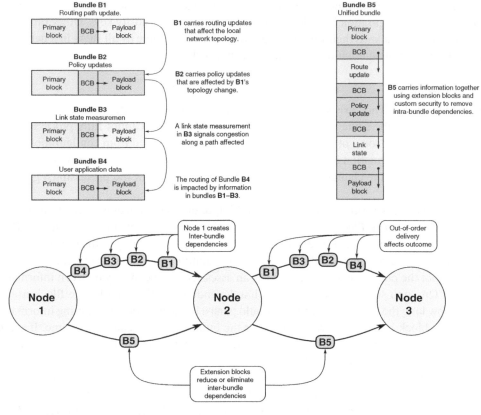

Figure 8.2 Inter-bundle dependencies impact bundle delivery. The inability of the DTN architecture to guarantee order or synchronization creates dependencies between bundles that can be addressed, in part, using BPv7 features.

will reach the BPA at Node 3 out of order. B4 may be received prior to any of the other information necessary help it reach its destination.

One way to mitigate certain types of inter-bundle dependencies is to collapse information into a single bundle. Just as extension blocks can carry session-like information to mitigate the difficulty of synchronization in the DTN architecture they can also mitigate certain kinds of inter-bundle dependencies. This is also illustrated in Figure 8.2 as bundle B5. In this case, the BPA at Node 1 passes a variety of information as extension blocks annotating the user application payload. Using bundle B5 in lieu of bundles B1, B2, B3, and B4 removes inter-bundle dependencies because there is only one bundle.

The practice of aggregating multiple information in a single bundle can be useful in certain circumstances. However, networks are rarely used to transmit a single bundle. If many bundles were being sent by the BPA at Node 1, it could become impractical to augment all of them with multiple extension headers to avoid any type of out-of-order delivery. Inter-bundle dependencies can be mitigated, but they cannot be eliminated.

BPSec mechanisms should carefully consider the impact of security operations as they relate to inter-bundle dependencies, particularly bundles carrying network-related information.

8.2.2.1 Network Information

A network information dependency is a particular type of inter-bundle configuration dependency where information about the BPv7 network is required in order to provide or update configuration of the network. The lack of end-to-end data synchronization has implications beyond the inability to synchronize information at messaging endpoints. While end-to-end communication might be significantly constrained, communication within certain localities (such as within 1- and 2-hop neighborhoods) might be beneficial. Generally, the BPAs that exist between a message source and destination rely on information gathered from multiple algorithms used to measure (or predict) congestion, route computations, link quality, and other network-related metrics. The distribution of network information in a BPv7 network may be accomplished either by placing this information in the payload block of a bundle dedicated to this purpose, or by adding extension blocks to other bundles.

Placing network information in a bundle as a payload implies that the information has a known destination – the bundle destination. This mechanism is useful when implementing algorithms between endpoints that need to communicate some network conditions. Since any kind of end-to-end mechanism in a DTN architecture might be limited to best-effort, this type of information exchange might yield poor results.

Alternatively, carrying this type of information as extension blocks has a few benefits, as follows.

- Extension blocks requires fewer resources to produce than whole bundles.
- Extension blocks are more appropriate for inspection by waypoint BPAs.
- In cases where bundle exchange is plentiful, extension blocks can reach all parts of the network.

Regardless of whether network information is carried in an extension or payload block, this information should be available for inspection by waypoint BPAs. Encrypting non-sensitive network measurement data in a BPv7 network would prevent the network from making best use of available information and resources. This type of inefficiency is much more problematic within the DTN architecture which often cannot supply on-demand access to connectivity.

8.2.2.2 Fragmentation Dependency

Bundle fragmentation, if allowed by policy, can create an inherent inter-bundle dependency where all bundle fragments are required at the block acceptor in order to reconstruct and decode the bundle. BPSec requires that if a bundle is fragmented, all security blocks must be contained within the same fragment as their target blocks. This avoids conditions where a security validator or a security acceptor may receive target blocks without the corresponding security blocks or security blocks without their target blocks, resulting in a security error. For this reason, fragmentation dependencies are not typically a security-related dependency, but there can be security challenges associated with handling fragments.

8.3 Security-Related Dependencies

The existence of dependencies within and among bundles informs recommendations for where security services should be used. A separate set of security-related dependencies discuss the mechanics of applying recommended security services. Security-related dependencies focus on four topics that govern how security services are implemented at a BPA: security operations within a security block, security blocks within a bundle, security configuration both in bundles and at processing BPAs, and policy at BPAs.

8.3.1 Operation Dependencies

As represented within a bundle, a security operation is captured with a security block and contains cryptographic material associated with the application of a security service to some other block within the bundle. Because these operations interact with two different types of blocks, there are two dependencies for every security operation: a dependency on the encapsulating security block and a dependency on the target block.

Structural dependencies between security operations and security blocks exist because the interpretation of the security operation, and its mapping of security results, relies on the structural elements of the security block. For example, security target results within a security block are ordered relative to security target identifier. BPSec design constraints define rules meant to simplify the interpretation of these types of dependencies, to include rules that ensure security operations remain unique and unambiguous and algorithms relating to how operations can be added, processed, and removed from security blocks.

Similarly, security operations have a structural dependency on the security target block. Not only must the information within the security target block be present in the bundle, but it must be represented in the bundle in the correct way to be correctly identified and input into the security operation's security context. BPSec provides additional processing rules associated with ensuring target blocks are not obfuscated in such a way that security operations applied to those blocks cannot be completed.

8.3.2 Block Dependencies

The construct of a security block and a security target block implies a dependency; the security block must be processed prior to processing the target block. For example, a target block might need to be decrypted or have its integrity verified prior to a BPA processing the contents of that target block. This type of dependency, shown in Eq (8.1), indicates that the processing of $Block_{target}$ is dependent on the prior successful processing of $Block_{security}$.

$$Block_{target} \rightarrow Block_{security} \tag{8.1}$$

Security block processing becomes more complex when dependencies exist between security blocks and among security target blocks. To handle this complexity, the BPSec intentionally limits certain features so minimize the occurrence of dependencies. In cases where inter-block dependencies might still exist, special processing rules have been added to the protocol to ensure the unambiguous processing of blocks.

Figure 8.3 illustrates one example of why security operation dependencies must be carefully considered within BPSec. In this figure, a bundle contains a BIB (BIB 1) with security operations providing *bib-integrity* over the primary block, an extension block, and the payload block. If some downstream BPA wishes to encrypt the extension block using one security context and the payload block using another security context, there exist security operation dependencies that must be addressed first. In this example, the security operations $OP(bib - integrity, Block_3)$ and $OP(bib - integrity, Block_1)$ both exist within BIB 1. BIB 1 cannot be encrypted twice – once using BCB 1 and once using BCB 2.

Figure 8.3 is an example of a *valid* security operation dependency that results in an *invalid* security block dependency. The information dependencies inherent in encrypting extension block 3 and its signature and the information dependency between encrypting the payload and its signature are both valid. However, the block dependency where a BIB would be the target block of two separate BCBs is not allowed.

8.3.3 Configuration Dependencies

The third security-related dependency is the information dependency between configurations at different BPAs throughout the network. Unlike other security and bundle dependencies,

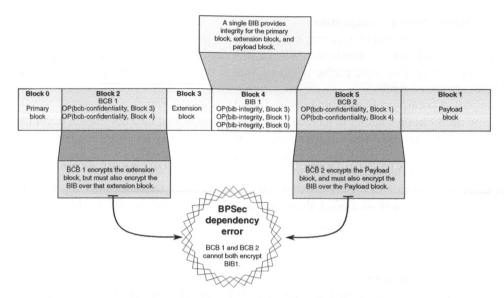

Figure 8.3 Security operations and blocks have different dependencies. Encrypting security target blocks and their *bib-integrity* mechanisms is allowed in BPSec. However, security operations can be packaged into blocks in ways that violate block processing dependencies.

these do not directly involve bundles or blocks at all. Because information in a BPv7 network might not be able to synchronize end-to-end data, individual bundles must either carry information or assume that information will be available at receiving bundles at some point in the future.

In cases where it is assumed that information will be present at a downstream BPA at the time of bundle receipt, there exists an information dependency between the security source BPA adding a security operation to a bundle and security verifier or security acceptor BPA processing that bundle. Since, in this instance, the information is not carried in the bundle itself, this is considered a part of the configuration of the various BPAs in the system.

Importantly, for effective processing this information must be consistent. If a security source adds a security operation which requires knowledge or implementation at downstream nodes and those knowledge or implementation does not exist, then the security operation will not be verifiable or acceptable in the network.

The set of configuration dependencies existing between and among BPAs can be categorized into security context support, security context configuration, and policy expressions.

8.3.3.1 Security Context Support

Every BPv7 implementation must understand and process BPSec security blocks, which means that every BPA must implement – at a minimum – some common set of security contexts and cipher suites. While security blocks may carry some configuration information related to security contexts, bundles cannot carry the information needed to wholly implement a security context at a node.

For this reason, the security source BPA must have some confidence that downstream BPAs implement the required security contexts in the network. If a BPA receives a bundle containing a security block referencing an unsupported security context, there is no BPSec-compliant mechanism by which the BPA could process the security operations resident in that security block, regardless of any policy statement requiring the BPA to do so.

8.3.3.2 Security Context Configuration

Part of the implementation of a security context includes the ability to customize the behavior of that context for different circumstances. When a BPA needs to generate or process cryptographic materials it is likely that some configuration relevant to the materials being processed is used to produce customized results.

Examples of security context configuration include shared secrets (e.g. symmetric long-term keys) and decisions on whether to include additional authenticated data beyond the target block. These configurations could be included in security blocks as part of security context parameters. However, receiving BPAs must also be able to process these parameters when present and use appropriate default values when parameters are not present in the security block.

In cases where both a BPA configuration and a security block contain competing versions of the same security context parameter, there may be an issue in understanding which parameter to use in what circumstance. This type of ambiguity is resolved through policy statements on the receiving BPA.

8.3.3.3 Policy Configuration

Security policy both at a local BPA and across a BPv7 network provides information about how events related to security processing should be handled. Unlike security context configuration, policy exists at a BPA to determine when security contexts should be applied. Specifically, policy determines what role a BPA should take relative to a security operation and what reactions should occur when security processing succeeds or fails.

Without this policy specification, no security processing can be done at the BPA. For example, a bundle carrying multiple security operations may be received by a BPA that has no relevant security policy statement for the operations in that bundle. In this circumstance, the BPA would perform no security processing and would handle the bundle as if no security were present.

The relationship between policy expressions and security roles is shown in Figure 8.4. In this figure, a BPA at Node 1 might determine, by analyzing local policy, that it should be encrypting and signing certain contents of a bundle. The BPAs at Nodes 2 and 3 may, similarly, have policy expressions that determine how existing security operations are verified and accepted, and when new security operations should be created.

An important observation from Figure 8.4 is that these policy expressions are made independent of any single bundle. They can be configured and communicated across a network and updated as a function of commanding, state-based autonomy, time, or other considerations.

8.3.4 Security Dependency Mappings

Security elements within the BPSec represent different types of dependencies, both in the nature of how the elements inform each other, and how they are communicated throughout a BPv7 network. The summary of this mapping is provided in Table 8.1. In this table, the BPSec elements of security operations, blocks, contexts, and policy are compared with informational and processing dependencies that apply within, and across, bundles.

As the manifestation of a security service in a bundle, the security operation is concerned with protecting information within a bundle using the structure of the bundle. Operations only apply within the context of their carrying bundle and any additional information is only available through their encapsulating security block or local configuration on the implementing BPA. However, policy configurations at local BPAs determine the security roles that impact how security operations are processed. Similarly, security context configurations and security context support determine

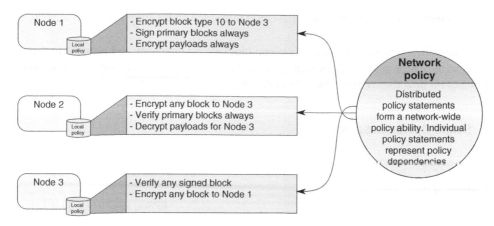

Figure 8.4 **Policy configurations determine security roles.** Security blocks do not mandate their security verifiers and acceptors in advance. Policy at individual BPAs determines the security role at the time of bundle reception.

Table 8.1 BPSec security element dependencies types.

	Information	Structure	Inter-bundle	Intra-bundle
Security operations	◑	●	○	●
Security blocks	○	●	○	●
Security configurations	○	●	●	○

BPSec security elements represent different types of dependencies. Individual topics can be always represented by a particular type of dependency (●), optionally represented by a type of dependency (◑), or never represented by a type of dependency (○).

how cryptographic materials are used in relation to these operations. Therefore, security operations have information dependencies on the local configurations of a BPA.

Security blocks may carry information related to block processing, and may themselves be processed as a function of adding, removing, or separating operations from the block. As extension blocks within a bundle, security blocks only contain information related to the processing of security within a single bundle. Security blocks are not impacted by policy and security context configurations because blocks are created to hold security operations, and it is the security operations that are affected by configuration settings.

Security configurations are dependent on the information within bundle structures – such as understanding what security blocks are present in a bundle and available for evaluation. Since these configurations determine what security information is added to, and removed from, a security operation for any traffic exchange in the network, they are also the mechanism by which inter-bundle information is applied.

8.4 Dependency-Related Constraints

Understanding the various dependencies that exist between elements of bundles, blocks, and BPAs is helpful to understanding the rationale behind constraints imposed by BPSec. These constraints reduce the complexity associated with block management and reduce the possibility of intra- and inter-bundle dependencies that could introduce ambiguities related to security block processing.

These constraints represent a balance between block-based flexibility and preserving deterministic processing. These protocol constraints, covered in other chapters, are summarized next with an emphasis on how they reduce block dependency.

8.4.1 Single-Operation Sources

> *…[a] BCB … MUST be used instead of adding both a BIB and then a BCB for the security target at the security source.*
>
> *BPSec, RFC 9172, §3.9, [2]*

BPSec security sources can add only a single security operation for a target block into a bundle. This constraint is enforced even if more than one security operation could be added to the target block and preserve security operation uniqueness. For example, the security operations of *bib-integrity* and *bcb-confidentiality* could both exist in a bundle as unique operations. However, these two operations cannot both be added by the same security source.

Because a security operation is represented in a bundle by a security block, this constraint implies that a security source cannot add more than one security block for a given target block to a bundle. This restriction reduces the number of possible security operations (and security blocks) in a bundle. Since this restriction prevents same-source duplication across security target blocks, it also reduces the likelihood of dependencies between security operations in the network.

Most security operations involving a given target block are likely to be applied either by the BPA adding the block to the bundle, or at special BPAs serving as gateway or other special functions in the network. Restricting the number of operations that can be added by a single BPA, therefore, is likely to restrict the number of operations added to a block for the life of the bundle.

8.4.2 Unique Security Services

> *A security target [of a BIB] listed in the Security Targets field MUST NOT reference a security block defined in this specification (e.g. a BIB or a BCB).*
>
> *BPSec, RFC 9172, §3.7, [2]*

BPSec prohibits situations where otherwise independent security operations would provide redundant security services.

One such situation is shown in Eq (8.2a), where a BIB is used to sign a BCB that, by definition, already contains an integrity mechanism.

$$Block_{target} \rightarrow Block_{BCB} \rightarrow Block_{BIB} \quad \text{(prohibited by BPSec)} \tag{8.2a}$$

This construct is prohibited because the service *bcb-confidentiality* provided by the BCB includes its own integrity mechanisms over both the generated ciphertext and identified additional authenticated data. Therefore, there is little value in adding a separate, additional integrity mechanism over the BCB.

It is important to note that allowing a BIB to target a BCB would have some minor benefit in certain circumstances, such as when a security verifier would like to verify the integrity of the BCB block without attempting to perform a decryption of the BCB.[2] However, allowing both a BIB to target a BCB, and a BCB to target a BIB can lead to circular dependencies which would result in a loss of determinism in security processing, as discussed in Section 8.4.3.

For this reason, BPSec prohibits this construction. Intermediate verification of integrity mechanisms could be accomplished through the definition of special security contexts that define both additional authenticated data via a cipher suite and dedicated integrity mechanisms that could be checked by a security verified without decryption.

A second set of examples of redundant security services, shown in Eq (8.3), are checking the integrity of integrity signatures and encrypting encrypted data.

$$Block_{target} \rightarrow Block_{BIB_1} \rightarrow Block_{BIB_2} \quad \text{(prohibited by BPSec)} \tag{8.3a}$$

$$Block_{target} \rightarrow Block_{BCB_1} \rightarrow Block_{BCB_2} \quad \text{(prohibited by BPSec)} \tag{8.3b}$$

Checking the integrity of an integrity mechanism is considered redundant because corruption of the first integrity mechanism is expected to manifest as a failure to verify the integrity of the target block. Therefore, detecting corruption of the first integrity mechanism by a second integrity mechanism would not result in a more robust reaction to corruption.

Encrypting already encrypted data (super encryption) can have utility in certain circumstances. However, encrypting the entire contents of a BCB will hide, among other things, the target of the BCB. This makes it impossible to determine if a target block is encrypted without decrypting every encrypted BCB block and examining its security target list. Therefore, every encrypted BCB would need to be decrypted at every hop prior to processing any target block to determine what is/is not encrypted in the bundle. Since the contents of a BCB do not need to be encrypted to preserve the confidentiality of their target block, this amount of confusion would provide no additional benefit to target block confidentiality and is prohibited.

8.4.3 Exclusively Linear Dependencies

The chaining of security blocks must not impact the determinism of security block processing; an acceptor or verifier must always understand the order in which security blocks, and their target blocks, must be evaluated. The simplest way to prevent processing ambiguity is to prohibit circular security dependencies.

Definition 8.3 *(Circular Security Dependency)* Any dependency chain of blocks that includes duplicate block types, such as shown in Eq (8.4).

The use of block type (rather than block identifier) is sufficient to identify duplication because a given chain of security blocks are all related to operations on the same target. For example, having two BIB blocks in a chain declares that an integrity mechanism depends on another integrity mechanism, just as two BCB blocks in a chain declares that a confidentiality mechanism depends on another confidentiality mechanism.

$$Block_{target} \rightarrow Block_{BCB_1} \rightarrow Block_{BIB_1} \rightarrow Block_{BCB_2} \quad \text{(prohibited by BPSec)} \tag{8.4a}$$

---

2 Some cipher suites only verify the integrity mechanism as part of the decryption process.

$$Block_{target} \rightarrow Block_{BIB_1} \rightarrow Block_{BCB_1} \rightarrow Block_{BIB_2}$$

$$Block_{target} \rightarrow Block_{BCB_2} \rightarrow Block_{BIB_3} \qquad \text{(prohibited by BPSec)} \qquad (8.4b)$$

Equation (8.4a) demonstrates the (prohibited) case where $Block_{BIB_1}$ (holding an integrity mechanism over block $Block_{BCB_1}$) is, itself, encrypted by block $Block_{BCB_2}$. This construction is already prohibited by BPSec according to Eq (8.2a). However, if it were not, the contents of block $Block_{BIB_1}$ would be encrypted and, thus, unknown which BCB was protected by it and which BCB was used to encrypt it. A verifier or acceptor would not know the correct order of processing blocks $Block_{BCB_1}$ and $Block_{BCB_2}$.

Equation (8.4b) demonstrates the (prohibited) case where $Block_{target}$ is first signed by $Block_{BIB_1}$, then $Block_{BIB_1}$ and $Block_{target}$ are each encrypted by $Block_{BCB_1}$ and $Block_{BCB_2}$, respectively. Then the results of the encryption are, themselves, protected by integrity mechanisms provided by $Block_{BIB_2}$ and $Block_{BIB_3}$. This construction is also prohibited by Section 8.4.2, but were it to exist, would cause processing ambiguities. In this instance, $Block_{BCB_1}$ cannot be evaluated until it is decrypted, which cannot happen until $Block_{BIB_2}$ is evaluated. This means that not all BIB results can be evaluated at once, which places a computational burden and possible ambiguity on how to process BIB blocks that contain multiple security results.

8.5 Special Processing Rules

BPSec constraints on the dependencies among security blocks reduces the combinations of security blocks that must be processed in a well-formed, secured bundle. This section describes the way in which properly formed, secured bundles must be handled. These processing rules cover three distinct types of processing: the order of evaluating security blocks, required combinations of security blocks, and special circumstances when handling block multiplicity.

8.5.1 Inclusive Confidentiality

When a security target block is encrypted any other security block that relies on the unencrypted security target block must also be encrypted. This processing rule prevents security information about the unencrypted target block from persisting after the target block has been encrypted. Because BPSec only defines two security block types (BIB and BCB), and because there exist constraints on dependencies between and among BIBs and BCBs, there is only one circumstance where this situation may exist: when a BIB exists for a target block that is, at a different BPA, encrypted by a BCB.

> *When adding a BCB to a bundle if some (or all) of the security targets of the BCB match all of the security targets of an existing BIB, then the existing BIB MUST also be encrypted.*
> *BPSec, RFC 9172 §3.9, [2]*

> *When adding a BCB to a bundle, if some (or all) of the security targets of the BCB match some (but not all) of the security targets of a BIB, then...[a]ny security results in the BIB associated with the BCB security targets MUST be removed from the BIB and placed in a new BIB. This newly created BIB MUST then be encrypted.*
> *BPSec, RFC 9172 §3.9, [2]*

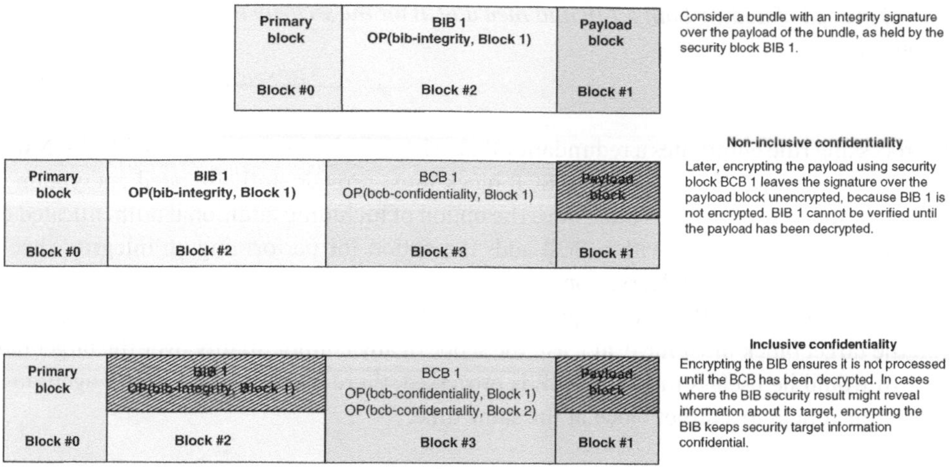

Primary block	BIB 1 OP(bib-integrity, Block 1)	Payload block
Block #0	Block #2	Block #1

Consider a bundle with an integrity signature over the payload of the bundle, as held by the security block BIB 1.

Non-inclusive confidentiality

Primary block	BIB 1 OP(bib-integrity, Block 1)	BCB 1 OP(bcb-confidentiality, Block 1)	Payload block
Block #0	Block #2	Block #3	Block #1

Later, encrypting the payload using security block BCB 1 leaves the signature over the payload block unencrypted, because BIB 1 is not encrypted. BIB 1 cannot be verified until the payload has been decrypted.

Inclusive confidentiality

Primary block	BIB 1 OP(bib-integrity, Block 1)	BCB 1 OP(bcb-confidentiality, Block 1) OP(bcb-confidentiality, Block 2)	Payload block
Block #0	Block #2	Block #3	Block #1

Encrypting the BIB ensures it is not processed until the BCB has been decrypted. In cases where the BIB security result might reveal information about its target, encrypting the BIB keeps security target information confidential.

Figure 8.5 BPSec requires inclusive confidentiality. Only encrypting the target block of a BIB would result in the inability to verify that BIB even though the target block integrity could otherwise be assured by an authenticating cipher suite.

This processing rule is required to avoid the situation where a BIB holding a plaintext integrity result over a target block cannot be verified because the application of a BCB has replaced the plaintext of the target block with ciphertext. In this situation, both the target block and the pre-existing BIB over the unencrypted target block must be encrypted with different confidentiality operations. The two instances of *bcb-confidentiality* might exist as two separate BCBs, as shown in Eq (8.5a, b), or (under certain circumstances) combined into a single BCB carrying multiple security operations. When this happens, the following security blocks would exist in the bundle.

$$Block_{target} \rightarrow Block_{BCB_1} \qquad \text{(allowed by BPSec)} \qquad (8.5a)$$

$$Block_{target} \rightarrow Block_{BIB} \rightarrow Block_{BCB_2} \quad \text{(allowed by BPSec)} \qquad (8.5b)$$

The differences between inclusive and non-inclusive confidentiality is shown in Figure 8.5. This figure first shows a bundle that has a primary block and a payload block, with the payload block's integrity protected by BIB 1. A BPA (separate from the BPA which added BIB 1) determines that the payload block must be encrypted. Were BPSec to allow non-inclusive confidentiality, then the payload would be encrypted and BIB 1 holding the *bib-integrity* service over the payload would not be encrypted. A downstream BPA trying to verify this *bib-integrity* service would fail if the payload wasn't first decrypted. This situation is avoided by using inclusive confidentiality, which encrypts both the payload and BIB 1. In this case BIB 1 cannot have its integrity checked until it is decrypted.

8.5.2 No Service Redundancy

Constraints imposed by BPSec prevent the syntactic redundancy of security operations – that the same security operation cannot be applied to the same security target block twice in a bundle. This constraint does not address the more subtle issue of semantic redundancy – in which different security services provide similar security capabilities. BPSec provides processing rules to reduce the likelihood of semantic redundancy provided by different security operations.

> *In cases where a security source wishes to calculate both a plaintext integrity-protection mechanism and encrypt a security target, a BCB with a security context that generates an integrity-protection mechanism as one or more additional security results MUST be*

> *used instead of adding both a BIB and then a BCB for the security target at the security source.*
>
> BPSec, RFC 9172 §3.9, [2]

This processing rule eliminates a redundancy that exists within the BIB and BCB blocks. Namely, that the cipher suites used to populate a BCB block must provide authenticated encryption and that the authenticated encryption must have the option of including additional authenticated data (AAD). The inclusion of AAD with a BCB adds the option for performing an integrity check of material prior to its attempted decryption.

Since a BIB cannot be verified or accepted if it is encrypted, then the scope of any integrity operation on the target block must have, at least, the scope of any confidentiality over the target block. Since a BIB would be useless until the BCB is processed, there is no need for a BPA agent to add both a BIB and a BCB to a target block at the same time.

8.5.3 Process Confidentiality First

> *When BIBs and BCBs share a security target, BCBs MUST be evaluated first and BIBs second.*
>
> BPSec, RFC 9172 §5.1, [2]

Specifying security block processing order helps to avoid ambiguity when applying security services. Because BPSec defines a number of protocol constraints meant to reduce ambiguities, there is only one processing rule required to ensure correct processing order between BIB and BCB blocks.

BPSec prohibits the $Block_{BCB} \rightarrow Block_{BIB}$ dependency, which means that no BIB may target a BCB. Similarly, BPsec prohibits a security target block from having the same security operation applied more than once in a bundle, meaning that multiple BIBs or multiple BCBs cannot target the same block.

Therefore, the only valid interaction between a BIB and a BCB is when a single BIB and a single BCB share the same security block. This condition can only occur when a BIB was applied to a target block and then a BCB was applied to the same target block by a separate security source. In this case, a *bcb-confidentiality* service must be placed over the target block and another *bcb-confidentiality* service placed over the original BIB holding a plaintext signature over the plaintext target block.

To properly handle this case, both $Block_{BCB_1}$ and $Block_{BCB_2}$ must be processed before processing $\rightarrow Block_{BIB}$. $Block_{BCB_1}$ must be processed to decrypt the target block and restore its plaintext and $Block_{BCB_2}$ must be processed to decrypt $Block_{BCB_1}$ so that the plaintext integrity result can be read. Therefore, all BCBs relating to this security operation must be handled before the BIB is handled.

8.6 Handling Policy Conflicts

Some of the dependencies discussed in this chapter involve configurations resident on BPAs in the network. These configurations are important because they can affect the processing of blocks and bundles.

One important type of configuration specifies the policies resident at every BPA and used to disposition bundles. Some of these policies relate to the handling of BPSec security blocks. Other policies might be used for handling quality of service, network management, routing, and other network functions.

However, the time-variant nature of some BPv7 deployments also make is unlikely that these configurations can be kept synchronized across the network. One way BPv7 mitigates this likely lack of synchronization is by carrying some policy statements in the bundle itself.

8.6.1 In-Bundle Policies

All BPv7 bundles define a set of bundle processing control flags and all extension blocks carry a set of block processing control flags. These flags allow individual bundles to carry their own policy with them, to mitigate the lack of synchronization among BPAs.

Bundle processing flags in BPv7 encode policy decisions such as whether a bundle can be fragmented and what types of activities should be reported to the bundle's Report-To EID.

Block processing control flags (BPCFs) associated with each extension block describe how the extension block should be handled when a bundle is fragmented and how to handle the case where a BPA fails to process an extension block. The failure-to-process policies specified in the BPCFs are discussed in more detail in Section 6.5.2.1. However, two important BPCFs are:

1. **Bit 2 (0x04)**: *Delete bundle if block can't be processed.*
2. **Bit 4 (0x10)**: *Discard block if it can't be processed.*

The use of in-bundle policy allows individual BPAs to omit certain policy decisions from the per-BPA local configuration. This simplifies BPA synchronization somewhat by reducing the amount of information that would need to be synchronized and also enables policy implementation interoperability with BPAs where local configurations cannot be applied. Further, it avoids ambiguities in processing because these policies are always carried with the bundle.

8.6.2 Security Versus Bundle Policy

BPA bundle-handling policy can be built to avoid conflict with in-bundle policies specified in processing control flags. In cases where a BPA might want to override bundle policies, that decision can be made locally without needing to synchronize that override with all BPAs in the network.[3]

BPA security policy, however, does not have the same independence with in-bundle policies. The block processing control flags associated with an extension block might result in the removal of a block from the bundle, or the bundle from the network. Similarly, security processing might result in the removal of blocks or the bundle.

This places a process ordering dependency between applying processing control flags and applying security operations. This dependency is best illustrated with a case study.

8.6.3 Case Study: Verify Unknown Block

Consider a case, illustrated in Figure 8.6, where some extension block ($Block_{ext1}$) has Bit 4 (0x10) set in its BPCF field, indicating that if a BPA does not understand how to process this block, it should be removed from the bundle. Further, consider that there exists an integrity signature over this target block, creating the dependency $Block_{ext1} \rightarrow Block_{BIB}$.

If the bundle carrying these blocks is received by a BPA which is both configured as a security verifier for the BIB and also does not know how to process $Block_{ext1}$ then there exists a policy

---

3 If policy were synchronized across the network, overrides would not exist because the desired behavior would have been codified at the bundle source.

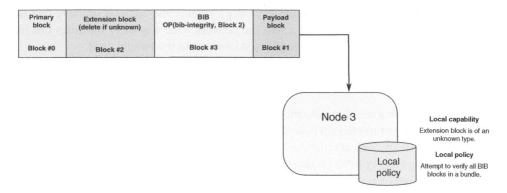

Figure 8.6   Block policy and security policy can conflict. A BPA receives a bundle containing an extension block it can not process and a security operation that it can process. This situation results in different behavior based on whether block or security policy is applied first.

conflict. The policy encoded in the BPCF of $Block_{ext1}$ requires that the extension block be removed from the bundle, but the local BPSec security policy requires that the BPA verify the integrity of $Block_{ext1}$ as part of security processing. In this case, we presume that the integrity of $Block_{ext1}$ has been preserved and that all security configurations are correct in the network – the security verifier should successfully verify the integrity of $Block_{ext1}$.

There are two options for how a BPA would proceed in this instance. Option 1 is that the BPA process the security block first and then the extension block, and Option 2 is that the BPA process the extension block first and then the security block. While both options will result in the removal of the block from the bundle they generate different policy behaviors and reactions.

The question asked by this case study is: which option should be the proper behavior at a BPA? Should security or block policy be applied first when evaluating an extension block in a bundle? The following subsections present each option, followed by a reflection on why there is a definitively correct answer to this question.

8.6.3.1   Option 1: Security Policy First

If security dependencies are resolved first, then the BIB block is processed before the BPCFs of $Block_{ext1}$. This means that the integrity of $Block_{ext1}$ would be assessed first. Then, the BPA would attempt to process the (now verified) $Block_{ext1}$ and realize that it does not know how to process blocks of that type. In accordance with $Block_{ext1}$'s BPCFs, the BPA would then remove $Block_{ext1}$ from the bundle.

Since security policy has already been verified, it is possible that the BPA would preserve the BIB in the bundle, even after $Block_{ext1}$ had been removed.[4]

This leads to a situation, illustrated in Figure 8.7 in which the following seemingly conflicting statements can all be true:

- $Block_{ext1}$ was recorded as having passed integrity at the BPA.
- The bundle was forwarded to another BPA without $Block_{ext1}$ but preserving the dependency $Block_{ext1} \rightarrow Block_{BIB}$.
- The next receiving BPA would not be able to distinguish the case where $Block_{ext1}$ was lost in transit or removed based on its BPCFs.

---

4 While a sophisticated implementation of a BPA should catch this and also remove the security operation from the BIB relating to $Block_{ext1}$ there is nothing in the BPv7 specification which requires this behavior.

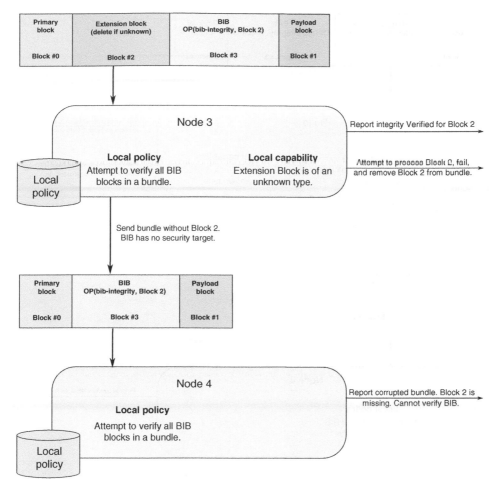

Figure 8.7 Prioritizing security policy over block policy. Processing security policy first could lead to the verification of a target block that would then be deleted by a BPA while processing block policy. This can lead to non-intuitive results such as a BIB in the bundle without a target block.

8.6.3.2 Option 2: Block Policy First

If block policies are resolved first, as illustrated in Figure 8.8, then $Block_{ext1}$ will fail to be processed and removed from the bundle. Once this has occurred, security policy will note that the BPA is a security verifier for the BIB, but also that the dependency $Block_{ext1} \rightarrow Block_{BIB}$ is not present in the bundle because $Block_{ext1}$ is no longer in the bundle.

In this case, a security error will be generated because the BIB block will fail to verify, and the block (and bundle) would need to be handled in accordance with security policy.

This leads to a more expected series of policy reactions:

- $Block_{ext1}$ was removed from the bundle.
- Local security policy determines that the integrity of $Block_{ext1}$ failed, because $Block_{ext1}$ was missing from the bundle.
- The bundle may be processed as per security policy, which may include generating appropriate log messages or status reports.

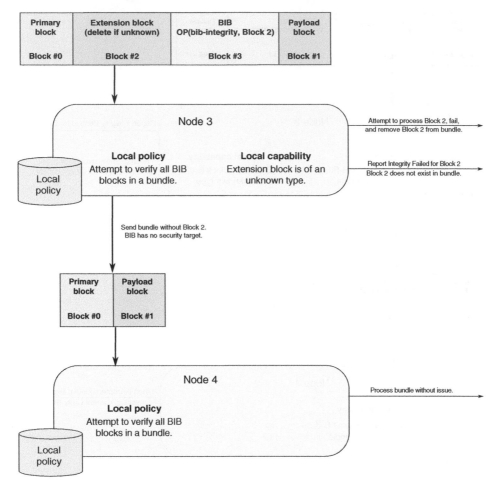

Figure 8.8 Prioritizing block policy over security policy. Reviewing block policy first would lead to the removal of a target block prior to processing security over that block. The resultant security failures would produce more intuitive reporting.

8.6.4 Reflections on Processing Order

The prior case study highlights an instance where BPv7 block policy and security policy yield different results based on the order in which policy dependencies are resolved by a BPA. Since BPSec only requires that security blocks be processed prior to bundle transmission/forwarding or payload delivery at a destination, both processing options are technically supported by BPv7 and BPSec.

Choosing to process security policy first ensures the proper interpretation of block policy for two reasons.

1. If an extension block has been encrypted, it is impossible to process that extension block until it has been decrypted, which requires processing security blocks.
2. If an extension block has its integrity verified, the block (including its BPCFs) should not be trusted unless the contents are shown to have preserved integrity, which requires processing security blocks.

Processing security policy first yields the unintuitive result that $Block_{ext1}$ had perfect integrity at a BPA which otherwise removed $Block_{ext1}$ from the bundle. Alternatively, processing BPv7 policy first

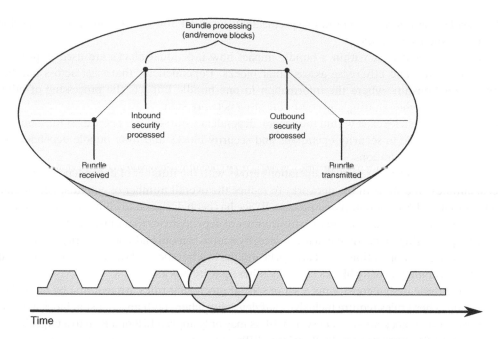

Figure 8.9   A time-based view of block processing. The processing of a secured block can be envisioned as a rising/falling function. The rising edge is not instantaneous, starting with block receipt and ending with successful security processing. The falling edge is similarly not instantaneous starting with security processing and ending with transmission.

avoids this unintuitive result, but allows a situation where a malicious node could alter a block's BPCFs to cause the block to be dropped prior to verifying that the BPCF's were altered.

8.6.5   Security Roles and Timing

The ambiguity that stems from whether to process security or block policy first can be resolved by viewing the processing of all policy as happening in one of three discrete time events:

1. Time T_1: The receipt of a bundle.
2. Time T_2: The processing of bundle contents.
3. Time T_3: the transmission/forwarding of a bundle.

A time-based view of block processing is illustrated in Figure 8.9. The actions taken by security verifiers and security acceptors occur at time T_1. At this time, security policy is used to confirm that, upon receipt, appropriate security processing was done on the inbound bundle. Any additional processing of the bundle, such as might occur as part of analyzing BPCFs, occurs during T_2. Finally, at time T_3 any new security blocks added by the BPA (acting as a security source) are applied and the bundle can be sent along its path.

8.7   Summary

There exists a complex set of dependencies among blocks within a bundle, bundles with a network, and the configuration and policies resident at BPAs across the network. These dependencies can be based solely on information or more specifically focused on how information exists in specific structural elements of bundles.

Dependencies that exist within a bundle impact how individual blocks are used to process, decode, configure, and otherwise assess other blocks. Dependencies that exist across bundles showcase relationships where the information in one bundle impacts the processing of other bundles – such as when sharing and communicating network state.

Security dependencies exist within the general dependency framework. There exist intra-bundle dependencies related to security operations and security blocks and inter-bundle dependencies relating to BPA configurations.

Processing complexity for security operations grows with the number of dependencies on other blocks, bundles, and BPAs in the network. To reduce the overall number of potential operational scenarios that a BPA must detect and process, BPSec has been designed to eliminate many combinations of information and, thus, reduce the number of dependencies that must be checked as part of security processing. In particular, a given security source can only add one security service on a target block, security operations must always be unique in a bundle, and there must only be linear dependencies among security blocks targeting other security blocks.

To enforce these design constraints a set of special processing rules are required by the BPSec specification. These rules require inclusive confidentiality where both an extension block and any signature over that block must be encrypted. BPAs may only apply a BIB or a BCB to a target block, and receiving BPAs must process BCBs prior to BIBs.

Even with protocol constraints, there exist certain scenarios where security processing order is either mandated by BPSec to assist with interoperable implementations or otherwise recommended to avoid challenges with other, non-security-related bundle policies, such as those encoded in block processing control flags. In cases where there exists ambiguity on whether security or block dependencies should be processed first, it is always correct to process security policy first and block processing control flag policy second.

References

1 Burleigh, S., Fall, K. and Birrane, E. J. [2022]. Bundle Protocol Version 7, RFC 9171. **URL**: https://www.rfc-editor.org/info/rfc9171.

2 Birrane, E. J. and McKeever, K. [2022]. Bundle Protocol Security (BPSec), RFC 9172. **URL**: https://www.rfc-editor.org/info/rfc9172.

9

Threat Considerations for BPv7 Networks

As a specification, BPSec defines the building blocks needed to protect and verify data transferred using a BPv7 network. The correctness of the design of the security context is critical in order to ensure that a BPv7 network can actually be secured against anticipated threats in the operating environment. In addition to BPSec, the policies governing the operation of the network play a role in the overall security posture. This chapter describes classes of potential threats that might be encountered in an operational environment and provides discussion on how those threats may apply to delay-tolerant networks.

The overall security posture of a BPSec-enabled BPv7 network can only be determined by evaluated specific aspects of the network deployment, such as chosen security contexts. This chapter provides a broad set of threats to consider when architecting or designing a BPSec-enabled BPv7 network. This chapter forms the basis for further study of threats that may apply to particular security contexts, security mechanisms, and security ecosystem implementations.

This chapter also discusses security implications inherent to the Disruption Tolerant Internet (DTN) architecture and compares these to the Internet Threat Model. Attacker objectives and capabilities are described in order to help understand attacker motivations for a particular delay-tolerant network application. Finally, several key classes of attacks are discussed in the context of delay-tolerant networks to help identify potential threat vectors when designing a delay-tolerant network or application.

After reading this chapter you will be able to:

- Consider attacker perspectives when applying BPSec.
- Describe the classes of network security threats to delay-tolerant networks.
- Understand the security challenges of delay-tolerant networks.

9.1   Security Implications of BPv7 Networks

The variety and flexibility of BPv7 networks lead to a set of interesting and challenging security circumstances for which BPSec must be able to provide protection. These circumstances span locations and environments anywhere from deep space to terrestrial to undersea networks. The processing platforms and configurations in these environments each pose their own security challenges outside of the scope of BPSec. While similar categories of threats may be encountered in any network, the special nature of the BPv7 transport affects the network security posture in ways that are notably different. It is because of these differences and because of the wide variety of use cases that the flexibility in the design of BPSec becomes helpful.

Securing Delay-Tolerant Networks with BPSec, First Edition. Edward J. Birrane III, Sarah Heiner, and Ken McKeever.
© 2023 John Wiley & Sons, Inc. Published 2023 by John Wiley & Sons, Inc.
Companion website: www.wiley.com/go/birrane/securingdelay-tolerantnetworks

The most stressing environments in which BPv7 can be deployed are those conformant to the Disruption Tolerant Internet (DTN) architecture. Therefore, focus on threats associated with delay-tolerant and disruption-tolerant networks illuminate many of the special threats encountered by BPv7 deployments. Unless otherwise specified, references to BPv7 networks in this chapter imply the use of BPv7 for DTN transport.

9.1.1 Network Topology

Similar to conventional networking architectures, BPv7 networks require that nodes along the path from source to destination will inspect, process, and route bundles. Security processing for BPv7 must protect against the same types of attacks as found in conventional networking, but must also address a few key differences. First, attackers may not be able to conduct certain attacks – or reach other network resources – if a return path is not available. Second, bundle lifetimes are much longer than in conventional networking. Attackers in a BPv7 network will have more time and opportunity to conduct attacks that may involve activities such as password cracking.

The defining features of delay-tolerant networks – operation in environments with high delay, high disruption, and/or potentially in a frequently disconnected state – pose security challenges that make BPSec an important facet of a DTN security architecture. In conventional networking, it is generally a safe assumption that while a path exists between two endpoints, then the bidirectional path between those endpoints exists with sufficiently low delay that they are able to exchange messages at machine speed. In environments with high latency or loss, these conventional protocols and services which were built without these design constraints may fail to operate as expected.

9.1.2 Timing and Key Management

Key management, expiration, and rotation are typically much easier in conventional networks. Given readily available forward and return paths with relatively low round-trip message times, this can easily be done. In delay-tolerant networks, key management can become much more difficult. In high-delay environments where nodes may infrequently come into contact, it may be difficult to apply any existing key management protocols such as the Internet Key Exchange Version 2 (IKEv2) protocol as defined in RFC 5996 [1] because the time required to complete all of the back and forth messaging may take an unreasonable amount of time.

As an example, Figure 9.1 shows the message exchanges necessary to establish a security association between two endpoints. The first message exchange, IKE_SA:INIT, establishes the cryptographic suites to be used, exchanges nonces, and conducts a key exchange. The second exchange, IKE_AUTH, authenticates the endpoint identities, exchanges certificates, and establishes the security association to allow for protected communications between the endpoints.

In a conventional network, this exchange of four messages could occur over a very short time – less than a second. In a BPv7 network, such an exchange could take a very long time or may not be possible if return paths for response messages are not available. While there are some BPv7 applications where these conventional key management techniques would still be useful, it can't be assumed to be the case for all applications.

9.1.3 Timing and Incident Response

Another challenge in delay-tolerant networks similar to key management is security monitoring and response at the node and network levels. Since nodes in the network may be disconnecting and reconnecting with other nodes in the network, it may not be feasible to implement a centralized monitoring solution that requires periodic updates. In high-delay environments, using centralized solutions means that any information received at a collection point may be quite old. Likewise, any attempted response may not be able to reach the node in a reasonable amount of time.

Similarly, it may be difficult to provide software or configuration updates as part of an incident response if a node is in a primarily disconnected state, unreachable, or has hardware or software limitations that prevent software updates.

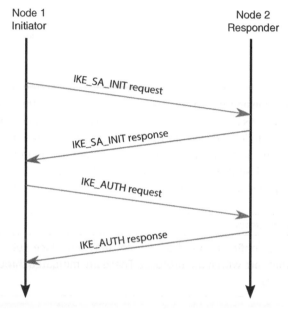

Figure 9.1 IKEv2 messaging. Establishment of an IKEv2 security association requires two message exchanges. This exchange presumes very low-latency round-trip communication time.

9.2 Threat Model and BPSec Assumptions

This section describes the considerations for security threats that went into the design of BPSec and is intended to provide a BPv7 network architect, implementer, or administrator a broader understanding of threats which might be posed to their networks. While it does not provide a complete guide to building a secure BPv7 implementation, the reader should have an understanding of the threats that the BPSec security extensions are intended to protect against and be aware of some of the broader security issues that might affect a particular implementation. The capabilities and objectives of an attacker are captured through a threat model; here, the Internet Threat Model described in RFC 3552 [2] is applied to BPv7 networks and serves as a useful baseline to consider DTN security and the design of BPSec.

A threat model describes the capabilities of, and actions taken by, an attacker to compromise a target system. A threat model tends to describe the threat capabilities from a worst-case perspective in order to rigorously evaluate the security of the system under consideration. In a real-world system implementation, a variety of security mechanisms may be used, including physical (e.g. access restrictions), policy (e.g. user verification and registration procedures), and technical (e.g. BPSec).

9.2.1 The Internet Threat Model

The overall security posture of a specific system is the combination of the security posture of all of the individual components and subsystems as well as the security of the integration of those

Table 9.1 Attacks defined by RFC 3552.

Passive attacks	Active attacks
Confidentiality violations	Replay attacks
Password sniffing	Message insertion
Offline cryptographic attacks	Message deletion
	Message modification
	Man-in-the-middle

subsystems. Here, we are considering BPSec and its implementation through security contexts, which is intended to be a means of technical security within a DTN system. So, for the purposes of BPSec, the Internet Threat Model described in RFC 3552 [2] proves to be a useful starting point to describe a BPv7 network attacker in order to discuss vulnerabilities and protection mechanisms. The Internet Threat Model describes classes of passive and active attacks that form the basis of most traditional attacks against networked hosts. These attacks are listed in Table 9.1.

Protocol Versus Implementation

The implementation and configuration of a protocol may introduce vulnerabilities that are separate from those which are within the protocol. These are mitigated through other software and security best practices.

These attacks are a general set of categories to consider. The passive attacks described by RFC 3552 apply slightly differently to a BPv7 network than they might for a conventionally networked system. For example, as described by RFC 3552, confidentiality violation attacks involve collecting information intended to be private that has not been cryptographically protected. This is because Internet security mechanisms are generally well-understood to provide protection when they are appropriately applied.

When evaluating BPSec, this attack is analyzed differently, to determine that the inherent design of BPSec can protect against confidentiality violation attacks, even when deployed in a delay-tolerant network. Even when assuming that all user and application data is protected by cryptographic mechanism, the packaging and longevity of those mechanisms must be considered. Of course, BPv7 networks are also subject to improper configuration or implementation of BPSec that may expose data intended to be private. However, those concerns would be for a particular implementation and not a weakness of the BPSec design itself.

9.2.2 BPSec Design Assumptions

To avoid an unbounded discussion on threats and vulnerabilities, it is important to separate the mechanisms and processes of BPSec from the implementation of the BPv7 application. This can be accomplished by making a few assumptions about the implementation. These assumptions largely focus on the software and configuration of the platform or platforms to be used within a BPSec-enabled BPv7 application. Additionally, there are a number of other considerations at the boundaries with other standards and systems which provide other security challenges outside of the scope of the BPv7 protocol and implementation.

9.2.2.1 Proper Implementation

A common attack vector is through weaknesses in software implementation. Considering an implementation of BPSec, flaws in input processing may allow an attacker to inject malicious code to either take over the host or subvert proper operation of BPSec. It is also important to consider the vulnerability of any external dependencies, libraries, or modules in the implementation as any vulnerability there becomes a vulnerability of the overall system.

> **Faulty Implementation**
>
> BPSec is not designed to protect against security software implementation flaws.

In order to combat these vulnerabilities, appropriate software development principles and dependency tracking become critical. For any implementation, it is important to follow best practices to avoid buffer overflows or other input processing vulnerabilities which can expose sensitive data or allow remote code execution. In addition, it is important to track the use of external dependencies within an implementation through a software bill of materials and stay aware of security updates for those dependencies, updating the BPSec implementation as needed and possible.

Instead, the security of BPSec within a BPv7 network relies on secure and correct implementation. The exploitation of a software vulnerability could severely compromise the operation of a secure DTN application and could be executed in such a way that it takes advantage of the complex interaction between a BPSec software implementation, the underlying operating system, and/or the hardware platform itself.

Even if a BPv7 application is well-written and resistant to input attacks, an attacker could take advantage of vulnerabilities of other software services on the same host to gain access to the host. From there, an attacker would be able to attempt other attacks against the BP application, perhaps through other privilege escalation or memory access attacks.

It is certainly no easy feat to create completely secure software. However, because an implementer has control over the implementation of the software and has the ability, with effort, to create secure software, BPAs associated with a controlled, secure subnet should be able to be resistant to compromise from an outside attacker. For this reason, BPSec is not be designed to protect against implementation or other host-based vulnerabilities.

9.2.2.2 Proper Configuration

Erroneous configurations (whether unintentional or malicious) are a common source of vulnerability for any implementation, and BPSec is not different in this regard. These errors can manifest within many areas of an information processing system, to include host-based configuration errors. Host-based configuration errors are particularly problematic as they can allow attackers access to the hosting compute platform. For example, an unnecessary web server may be

> **Configuration Error**
>
> BPSec is not designed to protect against configuration errors.

accidentally left running after troubleshooting an issue, exposing a network port for remote connections. If this web server is vulnerable, it could provide a foothold for an attacker to access the platform and attempt other attacks against the BPA on that host or other nodes in the BPv7 network.

Configurations associated with a BPSec implementation are similarly important. Depending on the complexity and parameters of the implementation, there could be opportunities for misconfiguration that might be unintentionally detrimental to security. One way this can happen is through the accidental selection of a vulnerable cipher suite for Block Confidentiality Blocks (BCBs) would mean that BPv7 block data would be insufficiently protected in transit and vulnerable to confidentiality attacks.

While these are serious to the overall security of an overall platform, these are issues that must be addressed for any security implementation. BPSec should be implemented in a way that simplifies configuration for an end user. For complex implementations that offer a wide variety of options, configuration validation tools and scripts can be useful to identify known issues.

9.2.2.3 Appropriate Security Contexts

> **Security Contexts**
>
> BPSec requires that deployed security contexts are available and, through security policy, properly used and configured.

The application of BPSec to a BPv7 network requires the selection of a BPSec security context which defines the cryptographic mechanisms, formatting, and processing to be applied to blocks in order to achieve intended security outcomes. Furthermore, policies are defined to govern the application of security contexts and, generally, the operation and use of the secure network. In order for BPSec to be effective, security contexts and policy must be both correctly designed and implemented. When the implementation differs from design intent, that can create vulnerability that an attacker may be able to exploit to compromise the network.

To avoid implementation errors, best practices relating to documentation and system description should be followed to ensure understanding between a network designer and a solution developer. Network designers and architects should make use of enough descriptive content, such as pseudocode and test vectors, to illustrate expected processing across a range of potential processing logic. Developers should ensure that implementations match these design specifications. See **Focus: CVE-2014-0160, Heartbleed** on next page.

9.3 Attacker Objectives and Capabilities

Before discussing different attacks on delay-tolerant networks, it is useful to describe an attacker in a bit more detail to set context for those attacks. Given the breadth of potential DTN applications, it is important to consider attackers generally. Objectives of an attacker relate to the motivation for attacking. The objectives and motivation can be useful in exploring potential avenues and means of access in order to attempt attacks and can help understand the most likely types of attacks. The capabilities of an attacker refers to the privilege level within the scope of the network.

9.3.1 Attacker Objectives

> **Design Note**
>
> Developing the attacker objectives and capabilities is an important exercise that must be considered during the design of BPSec security contexts and security policies.

When considering the security of delay-tolerant networks, and specifically BPv7, it is important to keep in mind that these networks are just as much a means of access for an attacker as they are for legitimate nodes on the network. When attempting to compromise a BPv7 network, an attacker's motivation and goals will drive the manner in which they attack the network. For example, an attacker may seek to gain access to other networks from across a delay-tolerant network. Alternatively, an attacker may be trying to gain access to an out-of-band management link by disrupting the network in an attempt to activate a recovery response from a network operator.

Focus: CVE-2014-0160, Heartbleed

The Heartbleed vulnerability affected a large number of software packages that used the open source OpenSSL library. Source: Markus Spiske/Unsplash.com

An example of the challenges of software implementation security can be found in the Heartbleed vulnerability. In the open source transport security protocol library, OpenSSL [3], Heartbleed allowed an attacker to take advantage of flaws in how particular packets were handled in the Transport Layer Security (TLS) and Datagram Transport Layer Security (DTLS) protocols to read out additional data from process memory. This could allow an attacker to retrieve sensitive information, including private keys. With this information, an attacker could then potentially decrypt intercepted traffic. Even though OpenSSL followed the specifications for TLS and DTLS and implemented the appropriate processing and cryptographic suites, this software implementation vulnerability had the potential to allow attackers to subvert these protections.

Besides being a critical implementation vulnerability, the impact was compounded by the proliferation of the OpenSSL software and highlights the importance of software supply chain security. Given the importance of transport security as an element of security on operating on the Internet, the completeness of the open source OpenSSL software library implementation made it an attractive solution to provide transport security within other software applications. This vulnerability therefore became a vulnerability of these other software applications. After updating to a later version of OpenSSL, these software applications were no longer vulnerable.

In DTN applications, there may exist design constraints which prevent timely software updates or may make them impossible. For this reason, it is important to consider and evaluate not only the security of software used to implement BPv7 or BPSec, but also any external software dependencies of those implementations. Implementation vulnerabilities have the potential to allow an attacker to subvert even a well-designed security policy and protocol.

Availability, integrity, and confidentiality form the basic building blocks for information security and likewise are useful when discussing attacker objectives.

BPSec provides mechanisms to help protect the network from compromise, but these mechanisms need to be used correctly to provide that protection. The *bcb-confidentiality* and *bib-integrity* services provide confidentiality and integrity, but other means must be used to ensure availability of the network. Network availability is provided by protecting processing resources and ensuring reliability of the nodes, which is typically implementation-specific, making it inappropriate to define protection mechanisms in a standard like BPSec.

9.3.2 Attacker Placement

> **On-Path Attacker**
>
> An On-Path Attacker is one which is in the middle of the flow of network traffic.

In the DTN architecture, it is possible that a network will be made up of a collection of federated subnetworks organized into separate security enclaves. These subnetworks may share common data transport and routing methods to increase the size and reliability of the collective network. However, depending on the operations and management model, these other subnetworks have the potential to be On-Path Attackers (OPAs), either through compromise or by accident (e.g. through misconfiguration).

Figure 9.2 shows an example of a BPv7 network with three independently managed security enclaves. The nodes across these networks all conform to BPv7 and share common link layer interfaces which allows them to pass bundles – to include bundles that did not originate within a node's own security enclave. While this increases the capacity and reliability beyond what each enclave could provide individually, it means that any node along the path could either intentionally or unintentionally prove to be a threat to the overall security of the network. Given the access to the blocks of the bundle, an attacker can attempt attacks on the confidentiality, integrity, and availability of other parts of the network.

In order to intercept protected bundles, an attacker needs to be either an OPA or a Near-Path Attacker (NPA). While OPAs and NPAs are both able to intercept bundles and send bundles, they differ in that an OPA is able to replace and modify traffic, whereas the NPA is only able to add new traffic into the network.

Definition 9.1 *(On-Path Attacker)* An attacker which has direct access to one of the nodes on a network, placing it in the middle of the flow of network traffic and allowing direct manipulation of that traffic

Definition 9.2 *(Near-Path Attacker)* An attacker which does not have direct access to a node on a network, but instead through physical proximity is able to receive bundles from and transmit bundles to the network but is unable to directly manipulate traffic on the network

An attacker is able to become an OPA or NPA in a few ways, including:

- Compromising an existing on path node.
- Become on path through topology attacks.
- Become near path by proximity access.

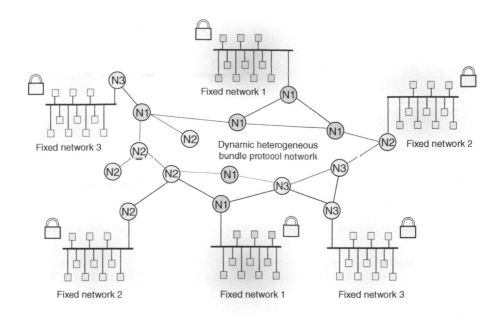

Figure 9.2 Vulnerabilities exist end-to-end in a network. Federating multiple security enclaves (notated as Fixed Networks) exposes traffic to vulnerabilities both from other enclaves and from the intermediate networks that connect them.

9.3.2.1 Node Compromise

An attacker may attempt to compromise the platform or BPA application of a node as a means to gain control of the BPA functionality in order to access the BPv7 network. To achieve this, an attacker would need to exploit some vulnerability in the system or the application such as in the underlying operating system, the BPA, or some other application or service. By gaining control of the BPA, an attacker may then have access to BPSec-protected communications.

9.3.2.2 Topology Attacks

Another means by which an attacker may become on path is through attacks on routing algorithms. By modifying routing advertisement messages, an attacker can influence the routing through the network, thus placing a compromised node on a path to become an OPA. In a wireless network, an attacker can also become a NPA if they have a node or nodes in proximity to other BPv7 network nodes and are able to intercept and inject bundles into the network by being within the wireless transmission footprint. Depending on the nature of the routing, an attacker may be able to transition from a NPA to an OPA if they are able to degrade or disrupt an existing, preferred link on the network by intentionally interfering with that link.

Figure 9.3 highlights several different attacker positions on an example BPv7 network with two separate security enclaves, networks A and B, two compromised nodes, $A2$ and $B3$, representing On-Path Attackers OPA_1 and OPA_2 respectively. There is also a passive attacker NPA in proximity to network nodes $A1$ and $B2$ that is able to receive radio transmissions and conduct passive attacks against those two nodes. The attacker OPA_2 which has compromised node $B3$ is on path and able to receive and inspect traffic between nodes $B2$ and $B1$ since it shares secret key material. The attacker OPA_1 which has compromised node $A2$, however, is able to intercept communication between nodes $B1$ and $B4$, but is unable to inspect bundles because it does not belong to the network B security enclave.

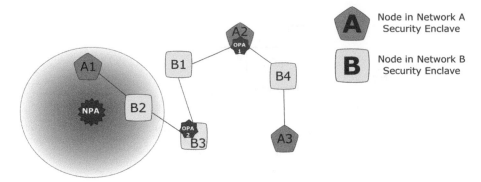

Figure 9.3 Topology impacts attackers capabilities. A sample network shows communications through two different security enclaves, *A* and *B*, both of which are subject to a Near-Path Attacker *NPA* and two nodes compromised by On-Path Attackers OPA_1 and OPA_2.

9.3.2.3 Proximity Access

Gaining proximity access to a BPv7 network is both easier and more difficult than in more static networks.

Proximity to a BPA is made difficult because BPv7 networks are often partitioned as a function of motion, propagation delay, link disruption, and power management on BPA host platforms. Just as BPv7 routing and topology must consider time, an attacker must achieve proximity in both space *and* time to be an effective NPA.

Proximity to a BPA is somewhat easier as BPv7 networks are typically wireless networks that communicate over longer distances, as long distances have greater signal propagation delays and likelihood for link disruption. Placing a node within this larger transmission footprint may be easier than in other wireless network deployments.

9.3.3 Attacker Privileges

The privileges associated with resources under an attacker's control significantly impact the threat posed by that attacker. Attacker privilege is associated with the level of access and set of capabilities of the compromised resource within the context of the network.

Rather that discuss all types of accesses, this section focuses on the compromise of a node in a network, such as the compromise of a BPA within a BPv7 network. From that perspective, three types of node privileges can be defined that become attacker privileges when an attacker takes control of that node.

- **Unprivileged Node**: Essentially an outside observer
- **Legitimate Node**: Node within the secure environment
- **Privileged Node**: Administrator-like capability with full access to the network

An unprivileged attacker is one that has control over an unprivileged node in the network. Protecting against these types of attackers is the most basic case of networking security. The fundamental capabilities of BPSec can provide inherent protection of the integrity and confidentiality of bundles on the network using the Block Integrity Block (BIB) and BCB when employed with an appropriate cipher suite.

When an attacker has some access to the network through a compromised node, it is important to be able to isolate the attacker as much as possible. This isolation is not part of the capabilities provided by BPSec security blocks. Protecting the network from compromised nodes requires additional considerations relating to the applied BPSec security context, network and key management architecture, and possibly active security response capabilities.

Consider the ways in which security context selection impacts the threat of a compromised node. The utilized security context determines, among other things, the types of keys that generate the cryptographic materials associated with security blocks. If the context uses symmetric cryptography and all nodes have shared key material, an attacker gaining access to such a legitimate or privileged node can compromise the confidentiality and integrity of the entire network.

Depending on the method, attackers may gain these privileges with or without knowledge of any passwords or other secret credentials. For example, an attacker may be able to access the system memory and recover a derived session key which may be temporary. This could allow an attacker to decrypt traffic until a new key is derived, and once that happens, they will lose access since they did not recover the key used to derive that session key.

9.4 Passive Attacks

Passive attacks are a class of attacks which have no discernible effect to the network or its operation. These attacks can be conducted by an unprivileged attacker outside the network or by a privileged attacker already within the network. These attacks attempt to gain information and insight about the operations of the network and the content and nature of the message therein.

Sometimes passive attacks enable subsequent, active attacks on the network, its nodes, or hosted services. Consider the case where an attacker wishes to compromise the confidentiality or integrity of a network. Passive attacks enable gathering useful information while avoiding detection. With this access, it may become possible to, later, inject fabricated messages thus providing a vector to attack other nodes or networked systems.

In order to conduct passive attacks, an attacker must observe and eavesdrop on latent traffic within the network. In the case of a network that uses a wireless radio link, an attacker would be able to observe traffic through proximity to transmitters and appropriate selection of hardware components such as antennas, amplifiers, and receivers.

Passive attacks are difficult to detect by the elements providing information security under the best of circumstances. Some physical security approaches can help prevent or minimize the opportunities for passive attacks. For example, motion detection sensors and secured perimeters limit opportunities for attackers to gain proximity to certain network resources. Similarly, placing computing access is otherwise challenged environments, such as deep space, make some proximity communications more challenging.

An attacker can also conduct passive attacks by gaining an on path position within the network. While it may require active attacks to achieve this on path position, it still enables the attacker to then begin passive attacks. Depending on how the on path position has been gained, the attack may not be a purely passive attack – it may be possible for the attacker to be detected. If on path status is achieved through code execution on a legitimate node, additional message processing delay, memory use, processor use, or other error or system logs may indicate the presence of an attacker.

When dealing with the storage, transmission delays, and topological changes that may be encountered by a BPv7 network, passive attacks become almost impossible to detect. For this reason, it is best practice to consider that a BPv7 network is always under threat of passive attack.

With the gathered data, an attacker can analyze and learn about the network. In this subsection, the passive attacks outlined here assume that the attacker does not have access to the key information used to generate BCBs within the network and therefore does not have access to the transported plaintext traffic. If an attacker does have this access, the passive attacks described here are still somewhat relevant, but the focus of passive attacks likely shift to the transported application data instead of the BPv7 network or subnetwork.

9.4.1 Cryptanalysis

An unprivileged attacker may attempt to attack the cryptographic protections of the cipher suite used for BCBs or BIBs to reveal the secret key material used to generate ciphertext (in the case of BCBs) or digital signatures (in the case of BIBs). These are cryptanalytic attacks which leverage computational power or weaknesses in cryptographic mechanisms.

Definition 9.3 *(Cryptanalysis)*   The analysis of cryptographic algorithms, typically with the intent of identifying weaknesses or proving resilience.

These attacks may be brute force, meaning that the attacker is attempting to guess the secret key material used by the cipher suite defined by the security context. Brute force attacks are the most basic type of cryptanalytic attack and are the primary consideration during the design and parameter selection for cryptographic systems. For many algorithms, a longer key means that it will take more attempts and therefore more time to correctly guess. In these cases, key sizes are typically selected so that it becomes completely computationally infeasible for an attacker to succeed with a brute force attack.

As an example, the Data Encryption Standard (DES) is a symmetric-key algorithm published in the early 1970s and designed to use 56-bit keys. By the 1990s, computational capability had greatly increased which thereby reducing the time required to conduct a brute force attack such that DES effectively no longer provided adequate protection. The Advanced Encryption Standard (AES) emerged in the 1990s as a variable-length symmetric key algorithm which could provide a higher level of protection.

Some cryptographic algorithms may have implementation weaknesses that allow an attacker to determine key information faster than a brute force attack. The nature of these vulnerabilities vary depending on the design. As an example, the Rivest Cipher 4 (RC4) stream cipher in early security mechanisms of Wi-Fi had vulnerabilities which, as implemented, allow an attacker to determine the secret key through observation of transmitted data [4]. Since the first publication of this vulnerability, improvements have been made to reduce the collection requirements to be on the order of one million packets that can be collected in minutes using active message injection techniques.

Similar issues can be present in BPv7 networks. Consider a small sensor network whose nodes pass raw data to a local processing node which in turn aggregates information and sends it to a centralized repository. A small unattended sensor node may have a requirement for long battery life that may restrict the use of sophisticated cipher suites or complex security processing due to limited compute resources. On the other hand, the local processing node may have more resources to be able to take advantage of more sophisticated cipher suites. Such an architecture could allow effective cryptanalysis of data into the local processing node which could then impact the more secure information sent from that node.

9.4.2 Network Profiling

Important information for an attacker is insight on the network topology or connection points. Given the potential for frequent topological change in BPv7 networks, there are some cases where opportunistic or planned connections and their dependence on time do not require a different way to describe the network topology compared with conventional networks. That is, a conventional network topology could be described as a static graph with nodes and edges, whereas a BPv7 network could be described as a graph with nodes and time-variant edges.

There may be different reasons or motivation to attempt to profile a BPv7 network. If an attacker is attempting to generate a "roadmap" to plan a series of attacks to maneuver through the network, that may not be possible. However, profiling the time-dependent nature of the network may be interesting to an attacker as it may provide information on the location of a node at some point in the past, which may be valuable in and of itself or it may provide insight on where a node may be at some point in the future.

Also, depending on the scale and overall distance a network covers, it could be difficult for an attacker to be able to collect data at multiple places within a BPv7 network and combine the data to be able to analyze the structure or behavior of the network. For a highly distributed network where delay is the primary constraint – such as in a deep-space network – an attacker may need to have an equally distributed monitoring capability. This would mean that the attacker would be subject to similar delays when trying to collect and correlate observations of the network. Additionally, an attacker may need to be aware or able to predict the movement of nodes in the network in order to predict the path by which bundles will travel.

9.4.3 Traffic Profiling

An attacker may attempt to understand the nature of network traffic, typically by measuring characteristics and statistics of observed data. While traffic content may be protected through encryption, the simple presence of traffic may be of interest to an attacker.

In conventional networks, observed traffic statistics are predominately a function of the application-layer data and loosely on channel characteristics. Passive attackers in these networks might draw conclusions on the nature of traffic through externally observable metrics such as packet rate, inter-arrival time, and length. Such an attacker might be able to identify a file transfer separate from a voice over IP stream, even when the contents of the data are encrypted.

In a BPv7 network traffic patterns might manifest differently. Time-based metrics such as rate and inter-arrival time do not have the same meaning since data is packaged in bundles, stored, and transmitted as connections allow. Depending on how data is batched into bundles, bundle size may not be indicative of the application-layer data frames. In addition to other BPv7 processing behaviors, the selection

> **Bundle Timing Analysis**
>
> Aggregating data into bundles, and store-and-forward operations, make attacker timing analysis more difficult.

and implementation of security contexts can also affect how higher-layer traffic is converted into blocks and bundles.

Observing traffic characteristics in BPv7 networks is difficult, but not impossible. For an attacker to successfully profile BPv7 traffic, they must have more information on connectivity and channel characteristics compared with conventional networks as well as an understanding of how block and bundle processing is implemented on the target BPv7 network.

In Figure 9.4, an attacker A combines observed bundle characteristics and statistics with knowledge about the network to attempt to draw conclusions about the protected traffic. In this case, an attacker may be able to determine that, in a sensor network, a sensor node ($BPA1$) communicates different types of data back to a processing node ($BPA2$). One type of data can sensed information while another type of data might be regularly produces health and status information (such

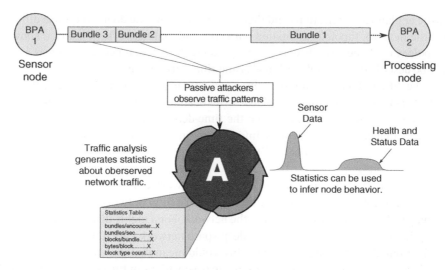

Figure 9.4   Traffic profiling can reveal network data. An OPA can infer network behavior from message statistics. Inspecting bundles to draw conclusions about the data being sent can be done even when bundles are protected with bcb-confidentiality services.

as remaining battery and storage). If an attacker knows when sensed data is being communicated, they can infer something about the network or the environment around the network.

From a design perspective, a particular application or implementation could create opportunities for an attacker to successfully profile traffic. For example, security contexts for different domains or traffic types may directly indicate different sets of data, processing domains, or logical functions within an application. In a BPv7 network, there may be a heterogeneous set of hardware platforms with different processing capabilities.

Different secure domains operating on a common BPv7 network may make use of different security contexts or security context configurations. If these separate domains are managed by different entities, then in the absence of any security context requirements imposed by the network operator, the designers of these secure domains could select or configure security contexts in such a way that may provide information to the attacker. In some use cases, this may not be a concern, but in situations where a single entity operates multiple secure domains, if configuration control is not correctly applied so that domains are not configured similarly and leak some information to the attacker.

Quality of service and bundle-processing behaviors can also provide information to an attacker about the nature of the traffic being transported on the BPv7 network. While an attacker may not be able to inspect the contents of a bundle, they may observe that bundles may be handled differently. It could be that blocks or bundles from a particular end user or application are given a higher priority and are transferred first. Additionally, any network or node behaviors that involve processing and responding to block contents could potentially inform an attacker on the nature of the traffic if the attacker knows the nature of that processing.

The objectives of an attacker attempting to profile traffic will likely vary depending on the nature of the BPv7 use case and implementation. For example, an attacker may be able to isolate network operations traffic on a BPv7 network. If an attacker attempts to maneuver through the network, they may be able to detect an increase in network operations traffic and may assume that an attempt to gain additional access has been detected. In that case, the attacker may use this information to halt operations, reduce their footprint, and attempt to avoid further detection or loss of access.

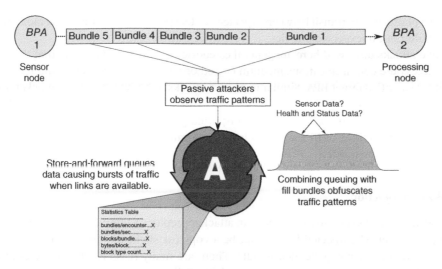

Figure 9.5 Queuing traffic impedes profiling. The store-and-forward nature of bundles separates bundle production from bundle transmission in cases where links come and go over time. This makes traffic analysis more difficult.

As illustrated in Figure 9.5, the store-and-forward nature of BPv7 can impede traffic analysis, as queued bundles are transmitted when the link is available which may be different than when the bundles themselves were generated. For delay-tolerant networks, this means that timing analysis of traffic may not correlate to application-level actions from networked platforms.

If traffic profiling is a specific concern, a BPv7 developer can consider adding processing techniques to obscure characteristics of traffic and potentially the BPv7 network itself. As an extreme example, a network designer could use mandatory periodic bundle transmissions of fixed-length to obscure characteristics of the data being transmitted over the network. In this example, even if there is no data to send, the node still creates and sends a bundle.

9.5 Active Attacks

Active attacks involve interaction in ways that can be detected and acted upon by the network, such as by directly injecting new messages or replaying old messages. The utility of defensive mechanisms can vary greatly depending on the nature of the BPv7 network being considered. In networks with good connectivity defensive mechanisms can detect and alert on potential attacks fast enough to enable timely responses.

In the DTN architecture, delays in response times create opportunities for an attacker that would not exist in a low-latency network. Attackers could attempt to execute a barrage of highly detectable attacks knowing that a network operator will not receive, much less act on, any indication of those attacks within the time frame of the network delay. This allows ample opportunity for an attacker to attempt a variety of attacks and observe any effects. Generally, BPv7 networks must rely on on-node defense mechanisms and not rely on external operators or other security services.

In order to protect against attacks in a high-delay network, it is important to consider protective measures and policies that are decentralized and afford protection to individual nodes within the network. The use of BPSec mechanisms such as the BCB and BIB through an appropriate security context can provide protection against many forms of active attacks. Additionally, hardware-based procedures can also help reduce the opportunity for attacks – in an example of an interplanetary network, if a platform is in transit between expected contact points, it can disable the radio interface or possibly filter out unexpected bundles.

The active attacks discussed here are general categories of attack and assume that the attacker is able to access the communications medium (wired or wireless) and transmit and receive messages to interact with a target BPA. Similar to passive attacks, we assume that the attacker does not have access to the key information used to generate BCBs within the BPv7 network and therefore is not able to generate ciphertext or access transported plaintext traffic. If an attacker does have this access, the protections of BPSec are no longer relevant and this elevates the concern to the transported application traffic.

9.5.1 Bundle Injection

Bundle injection attacks refer to cases where an attacker prepares a bundle and sends it to a nearby node for processing. The injected bundle can be a copy (or slight modification) of a previously observed bundle or a newly generated bundle. There are many reasons why an attacked would want to add bundles to a target network, to include the following.

- Attack application-layer services by providing them with malicious data
- Create a denial of service against a node or nodes to prevent bundle passing
- Retransmit old bundles or add copies of old blocks to bundles to disrupt operations
- Exploit bundle-processing software to gain access to a node.

Figure 9.6 shows an example of a bundle injection attack. An On-Path Attacker *OPA* can both replay stored bundles and add maliciously crafted bundles to the network targeting another node (*N*1). Injected bundles could exploit a vulnerability in the BPA of *N*2. This then allows the attacker privileged access to that node and the BPv7 network. The integrity of BPA software is both critical to the security of a BPv7 network and outside of the scope of what BPSec can protect.

In a BPv7 network, bundle status reports [5] may be a particular target for an attacker. These bundles are used to provide information on how other bundles are progressing through the system. Reports can be used to indicate that some bundle has finished transmission successfully or unsuccessfully. In addition, reports contain reason codes, such as "Lifetime expired" or "Forwarded over unidirectional link," which provide context for the status reports. By forging these bundles,

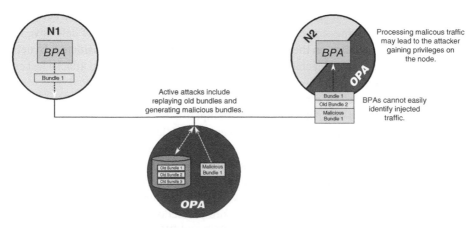

Figure 9.6 Injected bundles include both new and old bundles. An On-Path Attacker *OPA* can store and replay older messages, or use knowledge gained from observing the network to generate new, malicious bundles.

an attacker may be able to create disruptive behaviors by falsely indicating that transmission has failed or using an unexpected reason code for a given situation. The use of BPSec can help verify the authenticity of these reports, allowing nodes to ignore illegitimate reports.

9.5.2 Bundle Modification

Bundle modification attacks are very similar to message injection attacks from a technical perspective, but differ in that they do not add new bundles to the network. In a bundle modification attack, an attacker intercepts a bundle and attempts to manipulate the content or structure of that bundle. These attacks can include modification of the content of blocks in a bundle, addition of new blocks, replacement of blocks, removal of blocks, or reordering of blocks within the bundle. The objective of the attacker is to force a node to handle or process a bundle differently than intended.

An attacker might attempt to modify content and structures within bundles, such as the primary block and payload blocks. Source and destination identifier fields within blocks are of particular interest – by modifying endpoint identifiers, an attacker may send the bundle to a different location (potentially through a different route) or attempt to falsely indicate the source of a bundle. Modifying bundle and block Processing Control Flags allows an attacker to adjust node behavior. Consider a scenario where an attacker receives a bundle with the "Bundle must not be fragmented" flag set and changes that to instead be unset. In this case, the bundle may be inappropriately fragmented causing the bundle to at best be discarded or at worst provide opportunity for further attacks.

The *bib-integrity* service of BPSec can provide strong protection against modification attacks with appropriate selection of cipher suites and strong implementation. As long as BIBs target all critical portions of the bundle and any receiving nodes are designed to expect the presence of BIBs, then any security validation operations will be able to detect modified or missing blocks.

9.5.3 Topology

An attacker may also attempt to actively adjust the network topology or the way BPAs use that topology. Successful manipulation allows an attacker to force bundles to pass through specific nodes to which the attacker can intercept messages either directly or through proximity. Attackers may use one or more methods to influence bundle routing ranging from bundle transmission which could include forged blocks or bundles, replayed blocks or bundles, or jamming or other disruption of the links between nodes.

Figure 9.7 illustrates a federation of three networks ($N1$, $N2$, and $N3$) whose nodes cooperate to pass bundles. In this illustration, two nodes within network $N3$ wish to exchange data. The preferred path between the $N3$ source and destination is through $N2$ nodes. This path is delineated by the thicker black lines between the source and destination that omit any nodes from $N1$.

A topology attack executed by an attacker controlling nodes on $N1$ wishes to change this preferred path in an attempt to have traffic routed through $N1$ where it can be inspected. A $N1$ node uses a bundle injection technique to change the preferred path to instead pass through $N1$ nodes. This induced path is delineated by the thicker striped line between the source and destination that include nodes from $N1$. This allows the attacker to conduct passive attacks on traffic as it passes through and send that information outside the BPv7 network.

Topology attacks will have varied success depending on the nature of the routing algorithms in use within the network and will drive the nature of the attacks which may be used to accomplish this objective. For example, if a dynamic link-state routing algorithm is used which passes status within bundles, an attacker may attempt to forge link state messages to attempt to establish a better route through the network by indicating a particular link may have higher or lower performance

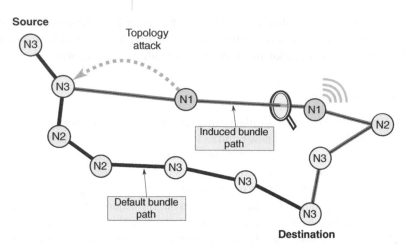

Figure 9.7 Topology attacks influence routing. Attackers can influence the routing of bundles by degrading or denying other links in the network. Here, a topology attack from a *N*1 node can cause an *N*3 node to route data through a series of *N*1 nodes.

characteristics. If a routing algorithm takes into account statistics or metrics derived from received bundles (or from the underlying link), an attacker may be able to manipulate those statistics by repeating prior bundles, perhaps modified in some manner.

The *bib-integrity* and *bib-confidentiality* services can help provide protection against messaging-related attacks on topology. In general, routing decisions should be made only on trusted data which was provided with clear non-repudiation and authentication. In a homogeneous network where all nodes are operated by a single entity, it may be more feasible to manage a common source of trust (e.g. cryptographic keys or a certificate authority) and apply BPSec mechanisms to all bundles and blocks in the network to provide uniform protection. If a network comprises nodes from multiple entities, it is still feasible, but may be more difficult to provide the same level of protection against routing attacks.

In some delay-tolerant networks, routing topology can be considered to be made up not only of existing links which connect end-to-end nodes, but also future links which have yet to be established. In Contact Graph Routing [6], known movement of nodes can be used to predict future links which can be planned for use in routing bundles. In addition to attacking the topology of the existing nodes, an attacker is able to disrupt a future link by impeding the movement of certain platforms. BPSec may be unable to prevent attacks like these that occur completely outside the operating domain of the network itself, but proper use of BPSec services can help protect the content as it passes through OPAs or other passive attackers.

9.6 Summary

When designing a BPv7 network, the threat environment is the primary driver for the selection of security mechanisms. BPSec provides integrity and confidentiality services that help to protect the messaging over a delay-tolerant network using BPv7, which is one part of the overall security design of an implementation, which would also include BPSec security context design, policy design, software security, and physical security. All of these aspects of system security should be developed in concert while considering the risk posed by various attacks. Given the nature of how delay-tolerant networks differ from conventional networks, there is not a "one-size-fits-all" approach to security design, so a thoughtful approach to security design must be done that meets the need of the anticipated use case of these unique networks.

Typically, there are many assumptions that are made about threat capabilities when considering the security design of a system. Either evolutionary or revolutionary improvements in threat capabilities over time can challenge these assumptions and have the potential to affect the security of a system. Increasing computational power or new operating principles can challenge assumptions previously held about the resilience of cryptographic mechanisms or the overall protections of a platform. For example, it has already been shown that Shor's algorithm offers a faster approach to factoring integers with quantum computers than is possible with classical computers. This challenges the assumption in many cryptosystems that factoring large integers is computationally difficult. As quantum computing hardware and systems continue to mature, this means that existing cryptographic algorithms will no longer be able to provide sufficient protection against a capable attacker. Other advancing capabilities such as artificial intelligence and even incremental computing power also have potential to challenge security assumptions made in system design.

While these future threats cannot always be predicted, it underscores the need to continuously reconsider the threat landscape even after design and deployment of a capability. Leaving opportunity to upgrade system elements or update the design of the system after deployment will allow a solution to remain flexible against emerging and unpredictable threats. While this would of course create a lucrative attack vector and open up the opportunity for system failures due to misconfiguration, this capability is very important to be able to keep a system protected against not only unforeseen threats, but also respond to the typical incremental security threats like software package vulnerabilities.

Finally, in addition to BPv7 nodes which source, sink, and facilitate data transmission, it is also important to consider nodes which inspect bundles and implement other security functionality as elements of security policy. For example, firewalls can be useful to manage the flow of blocks or bundles, and disallow unexpected traffic either within the network or at ingress or egress points. Likewise, intrusion detection systems can observe bundles within the network from a trusted position and identify behaviors that may be indicative of an attacker. The design of these appliances will differ from those found in conventional attackers, particularly due to the time-varying nature of networks and the potentially long delays which may present challenges in transferring status information or notifications across the endpoints.

References

1 Eronen, P., Nir, Y., Hoffman, P. E. and Kaufman, C. [2010]. Internet Key Exchange Protocol Version 2 (IKEv2), RFC 5996. **URL**: https://www.rfc-editor.org/info/rfc5996.

2 Rescorla, E. and Korver, B. [2003]. Guidelines for Writing RFC Text on Security Considerations, RFC 3552. **URL**: https://www.rfc-editor.org/info/rfc3552.

3 *CVE-2014-0160* [2013]. **URL**: https://www.cve.org/CVERecord?id=CVE-2014-0160. [Online; accessed 3-March-2022].

4 Fluhrer, S., M. I. and Shamir, A. (2010). *Weaknesses in the Key Scheduling Algorithm of RC4*.

5 Burleigh, S., Fall, K. and Birrane, E. J. [2022]. Bundle Protocol Version 7, RFC 9171. **URL**: https://www.rfc-editor.org/info/rfc9171.

6 Burleigh, S. [2010]. Contact Graph Routing, *Internet-Draft draft-burleigh-dtnrg-cgr-01*, Internet Engineering Task Force. Work in Progress. **URL**: https://datatracker.ietf.org/doc/html/draft-burleigh-dtnrg-cgr-01.

10

Using Security Contexts

The Security Context is a novel mechanism used to customize the behavior of security services without needing to define new security blocks or otherwise codify alternate representations of security operations. This mechanism is an integral part of BPSec because BPv7 networks may be deployed in varied networking environments each with special constraints on information exchange.

Security contexts are used by BPSec to insulate the handling of cipher suite materials from the varieties of network environments. It is expected that multiple security contexts will be developed over time. Understanding the purpose of these contexts is important to selecting what contexts to implement in a given network, and when new contexts must be developed and standardized.

After reading this chapter you will be able to:

- List the five services of a security ecosystem.
- Explain why traditional cipher suite definitions have limitations in certain networking environments.
- Evaluate security contexts for completeness against a set of necessary considerations.
- Demonstrate different approaches to handling the syntactic processing of data into and out of a security context.

10.1 The Case for Contexts

There are a few ways to discuss how security contexts are used in BPSec. One way is to focus on the definition of the security context to differentiate the concept from the concept of a cipher suite, as was done in Section 6.3. Another way is to outline the way in which security context information is represented within a BPSec security block, as is outlined in Section 6.4.4.1.

Distinct from the definition and representation of security context information, this chapter discusses the rationale for including contexts in BPSec. By describing how security contexts fit into the overall BPv7 security ecosystem, network architects and operators can determine when to apply what contexts, and in what circumstances new security contexts should be created.

10.1.1 A BPv7 Security Ecosystem

The term ecosystem, applied to information technology, encompasses more than a sense of design cohesiveness. Much like biological ecosystems, information ecosystems imply self-sufficiency that can grow and adapt over time. The ability of a system to evolve over time is what makes it akin to nature. Using this stricter sense, fixed Application Programming Interfaces (APIs) and fixed

Securing Delay-Tolerant Networks with BPSec, First Edition. Edward J. Birrane III, Sarah Heiner, and Ken McKeever.
© 2023 John Wiley & Sons, Inc. Published 2023 by John Wiley & Sons, Inc.
Companion website: www.wiley.com/go/birrane/securingdelay-tolerantnetworks

implementations might (or might not) enable the development of an ecosystem, but are not themselves ecosystems.

Because BPv7 networks may be used in a variety of networking environments – including but not limited to the DTN architecture – the sustainability and adaptability of an ecosystem are needed to secure bundles.

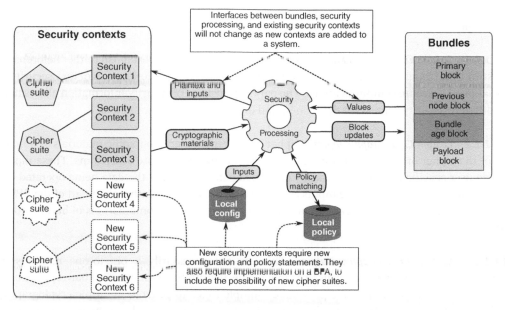

Figure 10.1   Security contexts evolve over time. Future block types might use existing security contexts or new security contexts as they are developed. Adding future security contexts does not change the existing APIs of security blocks, operations, or cipher suites.

Definition 10.1   *(Security Ecosystem)*   The set of algorithms, implementations, policies, configurations, and protocols that work together to generate, communicate, and process cryptographic materials associated with BPv7 bundles.

The relationships amongst various elements of the BPv7 security ecosystem are illustrated in Figure 10.1. Within this ecosystem, BPSec extension blocks define the behaviors that generate and process cryptographic information and the structures used to carry those data in the network. The BPSec definitions of security blocks and security operations, by necessity, must be fixed so that independent Bundle Protocol Agent (BPA) implementation can interoperate. The security context is used by BPSec as the method of sustainable adaptation.

10.1.1.1   Adaptation Properties
The security context concept provides an interface between a security operation and a set of cryptographic algorithms. Some of that interface is resident on the BPAs taking on the roles of security sources, verifiers, and acceptors. Some of that interface is resident in the security blocks carrying security operations.

> **BPSec Adaptations**
>
> BPSec uses security contexts to adapt to new environments and security capabilities over time.

There are five characteristics of the security context interface, each of which support the implementation of BPv7 security as an ecosystem.

1) Custom Configuration Parameters.
2) Custom Cryptographic Inputs.
3) Custom Cryptographic Outputs.
4) Parameter Binding.
5) Custom Policy.

10.1.1.1.1 *Custom Configuration Parameters* Security endpoints must establish (either beforehand or just-in-time) unique parameters and configurations to determine the specifics of how security is to be applied to secure communications. The parameterization of security information allows new and different security contexts to define new and different security parameters.

10.1.1.1.2 *Custom Cryptographic Inputs* Cipher suites generate cryptographic materials as a function of the cipher or other algorithms selected and the inputs to those algorithms. These inputs include user-specified parameters, locally generated parameters, and the data being operated on. By defining the set of cipher suite inputs as part of a security context rather than part of the BPSec specification itself, security blocks can be made to protect creative combinations of bundle information.

10.1.1.1.3 *Custom Cryptographic Outputs* There is not necessarily a one-to-one relationship between security operations and calls to a cipher suite algorithm. As an interface between the two, security contexts may call into a cipher suite multiple times to generate multiple cryptographic results. These results can be carried in security blocks or in target blocks at the discretion of the security context. This allows significant flexibility in how base services such as *bib-integrity* and *bcb-confidentiality* can be deployed in a BPv7 network.

10.1.1.1.4 *Parameter Binding* Once generated, cryptographic material must be communicated to other security nodes in the network for processing in such a way that the material is both understandable and secure. The extent to which custom parameters are also communicated with the materials they helped generate is also controlled by the security context. The ability to define new information to be carried with a security block (and what custom parameters are not to be included in a block) allows contexts to adapt to different end-to-end capabilities in different BPv7 networks.

10.1.1.1.5 *Custom Policy* Policy at BPAs allows different security operations to use different security contexts, even in the same bundle. This capability allows security capabilities to evolve over time without breaking backwards compatibility with existing security policies. The inclusion of new security contexts does not need to break the implementation of existing security contexts.

10.1.2 Cipher Suites

As with any other set of security extensions and security protocols, BPSec does not define its own algorithms for the generation and processing of cryptographic materials. This extends to the concept of security contexts, where custom behaviors deal with the construction of cipher suite inputs and the packaging of cipher suite outputs, but not the definition of new cipher suites.

As discussed in Section 2.3, cipher suites are those algorithms (and associated restrictions on those algorithms) used for the generation and processing of cryptographic material. To understand the conceptual difference between a security context and the cipher suite(s) used by that context it is useful to examine common components of a cipher suite.

10.1.2.1 Cipher Suite Terms

The terms cleartext, plaintext, and ciphertext are defined in the context of cipher suite inputs and outputs. While these terms are commonly used in the literature, there can be some conflation between their specific definitions. For the purposes of understanding how security contexts provide information into, and capture results from, cipher suites these terms are defined as follows.

Definition 10.2 *(Cleartext)* Any information that is not the output of a cipher algorithm. Data elements described as being in cleartext imply that they have not been encrypted and that their meaning can be inspected by anyone with access to the data without needing any special processing.

Definition 10.3 *(Plaintext)* Any information that is input into any cipher algorithm. The use of the term plain indicates that the data has not yet been modified by the current cipher algorithm operating on it.

Definition 10.4 *(Ciphertext)* The result of applying a cipher algorithm to some given plaintext. The resultant ciphertext is encrypted such that a decryption algorithm must be run to recover the original plaintext.

There is an important distinction between the terms cleartext and plaintext. Plaintext can also be ciphertext, whereas cleartext can never be ciphertext. These relationships are illustrated in Figure 10.2 as a function of cipher suite inputs and outputs.

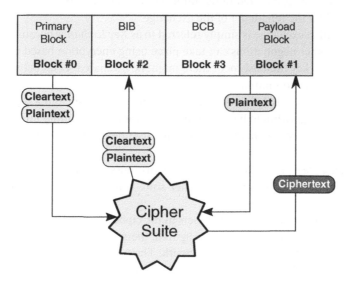

Figure 10.2 Cipher suite inputs and outputs. The terms cleartext and ciphertext are precisely defined relating to whether their contents have been encrypted. The general term plaintext refers to any input to a cipher suite algorithm.

It is particularly easy to conflate the terms plaintext and cleartext, as the act of encrypting is usually considered to be run on unencrypted data, thus converting cleartext to ciphertext. However, there are a few cases where the inputs to a cipher algorithm are not cleartext, to include the following.

- When an application provides encrypted data to a BPA which, subsequently, encrypts that data again using $OP(bcb\text{-}confidentiality, Block_{Payload})$. In this case, the input to the security context used by the *bcb-confidentiality* service is plaintext and not cleartext.
- When an application generates a signature over some encrypted data, as can occur with authenticated encryption, then the plaintext input to an integrity algorithm is ciphertext and not cleartext.

Similarly, the term ciphertext is only used to represent material that has been encrypted using a cipher. Other cipher suite operations, such as generating a signature, is not considered ciphertext. When discussing the generic output of a cipher suite algorithm, the more generic term *cryptographic materials* can be used.

Consistent and proper use of these terms is particularly important in cases where security contexts use multiple cipher suites or otherwise perform multiple cryptographic operations on data as part of applying a security operation.

10.1.2.2 Cipher Suite Algorithms

cipher suites must provide three capabilities to securely generate and process cryptographic materials. These capabilities include one or more algorithms negotiation information, apply ciphers, and authenticate information. As a suite of algorithms, not every algorithm will be used in every situation. Sometimes algorithm selection is a function of the parameters given to the cipher suite implementations. Other times, the security context itself decides which algorithms to use.

10.1.2.2.1 *Information Exchange*

An important first step in any secure exchange is the negotiation of special parameters. Most often, this includes the establishment of some ephemeral shared secret (e.g. a session key) and sometimes additional information relating to the kinds of ciphers and key sizes that can be used for the remainder of the exchange.

Often, this information exchange is simply referred to as *Key Exchange* because once ephemeral keys are established, other negotiations can take place using encryption based on the established keys. Prior to ephemeral key establishment there might be no shared secret between security endpoints and, thus, no secure communications.

Sometimes, cipher suites will define both a key exchange algorithm and a separate authentication and/or encryption algorithm to be used specifically for any remaining information exchange (handshaking) prior to the commencement of secured user data communications.

10.1.2.2.2 *Transport Cipher*

The transport cipher algorithm(s) is considered the primary responsibility of the cipher suite (the goal of the cipher suite being the implementation and support associated with a cipher).

These algorithms comprise the set of all ciphers in the cipher suite and that generate any ciphertext coming from the suite. When properly configured, these algorithms produce ciphertext that is appropriate for transport over an untrusted network. These algorithms also implement the transformation of ciphertext back to plaintext as part of the decryption process at a security acceptor.

Suites may contain different ciphers or the same cipher used in different modes. For example, the Advanced Encryption Standard (AES) has multiple modes of operation, to include the Galois Counter Mode (GCM) [1] and supports symmetric keys of three different key lengths: 128, 192, or 256 bits.

10.1.2.2.3 Authentication Codes Recent security ecosystems require that ciphertext generated by a transport cipher have an authentication mechanism applied to ensure that the ciphertext remains unchanged from source to destination. Authentication algorithms are used to produce these mechanisms, which can be carried with the ciphertext for evaluation at a security destination.

10.1.2.3 Partial Suites

From these definitions, it could be inferred that the algorithms within a cipher suite are dedicated solely to the encryption and decryption of user data. This is not necessarily the case, for two reasons.

- Some of the algorithms defined in a cipher suite describe how to check the authentication of encrypted materials.
- Authentication algorithms defined by a cipher suite might be used independently of any encryption to check the authentication of information that may otherwise not require confidentiality.

There exist certain cipher suites that intentionally omit ciphers as they are only used for message authentication and integrity protection. These *NULL-encryption* suites have found utility in resourced-constrained devices where encryption might be difficult to implement correctly and, generally, not needed for operations if all data in the network can be sent without confidentiality.

One example of such a *NULL cipher suite* is the *Pre-Shared Key (PSK) cipher suites with NULL Encryption for Transport Layer Security (TLS)* defined in RFC 4785 [2]. The three NULL ciphers defined in this cipher suite are listed in Table 10.1. See **Focus: IANA Cipher Suite Names** on next page.

Table 10.1 Pre-Shared Key cipher suites with NULL encryption for transport layer security.

Cipher suite	Key exchange	Cipher	Hash
TLS_PSK_WITH_NULL_SHA	PSK	NULL	SHA
TLS_DHE_PSK_WITH_NULL_SHA	DHE_PSK	NULL	SHA
TLS_Rsa:PSK_WITH_NULL_SHA	Rsa:PSK	NULL	SHA

Source: RFC 4785 Cipher Usage [2].

10.1.3 Security Configuration

Cipher suite algorithms are often parameterized beyond their plaintext input to allow for customized behavior. Some of these parameters are expected to be supplied by the user of the cipher suite while others may be generated internally (or negotiated remotely) by the algorithms in the cipher suite. Security contexts should be parameterized for similar reasons.

The many-to-many relationship between security contexts, cipher suites, and bundle data is illustrated in Figure 10.3. In this figure, concept of parameterized security contexts is shown as an API into the security context itself. Information from the local BPA, to include those data taken from a bundle, are processed by a security context through its unique API. This API is different from the inputs to cipher suites, which may be used and reused by different contexts over the life of the network.

To understand the role of security contexts and how parameters can customize their behavior and the behavior of cipher suites used within those contexts, some discussion of the types and sources of configuration data is warranted.

Focus: IANA Cipher Suite Names

The Internet Assigned Numbers Authority, among other duties, is responsible for maintaining a central repository for Internet-related protocol number registries.

The Internet Assigned Numbers Authority (IANA) maintains a set of cipher suites that can be used with various security protocols, such as TLS. These suites are identified both by a unique enumeration and a string name.

While no formal standard for cipher suite names exists, there are informal naming conventions used to quickly identify the algorithms contained within a suite, and to what purpose they may be applied. A common convention used is:

$$PROT\_EXC_{Key}\_EXC_{Cipher}\_WITH\_MSG_{Cipher}\_MSG_{Auth}$$

For example, the cipher suite:

TLS_ECDHE_Rsa:WITH_AES_128_GCM_SHA256

is used by the **TLS** Protocol, uses **Elliptic Curve Diffie-Hellman Exchange (ECDHE)** to establish keys, and protects handshake information using **RSA** encryption. **128-bit AES in Galois Counter Mode (GCM)** is used as the transport cipher and **256-bit Secure Hash Algorithm (SHA)** is used to authenticate ciphertext and, optionally, other information.

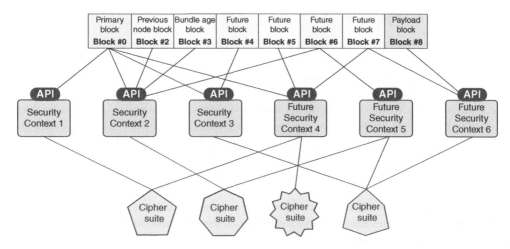

Figure 10.3   Security contexts interface BPSec to cipher suites. Security contexts support a many-to-many relationship between bundle data and cipher suites.

10.1.3.1 Configuration Sources

The sources of configuration data at any given BPA comes from the BPA itself, from bundles that the BPA processes, and from applications co-resident on the same node as that BPA. In unchallenged networks, these sources may be located anywhere in the network because of the network's ability to perform low-latency, round-trip data exchange.

In certain BPv7 networks, other information sources might be necessary, such as more responsive local autonomy, neighborhood oracles, and data caches.

- **Local Autonomy**: BPAs may choose to alter their own configurations (such as expiring keys or updating routing tables) as a function of state prediction of the BPA and/or the network. In simple cases, planned motion and schedules of events can be used to provide high-quality estimates of the state of the network and the ways in which BPA must adapt to those changes over time.
- **Neighborhood Oracles**: The neighborhood of a network refers to a series of nodes that are expect to be within some kind of *regular contact* with a node over some useful period of time. These nodes carry information both for themselves and for nodes that either always exist within their sphere of influence or (through mobility) pass through their sphere of influence. While end-to-end data exchanges might not be possible, node-to-node exchange may be possible within a given neighborhood.
- **Data Caches**: BPAs may choose to store data relevant to the state of the network for use in its own future bundle processing. In these cases, BPAs may build a cache of useful network data, to include configuration information. These caches may persist even after the bundle which populate them has been forwarded, delivered, or otherwise removed from the BPA.

There are multiple potential sources for security configuration data, and what types of sources provide what types of information are governed by the capabilities of the network itself. For example, over the terrestrial Internet there may not be a concept of a neighborhood oracle as end-to-end connectivity is considered ever-present. In a deep-space constellation, a planetary lander may have regular contact with an orbiting communications relay, but less contacts back to Earth.

While some thought to where configuration data comes from is important to ensure that security contexts are not designed with unrealistic expectations, the best way to do that is to focus on the types of configuration data that exist within a security ecosystem. Configuration types provide a logical overview of where data in a system comes from, independent of the sources of that data which might be different between different network implementations and even different BPv7 deployments.

10.1.3.2 Configuration Types

A BPv7 security ecosystem will have four types of information that all need to be present to correctly configure cipher suite algorithms. Together, these four types of inputs are required to successfully generate cryptographic materials associated with any cipher suite used by any BPSec security context.

1) Static values
2) Negotiated values
3) External values
4) Message data

10.1.3.2.1 Static Values A static value is an immutable property of the security context or cipher suite algorithm being configured. They are typically codified as named constants or hard-coded literal values in implementations.

These values are useful for selecting modes within security contexts and cipher suite algorithms that provide choices. In those cases, security policy and other configuration sources can use these values to signal modes of operation and select amongst pre-defined behaviors. Similarly, implementations use these values to verify configuration inputs.

Consider the case of the AES transport cipher, which is defined for key lengths of 128, 192, or 256 bits. Each of these values represent a static value for cipher suites including support for AES. Policies can use these constant values to specify which variant of AES to use and security context and cipher suite implementations can use these values to ensure that they have received a valid key length.

10.1.3.2.2 Negotiated Values
A negotiated value is one that is cooperatively calculated across a network – typically between the endpoints of a secure communications channel. In this usage context the cooperative exchange occurs in "real-time" when the negotiated values are needed and not pre-determined in advance of future needs. Therefore, the ability to support negotiation depends on the ability of the underlying transport network to support end-to-end data exchange.

Negotiated values are the norm for information exchange over the terrestrial Internet as part of establishing communication sessions. In that use case, negotiations are often called handshakes. They are the preferred method of calculating shared information because, when supported, the provide the most up-to-date information possible.

A common use of negotiation is to use asymmetric-key encryption to negotiate a shared symmetric key that can be used for a single communications session.

10.1.3.2.3 External Values
External values are those that are specified external to the security context, cipher suite algorithm, and target message data. Like static values they exist prior to a given security operation. Like negotiated values they customize how a security operation is applied. All values represented as policy and configuration in a system are of this type.

Because external values is an encompassing set, they can be further subtyped based on their origin as asserted, self-determined, or pre-negotiated.

Asserted Values Configured values that are asserted are placed into the configuration of a system by some authoritative source, such as a network operator. These values are inherently trusted and represent the bulk of configuration information in manually operated networks.

An asserted value for a security system would be a policy statement related to the security role of a BPA as either the security source, verifier, or acceptor of a given security operation.

Self-determined Values A self-determined value is any value that is calculated by the BPA. This type of self-configuration is usually a function of some time of local autonomy running on the BPA. Local autonomy can be used to achieve increased self-reliance and self-management in any kind of network, but has a particularly enabling utility in DTN environments.

One way to self-determine values is state estimation, where a BPA chooses from one of several existing configurations as a function of local time or other state changes. Consider the case of key rotations, where a given key expires at a certain time or after a certain number of uses. In this case, the decision to rotate to a new key for some security operation is self-determined by the BPA.

Another way to self-determine values is to perform future state prediction on a BPA. Continuing with a key example, a BPA acting as a security source might predict what long-term key a future verifier or acceptor would be using at the time of bundle receipt. Such a predictive security source could either select a single appropriate key or otherwise calculate security results using multiple potential long-term keys based on its predictions.

Pre-negotiated Values Negotiating values in real-time is problematic in the DTN architecture, but that does not mean that BPv7 networks can not negotiate data. Pre-negotiation is the process of negotiated data *before* that data might be needed. Pre-negotiation is preferable because it can be done far enough in advance that round-trip exchanges can complete prior to needing the negotiated value.

Pre-negotiation is still problematic in that values provided too far in advance may expire or otherwise no longer be appropriate for use by the time they are needed. One way to avoid this consequence is to pre-negotiate multiple sets of values and then use local autonomy or timing information to self-determine which set of values to use at a given time.

In this way, as long as pre-negotiation algorithms can stay ahead of actual use cases security systems can remain configured.

10.1.3.2.4 *Message Data Values* Message data values are those present in the bundle being processed by a security operation. These values include the plaintext or ciphertext of target block contents and the security context parameters and results held in BPSec security blocks. Plaintext and ciphertext are clearly important to security operation processing.

Separately, the ability to keep parameters and other results associated with this plaintext and ciphertext provides an in-band mechanism for data communication. The data associated with a security operation does not need to be negotiated or build external to the security ecosystem when it can otherwise be specified and carried in the security block itself.

10.1.3.3 Limitations of Current Approaches

Network architects must carefully consider how security configurations (to include security policy settings) are communicated across a network. This consideration is difficult even in unchallenged networks, but becomes significantly harder in challenged networking environments such as the ones found in the DTN architecture.

Some limitations that may be present in a BPv7 network include the following.

10.1.3.3.1 *No Timely Access To Endpoints* Cipher suites the define key exchange algorithms that presume negotiation between security endpoints might not be useful in certain BPv7 networks. In cases where these cipher suites are required, security contexts might be required to bypass their key exchange algorithms in lieu of some other mechanism more appropriate for challenged networks.

Alternatively, security contexts might choose to use multiple cipher suites, with one cipher suite used for its key exchange algorithms and another cipher suite used for its cipher. In certain cases, the use of different existing cipher suites may require less implementation and configuration than defining new cipher suites with different combinations of existing algorithms.

10.1.3.3.2 *Out of Date Oracles* The concept of well-informed nodes in a network neighborhood is useful in challenged network design – particularly in cases where end-to-end data synchronization is unlikely. Pre-placing cached information at neighborhood oracles increases the change that a given node in the neighborhood would have access to those configuration data.[1]

Neighborhood oracles, themselves, may receive their information from authoritative endpoints making their local cache of information potentially just as out-of-date as a local node. When this happens, secure communications based on out-of-date information may lead to failures to authenticate or otherwise process cryptographic materials at security endpoints. This is particularly true when secure endpoints in the network exist in different neighborhood being configured by different oracles.

---

1 This is not guaranteed, especially when the local node is a node that has left the network.

Security context configuration should, where possible, place as much configuration in the security blocks that carry security operations as possible to minimize the reliance of downstream security verifiers and security acceptors on their own neighborhood oracles.

10.1.3.3.3 ***Reliance on Local Autonomy*** Without reliable access to external configuration data, BPAs have a higher reliance on local autonomy to both predict network changes and apply local configuration changes. This implies the need to develop new algorithms that can perform security policy and configuration functions at local nodes with little to no synchronization with other nodes in a system.

Security contexts should specify multiple parameters to include not only configuration data that was used in generating cryptographic materials, but should also include associated local state information to provide some context to downstream nodes on why a local BPA chose a particular configuration datum.

10.1.3.3.4 ***Required Out-of-Band Configurations*** To seed local autonomy, and to help bootstrap BPAs, out-of-band configuration mechanisms are usually required. These pre-place security configurations under the assumption that periods of connectivity, while relatively rare, are sufficiently to allow for the long-term pre-configuration of node resources.

For example, if a sensor node is in contact with a configuration oracle once every three weeks, but during those contacts a months worth of key and schedule information can be sent to that sensor node, then this out-of-band configuration can be used in lieu of negotiated, in-band configuration as part of session establishment.

Security contexts should be written to include sensible default values based on either static values or pre-negotiated values to take advantage of these out-of-band configuration mechanisms.

10.2 Using Security Contexts

The *use* of security contexts refers to the ways in which network architects and operators choose to apply security contexts to securing blocks within bundles. This use is separate from the definition of a security context and its representation in a security block, which is covered in more detail in Section 6.3. This use is also separate from the design and specification of new security contexts, which is covered in more detail in Chapter 11.

There are four ways in which existing security contexts are used to impact the security of blocks within a bundle.

1) Identifying existing security contexts for use in security policy.
2) Determining what blocks should be security using what contexts.
3) Providing context parameters to security block processing
4) Determining context options and results.

The remainder of this section discusses how network architects and operators can make decisions and apply configurations associated with each of these uses.

10.2.1 Identifying Contexts

Security contexts provide BPSec with the adaptability it needs to function given the breadth of envisioned BPv7 network deployments. This adaptability allows new security contexts to be defined over time as new network deployments, new cipher suites, and new kinds of BPv7 blocks are defined. This flexibility, however, does not imply that the identification or use of security contexts is arbitrary. In order to build interoperable devices – to include those devices that generate and process security blocks – security contexts must be easily identifiable and perform in the same way across the networks in which they are used.

BPSec processing nodes need to understand which security context was used to generate a security operation so that they can determine what security context to use when processing that operation. Since support for these contexts must be present in the implementation of the node itself, and since these nodes may exist in challenged networks where software changes might be difficult to implement, security context definitions need to not change and be easy to identify. These requirements imply that security contexts require moderation and enumeration.

> *Users may select from registered security contexts and customize those contexts through security context parameters … [s]ome users might prefer a SHA2 hash function for integrity whereas other users might prefer a SHA3 hash function.*
>
> BPSec §2.4, [3]

Security contexts support is not required to be uniform throughout the network. Different nodes may implement different cipher suites and different security context behaviors. This concept is illustrated in Figure 10.4. In this sample network, Node 4 is the only node supporting all network security contexts (identified as *ID*1, *ID*2, *ID*3, and *ID*4). Other nodes support different combinations as a function of their compute resources and expected traffic requirements.

As with other standards published by the Internet Engineering Task Force (IETF), security contexts are registered with, and moderated by, the Internet Assigned Numbers Authority (IANA). Since March 24th, 2011 IANA has maintained a registry for items associated with the Bundle Protocol (both BPv6 and BPv7). With the publication of BPSec as RFC 9172 on February 1st, 2022,

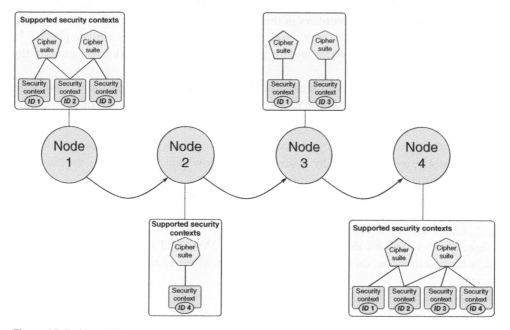

Figure 10.4 Not all BPAs support the same security contexts. While every BPA in a network must be able to parse BPSec security blocks, not every BPA must implement every security context. Security policy should only specify that BPAs are security verifiers and security acceptors for security contexts that they can support.

BPSec security context identifiers

Registration Procedure(s)
Specification Required
Expert(s)
Ken McKeever, Edward Birrane
Reference
[RFC-ietf-dtn-bpsec-27]
Available Formats

csv

Value 🔽	Description 🔽	Reference 🔽
-32768 to -1	Reserved	[RFC-ietf-dtn-bpsec-27]
0	Reserved	[RFC-ietf-dtn-bpsec-27]
1	BIB-HMAC-SHA2	[RFC-ietf-dtn-bpsec-default-sc-11]
2	BCB-AES-GCM	[RFC-ietf-dtn-bpsec-default-sc-11]
3-32767	Unassigned	

Figure 10.5 Initial BPSec Security Context Definitions. The initial set of security context definitions included a single integrity security context for block integrity and a single security context for block confidentiality. https://www.iana.org/assignments/bundle/bundle.xhtml\#bpsec-security-context.

this registry was appended to include a registration for *BPSec Security Context Identifiers* and initial definitions for two such security contexts, as illustrated in Figure 10.5.

IANA registries are *moderated* meaning that new entries can only be placed in them by IANA and with concurrence from identified experts overseeing the allocation of enumeration values. Standardizing security context identifiers in this way guarantees developers that security context definitions will not be renumbered or otherwise redefined. IANA provides a single place to look for new security context enumerations and for guidance on where to find the definitions of those contexts.

Values (or value ranges) in an IANA registry can be defined as reserved, unassigned, or otherwise given an assigned value. The meaning of each of these value descriptions determines how registry items can be enumerated in the future.

- **Reserved** values are those explicitly not assigned by IANA. These values will not be standardized and should not be used operationally. Reserved values allow implementations to use these values for specific, local purposes such as unit testing.
- **Unassigned** values are those that have not yet been assigned, but are expected to be used as part of future enumerations. Implementations should not use unassigned values, as those values might be standardized at a future time.
- **Assigned** values are those that have been assigned by IANA and are now considered standardized. In the case of BPSec, the security contexts BIB-HMAC-SHA2 and BCB-AES-GCM have been assigned.

10.2.2 Selecting Contexts

Network operators must determine what security contexts should be used for what kinds of data in a BPv7 network. Beyond simply understanding that contexts have been defined by IANA, operators must consider the capabilities of a security context and the nature of the data they protect. This

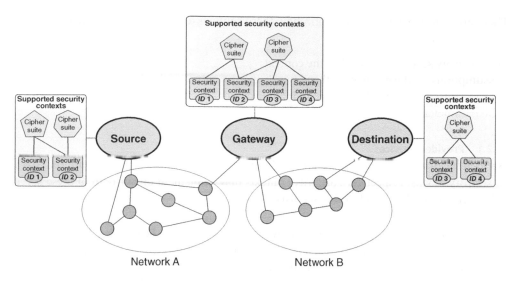

Figure 10.6 Security contexts can follow network topology. BPSec does not encode routing or topology information into security policy at a security source. But, policy can associate security contexts with topologically distinct portions of a BPv7 network.

informs both the expectation of what contexts might be encountered in the network and at what types of nodes those contexts might appear.

In cases where the same security contexts are supported across a network, and particularly at messaging endpoints, then BPSec can provide end-to-end security services. However, even in cases where end-to-end security is not possible, security contexts can provide some level of protection in segmented networks. Careful selection of contexts and their deployments consistent with network topology allow for this type of specialized usage.

Consider the case of bundles traversing an internetwork consisting of two otherwise administratively separate domains, as illustrated in Figure 10.6. In this figure, *Network A* uses security contexts *ID*1 and *ID*2, whereas *Network B* uses security contexts *ID*3 and *ID*4. Were a bundle to be sourced in *Network A* and delivered in *Network B*, a gateway node might serve as the security acceptor of *Network A* services and the security source of *Network B* services. This usage is similar to the concept of secure tunnels, without requiring the information hiding inherent in the security tunneling approach.

Even when similar processing actions can be defined across multiple security contexts, these contexts might require different responses from a policy perspective. The response to some processing error in a context which assumes operators-in-the-loop might be different from the response to the same error in a context which assumes completely autonomous operation.

A security context definition provides normative instruction on how to implement the algorithms, configurations, and policies associated with the generation, transport, and processing of cryptographic materials. This includes explaining how nodes should implement context processing, how information should be extracted from bundles, and how bundles should be modified as a result of applying security to them. In this way, a security context definition functions as its own security standard.

It is expected that security context definitions be generated under the auspices of standards organizations and that these definitions be seen as prescriptive guidance for developers. Were different vendors to implement different security contexts in different ways the overall security (and interoperability) of the system might be compromised.

There are four capabilities that inform how these contexts are deployed to secure data in a network, as follows.

1. The security services provided by the context.
2. Assumptions on the behavior of the network and the nature of the data.
3. The algorithms used by the context, to include cipher suites.
4. The parameters that can be given to customize the context

10.2.2.1 Provided Services

BPSec defines two security services, *bib-integrity* and *bcb-confidentiality*, aligned with the BIB and BCB security blocks, respectively. Security contexts may implement one or both of these services or other services if identified by other security blocks.[2]

> *[a] security context identifier MAY be used with a BIB, with a BCB, or with any other BPSec-compliant security block. The definition of a security context MUST identify which security services may be used with the security context, how security context parameters are interpreted as a function of the security operation being supported, and which security results are produced for each security service.*
>
> *BPSec, RFC 9172 §9.2, [3]*

Operators should select security contexts as a function of the services provided by that context. If a target block requires confidentiality, then the selected context must provide a confidentiality service, such as *bcb-confidentiality*.

10.2.2.2 Assumptions

Security context assumptions document the presumed capabilities of the network, information resident at security-processing BPAs, and what mechanisms are expected to be available for end-to-end communications.

For example, security contexts built for deployment over the terrestrial Internet might presume low-latency, high-reliability data exchange, access to certificate authorities, and other capabilities associated with Public-Key Infrastructure (PKI). However, security contexts built for low-power, infrequent communication devices might restrict round-trip data exchange and require pre-placed keys and symmetric-key cipher suites.

Operators should evaluate the documented assumptions of a security context prior to determining that the context should be used in a given BPv7 network. In cases where there is ambiguity on how network constraints might manifest, operators should restrict usable contents to those that make the least number of assumptions on the capabilities of the network.

10.2.2.3 Algorithms

Security context algorithms refer to both the cipher suite(s) used to generate cryptographic materials and the set of context-specific algorithms that determine how these materials interact with bundle information.

There are several cases where a security context might support multiple cipher suites. A security context might wrap a more comprehensive security interface, such as the Cryptographic Message Syntax (CMS) or the Concise Binary Object Representation – CBOR Object Signing and Encrypting

---

2 BPSec allows for the possibility of other security blocks to be defined in a bundle – these blocks are collectively termed Other Security Blocks (OSBs).

(COSE) interface. These interfaces provide a way in which parameters can be used to select from amongst multiple supported cipher suites.

Context-specific algorithms are those that determine how plaintext is constructed and what cryptographic materials are generated, and how those materials are stored within a bundle. While the BPSec provides some initial guidance on this in the form of security block structure, there may be additional dependencies, sanity checks, and other considerations to ensure that bundles remain well-formed through this process.

Operators should understand the cipher suites that a context implements and the capabilities of the context-specific algorithms. Contexts that support CMS or COSE may better integrate with existing security policies and tools. Contexts that provide unique constructions for plaintext allow operators to enforce dependencies and other relationships between and amongst blocks in a bundle.

10.2.2.4 Parameters

Different security contexts support different combinations of parameters that can be used to customize the behavior of the context. While the assumptions of a context imply certain hard-coded values associated with algorithms, parameters allow the same context to be used in different ways where necessary.

Operators must understand the ways in which security contexts can be parameterized. Because security policy specifies required behavior in response to defined events it is unlikely that the same policy can be specified for dramatically different security contexts. Different contexts might need to react to different processing events – for example, if different contexts establish keys differently (negotiation versus pre-placement versus key-wrap) then different policy statements may be needed for these different use cases.

Selecting contexts with maximum configurability allows for fewer individual contexts to be deployed (and thus supported) in a network.

10.2.3 Selecting Parameters and Results

While security context definitions act as standards in and of themselves, they must be customized to the data they operate on, the data they produce, and what processing options should be used as a matter of policy. Each security context defines one or more security context parameters and must define at least one security context result.

Every aspect of a security context's operation may be customized through parameters. For example, BPSec presumes that all security contexts operate on the block-type-specific data of the target block. Some security contexts might expand this behavior by including the target block's extension block header or allowing multiple target blocks to be catenated together to form complex plaintext. Some security contexts might constrain this behavior by specifying only a portion of the target block's block-type-specific data be adjusted.

Security results include the cryptographic materials generated by cipher suite algorithms, as customized by relevant parameters and applied to target blocks from the bundle. Contexts may generate multiple different types of results as a function of their regular operation and/or the parameters passed to the context.

Operators must understand how to select parameters and the ways in which these parameters will be used to populate security blocks and impact security processing at downstream BPAs.

10.2.3.1 Parameter Encoding

Security context parameters can represent any information needed to accomplish processing within the context. They may be encoded as primitive data types (e.g. integers and strings) or more complex data types such as serialized structured data only meaningful internal to the security context implementation. Parameters may be optional, may have pre-defined default values, and in some cases may be dependent on (or exclusive to) other parameters.

Operators must be aware of how parameters are encoded in the system to avoid misconfiguration. This is particularly important when security contexts define default values for parameters. In such a case, the absence of an operator-specified parameter does not mean that the parameter is unset when using the context.

10.2.3.2 Parameter Types

Operators must determine what parameters should be set as a function of the role of the BPA in a security operation's lifecycle. Not every parameter of every security context must be set. While the set parameters are different for any given security contexts, parameters can be discussed in general by the types of information they customize: security information, algorithm configuration, and processing options.

Information from the bundle (such as the context of a target block) is also an input into a security context. This information is not considered a parameter in this discussion because it is calculated in real time as bundles are generated and received in the network. Operators cannot specify bundle contents as part of system configuration.

10.2.3.2.1 *Security Information*   These parameters specify user-supplied information that is necessary for the processing of user data by the cipher suites and other security algorithms contained by the security context. Most cipher suite algorithms require some external input to help customize computed results. For example, when applying an authentication service, some information (such as a key) associated with the user must be input to the algorithm so that generated security results can be authenticated as coming from that user.

10.2.3.2.2 *Algorithm Configuration*   In cases where a security context supports multiple cipher suites, parameters may be used to determine algorithm selection. It is possible to define a security context that supports both plaintext integrity (such as would populate BIB security operations) and confidentiality (such as would populate BCB security operations). Security context parameters could be used to specify which operation is being requested by the security context.

10.2.3.2.3 *Processing Options*   Security contexts may define their own processing options independent of the cipher suites that encompass. These options include how to produce the plaintext fed to cipher suites, and what security results should be generated by which cipher suites. Some cipher suites allow for additional, unencrypted user data to be authenticated along with the user data to be encrypted. In this case, the option to include this *additional authenticated data*, and what parts of the bundle would be included, would be processing options.

10.2.3.3 Parameter Sources

Operators must be cognizant of the various sources of parameters given to a security context. The information that a network operator encodes into the local configuration and policy at a BPA might not be the only source of parameterization available when the security context processes that operation at a verifier or acceptor BPA.

The choice of what parameters an operator might specify should be based on an understanding of the multiple potential sources of values for that parameter, which may differ based on whether the security course is being created at a security source or processed at a security verifier or security acceptor.

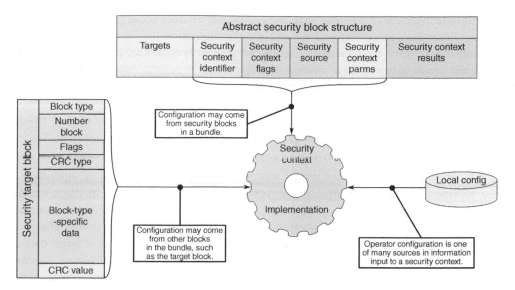

Figure 10.7   Security contexts fuse parameters from multiple sources. Security context behavior can be used to prioritize and fuse data from multiple sources. When the local BPA, security block information, and other bundle information present conflicting information, different security contexts can be used to deconflict based on the assumptions built into the context itself.

Figure 10.7 illustrates some of the potential sources of security context configuration. When a security context is being used as a security verifier or acceptor, the contents of the security block being processed might hold parameters associated with the context. At any security-processing BPA, other elements of a bundle, such as the target block, are used by the context. Similarly, and in all cases, local configuration is used to select options on how to apply the context.

10.2.3.3.1   Node Values   These are values resident at the local security-processing BPA. These values are selected by policy as a function of bundle, security operation, and security context. These values are placed at BPAs and may represent static, negotiated, or external values.

10.2.3.3.2   Security Block Values   In certain circumstances, parameter values from a prior BPA need to be used at a downstream BPA when processing a security operation. In these cases, node values are not helpful because they cannot reflect the state of the prior node. Values known only to the prior node are expected to be contained in the security blocks carrying security operation information in a bundle. Examples of block information include algorithm parameters such as wrapped keys and initialization vectors.

10.2.3.3.3   Default Values   While security contexts cannot customize themselves, they can define default values to be used when there is no other node or security block value. Certain parameters (such as user keys) should not have default values because hard-coding such values would reduce the security of the system. Other values, such as those associated with processing options, can have default values. Using default values in this way can reduce the overhead associated with parameter usage in a security block because parameters with default values can be omitted from the block entirely.

10.2.3.4   Result Types

The results of a security context are the results of applying a security service to the target blocks of a given security operation in the context of a security block. It is possible for a security context to output multiple instances of the same result type or to only produce some types and not others as a function of user data, processing options, and other configuration.

Operators may be given the option to customize what security results are produced by a context. These results would be carried in a bundle, either as part of the security block holding the associated security operation or placed in other parts of the bundle (such as replacing plaintext with a ciphertext result).

There are three types of security results that can be generated from a security context: primary cryptographic materials, secondary cryptographic materials, and downstream information.

10.2.3.4.1 *Primary Results* Primary results contain the outputs of cipher suite algorithms as applied to target blocks. When used in the context of a *bib-integrity*, this includes the data associated with the applied integrity mechanism, such as a signature. When used in the context of *bcb-confidentiality*, this includes the ciphertext that will replace some plaintext in a bundle.

The concept of multiple results is illustrated in Figure 10.8, where a single set of target data can generate multiple results based on information from local configuration. In this instance, a security context generates multiple results from multiple keys, which would be the case in a security context that allows for calculating multiple signatures.

The validation of these results is the function of a security verifier in the network. If some primary result fails to be validated at either a security verifier or a security acceptor, the security operation can be considered to have been corrupted. Operators should not configure contexts to omit primary results – and security contexts should always ensure that at least one such primary result is always produced as part of successful creation of a security operation at a security source.

10.2.3.4.2 *Secondary Results* Secondary results contain information that supports primary results. This may include additional signatures over non-target elements of a security operation. For example, when using a confidentiality cipher suite, in addition to producing ciphertext a security context may also produce an integrity result over additional authenticated data. Alternatively, a security context might generate more than one set of results for a given operation, perhaps using

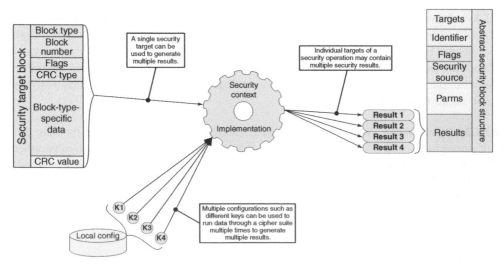

Figure 10.8 Multiple results provide security resiliency. Security services provided by a context might generate multiple results to handle different types of processing at different nodes.

different keys such that any single result must be acceptable, but not all results need to be accepted to consider the operation valid.

The validation of secondary results might be considered optional by a security context verification algorithm, as a function of the intent of the security context.

10.2.3.4.3 *Informational Results* These results may not be associated directly with cryptographic materials. This could include information relating to the use of various elements of a cipher suite, such as utilization information. This information, if present, provides some insight into the operation of the security context at the security source. Care should be taken to ensure that any such information would not be useful to an attacker able to observe bundle contents. With information about the underlying encryption algorithms or other security systems, an attacker may be able to compromise protection. Any information which may be beneficial for an attacker should be protected through a *bcb-confidentiality* service.

10.3 Summary

Security contexts were designed into the BPSec approach to allow for flexibility and adaptability. The flexibility of these contexts allow for different kinds of security results to be carried in BIB and BCB blocks in a bundle. The adaptability allows for new security contexts to be developed over time. Together, this provides a powerful, featured way to evolve a security ecosystem within BPv7 networks.

Security contexts handle all of the interfaces between bundles and cipher suites, to include passing cipher suite parameters, packaging cipher suite outputs, and determining how plaintext is generated from the contents of a bundle. The boundary between security context and cipher suite remains the generation of cryptographic material. Cipher suite algorithms that contain negotiation strategies, transport ciphers, and authentication codes may all continue to operate internal to the security context.

In special cases, algorithms from different cipher suites might need to be used – such as using negotiation algorithms from one cipher suite and transport ciphers from another. In this case, a brand new cipher suite can be defined using these algorithms or the security context can be designed to use multiple cipher suites. The use of multiple cipher suites by a context being the preferred approach with a BPA is resident on a node that includes existing implementations (possibly in hardware) of cipher suite algorithms.

The flexibility of the security context concept allows for new contexts to be defined to cover information exchange in a variety of environments. While developing new contexts is a strength of the BPSec approach, it can also lead to confusion if too many contexts are defined. To limit the number of operational security contexts, all new security contexts should incorporate as much configurable behavior as possible. This includes defining static, negotiated, and external values that customize the processing performed in the context.

Network operators use security contexts by identifying contexts, selecting appropriate contexts based on the nature of the network, and determining what parameters to use with selected contexts.

Security contexts are referenced by a globally unique identifier assigned and moderated by IANA. Part of registering new security contexts requires that the description and behavior of these contexts be standardized in an appropriate standards body (such as the IETF or CCSDS) and approved by the IANA designated experts for that registry.[3]

---

3 At this time of this writing, the designated experts to determine new security contexts are Ed Birrane and Ken McKeever.

The selection of what security contexts should be deployed in a network must include considerations for the security services, assumptions, algorithms, and parameters used by that context. Once the set of deployed contexts are selected, operators use these contexts by configuring security policy on BPAs that specify what services are applied to what parts of a bundle using what configurations.

References

1 Dworkin, M. J. [2007]. SP 800-38D. Recommendation for block cipher modes of operation: Galois/counter mode (GCM) and GMAC.

2 Blumenthal, U. and Goel, P. [2007]. Pre-Shared Key (PSK) Ciphersuites with NULL Encryption for Transport Layer Security (TLS), RFC 4785. **URL**: https://www.rfc-editor.org/info/rfc4785.

3 Birrane, E. J. and McKeever, K. [2022]. Bundle Protocol Security (BPSec), RFC 9172. **URL**: https://www.rfc-editor.org/info/rfc9172.

11

Security Context Design

BPv7 networks may be deployed in a variety of networking environments and BPSec security blocks require a way to adapt to the characteristics of these environments. Security contexts provide this adaptability as they act as an interface between cipher suites bundle information.

Designing security contexts is a complex activity whose success is fundamental to the correct and secure application of BPSec. This chapter explores concepts related to this design.

After reading this chapter you will be able to:

- Identify the ways in which the networking environment impacts security context design.
- Analyze how security context behaviors change the security implementation of a BPv7 network.
- Choose the syntactic representations of security context data.
- Evaluate how cryptographic binding can adjust the scope of BPSec security.

11.1 Overview

A security context represents a standard way of combining Bundle Protocol Agent (BPA) policy and configuration, bundle information, and cipher suite algorithms to process cryptographic materials appropriately for a particular network environment and/or information type. Just as there are several networking environments in which BPv7 bundles may be used, there will be several BPSec security contexts defined to enable securing those environments.

The need to design new security contexts might come from the need to interface with a new cipher suite or to combine algorithms from multiple cipher suites in unique ways. New security contexts might also be developed to use existing cipher suites, but adapt to new limitations or features of a particular transport environment. Contexts might also be defined to enable new behaviors associated with the cryptographic binding of bundle information or other unique scoping requirements of security results to enable more complex security policies.

This chapter discusses those considerations necessary for effective security context design, to include the nature of the BPv7 network, options for context behavior, and how information is encoded when placed in a bundle. Special attention is given to the important concept of cryptographic binding, which allows novel constructions of plaintext to associate target block data with other bundle data.

Securing Delay-Tolerant Networks with BPSec, First Edition. Edward J. Birrane III, Sarah Heiner, and Ken McKeever.
© 2023 John Wiley & Sons, Inc. Published 2023 by John Wiley & Sons, Inc.
Companion website: www.wiley.com/go/birrane/securingdelay-tolerantnetworks

11.2 Novelty

The first consideration taken when designing a new security context is whether or not a new context needs to be created.

> *Developers or implementers should consider the diverse performance and conditions of networks on which the Bundle Protocol (and, therefore, BPSec) will operate. Specifically, the delay and capacity of DTNs can vary substantially. Developers should consider these conditions to better describe the conditions in which those contexts will operate or exhibit vulnerability, and selection of these contexts for implementation should be made with consideration for this reality. There are key differences that may limit the opportunity for a security context to leverage existing cipher suites and technologies that have been developed for use in more reliable networks…*
>
> BPSec, RFC 9172 §9.3, [1]

As a practical matter, security context authors should first determine that a new security context is necessary and that the same capability cannot otherwise be achieved using the parameters of an existing security context. Limiting the overall number of defined security contexts has multiple benefits for network architects, users, and developers, to include the following.

- **Easier Context Selection**: Well-differentiated security contexts enable security policy makers to select the correct contexts for the correct deployments. Having many, similar security contexts increases the chance of conflating security contexts when configuring policy and, thus, increase the chance of a security misconfiguration within a network.
- **Efficient Implementations**: BPAs must implement the algorithms within the cipher suites of security contexts, to include processing associated with parameters and policy. If many security contexts are defined, then vendors must support multiple contexts. In cases where contexts are defined that use the same cipher suite, but with different encodings, options, or other processing rules BPAs might need to implement the same security algorithms in multiple ways, which would increase the cost and complexity of software while potentially reducing processing speed.
- **Better Policy Definition**: Supporting fewer security contexts also implies fewer security policy statements, especially in cases where a set of policy statements would need to be made for each defined security context. Reducing the number of policy statements both speeds up policy processing, but also reduces the likelihood of policy misconfigurations.

For these reasons, security contexts should only be created when they represent a sufficiently motivating departure from existing context definitions. When new contexts do need to be developed, careful consideration must be given to how these contexts might be customized to reduce the likelihood that similar deployments would need to develop even more contexts.

Limiting New Security Context Definitions

Security context authors should seek to limit the number of new security contexts they define. A smaller set of distinct and efficient security contexts is beneficial to network operators who must select and write security policy for these contexts.

Security contexts should only be standardized when there is sufficient motivation for interoperability. For some use cases with a single organization designing, implementing, and operating a private BPv7 network, a self-built security context utilizing a reserved security context identifier may be sufficient.

In the event that a revision or enhancement to a standardized security context is needed, it should be considered to be a new security context. Because standardized security contexts are registered with the Internet Assigned Numbers Authority (IANA), they cannot be modified without disrupting interoperability in networks which already implement the existing security context.

11.3 Network Considerations

The envisioned characteristics of a given network plays an important role in security context design. Understanding the cope of network characteristics – to include data, links, and access properties – reduces the likelihood that new security contexts would be narrowly scoped or otherwise be of limited utility. Further, by annotating security context descriptions with their supported environments allows network operators to know when to use which security contexts.

Three network considerations that designers should evaluate are data lifetimes, the presence of one-way traffic, and expectation of on-demand access to BPAs.

11.3.1 Data Lifetime

Bundles may persist in a BPv7 network for extended periods of times.[1] The relevant lifetime of bundle information is an important security consideration because the strength of cryptographic algorithms and the way they are applied in security processing is correlated to the expected amount of time an attacker would need to overcome, reverse, or otherwise defeat a security service.

Decisions such as the use of ephemeral keys, key length, and algorithm complexity must consider the nature of the data being protected. Designers must consider the presence of long-lived data and whether network operators should avoid the use of certain contexts as a function of expected data lifetimes.

Some relevant questions to consider relating to data lifetime include the following.

- Do cipher suite key sizes support long-term protection?
- Will exchanged keys stay relevant over the data lifetime?
- Will protected blocks be stored with the bundle?
- Are context parameters needed to carry other long term data?

Consider the data lifetimes of two different types of blocks: the payload block and the Previous Node Block (PNB). A bundle might need security protection of its payload block from the bundle source to the bundle destination over some extended period of time. Separately, a PNB in that bundle only needs protection for a single bundle transmission because this type of extension block is added just before bundle transmission and processed and removed by the next receiving BPA. Even when a bundle stays in a network for hours, days or weeks a PNB may only exist for slightly longer then the signal propagation time of the bundle.

For a bundle whose source and destination BPAs are not adjacent, the data lifetimes of the payload and the PNB are very different. A security context appropriate for short-lived data may be fine to use to protect the PNB but may not be appropriate for a longer-lived block such as the payload. Similarly, a security context appropriate for a payload might require using more processing, storage, and bandwidth than is needed for shorter-lived block.

---

1 In extreme cases, bundles could exist in the network for weeks, months, or even years.

11.3.2 One-Way Traffic

Some links in a BPv7 network might be unidirectional. Even when a link nominally supports bidirectional connectivity, changes in network characteristics over time may prevent effective round-trip information exchange, making these links effectively unidirectional.

Security context designers must consider whether the processing inherent to the context requires bidirectional communication as part of the actions taken at a security source, verifier, or acceptor. Attempting to secure data with a security context that requires bidirectional activities such as negotiations or other handshaking will cause security failures in the presence of one-way traffic. Where applicable, designers should explicitly state requirements for bidirectional communications.

One-way traffic elements to consider as part of the design of a security context are as follows.

- Does the context require negotiations or handshakes?
- Will different paths affect timeouts, key expiration, or other time-based elements that could prevent effective security verification or acceptance?
- Can local state be fully captured in parameters if it cannot be synchronized end-to-end?

Many cipher suites include protocols for negotiating security parameters and defining or communicating keys to be used for a particular security exchange. Security contexts that are built for use in networks with unidirectional links might not be able to use these algorithms. In these cases, contexts must either override the behavior of the cipher suites they define or use cipher suites that rely on out-of-band configurations.

There are several ways in which a link may only support one-way traffic, to include the following.

11.3.2.1 Long Signal Propagation Delays

Long propagation delays are effectively unidirectional when they prevent round-trip communication. This happens when the round-trip time is so long that either (i) the network topology changes so as to disrupt the communications link or (ii) message traffic is so delayed that applications waiting on data time-out.

11.3.2.2 Frequent Disruptions

Disrupted links can appear as unidirectional if they are unable to stay active for the duration of a round-trip exchange. BPAs operating in challenged environments might require multiple attempts to get a message across a link and responses to those messages may take multiple attempts to receive back again. In these cases, it may be difficult or impossible to associate messages with their responses.

11.3.2.3 Opportunistic Links

Links resulting from transient connections between BPAs can appear as unidirectional based on their short lifetimes. While there may be some short-term bidirectional communication, these links might not persist long enough to effectively carry bidirectional traffic.

11.3.2.4 Hardware Limitations

As a function of power, hardware design, pointing, or other physical considerations, telecommunications systems on some platforms may enforce a transmit-only or receiving-only operating mode. In these cases, links are unable to support bidirectional communication regardless of other network characteristics.

Round-Trip Paths

Lacking bidirectional links, two BPAs may still communicate with each other using different networking paths. Sometimes these alternate return paths can be used in lieu of bidirectional links. However, such alternate paths may experience different delays, data rates, and other characteristics that may prevent timely end-to-end data synchronization.

11.3.3 On-Demand Access

The existence of non-contemporaneous paths through BPv7 networks means that on-demand access to nodes cannot be assumed. Security contexts must consider whether on-demand access to a security oracle is available.

Certain security contexts might presuppose the existence of oracles to validate or provide security information related to a particular security session. One example of such an oracle is a certificate authority that can be used to retrieve or verify public keys and assign private keys.

Some questions a security context designer or network operator could ask to understand their reliance upon on-demand access to other nodes include the following.

- Does this context assume on-demand access to other nodes?
- Can this context capture assumptions in parameters if an oracle is not available?

Security contexts that operate using only on-demand access to nodes might need to capture assumptions made in the context parameters themselves. For example, contexts might carry information related to the expiration and exhaustion information of long-term or other public keys rather than wait for access to a security oracle to verify this information.

11.4 Behavioral Considerations

The behavior of a security context refers to the processing done by the context. Since the use of security contexts is to add behavior to security processing for different BPv7 networks, behavioral considerations are an important part of context design.

Designers should consider the behavior of their contexts as they relate to parameterization, authenticated encryption, key management, and target block associations.

11.4.1 Parameterization

Parameters provide a way to customize the behavior of a security context. This customization can reduce the overall number of contexts active in a network, particularly when multiple contexts would otherwise utilize the same underlying cipher suites.

> *To reduce the number of security contexts used in a network, security context designers should make security contexts customizable through the definition of security context parameters...*
>
> BPSec, RFC 9172 §9.2, [1]

This concept of parameterization is illustrated in Figure 11.1. Defining multiple, non-parameterized security contexts, as shown in Figure 11.1a, results in the definition of multiple security contexts when using different cipher suite modes, or when mixing-and-matching algorithms across cipher suites. Alternatively, equivalent behavior can be accommodated in a single security context as shown in Figure 11.1b by adding parameters to select and configure cipher suites.

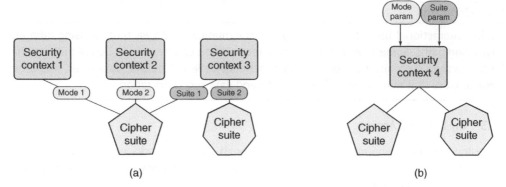

Figure 11.1 Parameterization reduces duplication. (a) Failing to parameterize security contexts leads to developing multiple contexts for minor behavioral differences. (b) One parameterized security context can provide similar flexibility as multiple non-parameterized contexts. Parameterization is a powerful tool to reduce the number of security contexts defined for BPv7 while allowing for expressive behavior.

11.4.2 Authenticating Encryption

Security contexts that encrypt information must, when used as part of the *bcb-confidentiality* security service, also provide authentication. There are a few ways in which an integrity mechanism (such as a Message Authentication Code [MAC]) can be calculated over the data being encrypted.

Some variations of authentication and encryption, summarized from a more expansive work published in [2], are shown next. It is recommended that security contexts prefer Encrypt-then-MAC behaviors where possible.

11.4.2.1 MAC-then-Encrypt
This behavior calculates a MAC over some plaintext, and appends that MAC to the plaintext prior to encrypting the set of data. Security contexts that use this approach will not protect the integrity of the ciphertext and, instead, can only check integrity after the plaintext and MAC have been decrypted.

11.4.2.2 Encrypt-then-MAC
This behavior first encrypts plaintext and then calculates a MAC over the generated ciphertext. This allows for checking the integrity of the ciphertext and, because it does not carry the MAC for the plaintext, avoids any cases where knowledge of the plaintext structure could enable attacks on the MAC of the ciphertext.

11.4.2.3 Encrypt-and-MAC
This behavior involves calculating a MAC over the plaintext, and then including the plaintext MAC with the ciphertext. This approach does not provide integrity over the ciphertext.

11.4.3 Key Management

The effectiveness of a security ecosystem is based on the effective compartmentalization of shared secrets across the network. The management of these shared secrets is both critical to the assumptions made by cipher suite algorithms and the activity most likely to be impacted by limitations in the transport abilities of a network. The way in which keys are determined, used, expired, and replaced must be part of any security context definition.

Security contexts should discuss how keys will be distributed and otherwise managed in a network. This includes discussions associated with the following.

- Whether long-term keys are used to produce session keys.
- Whether keys are pre-placed or negotiated beforehand or calculated just-in-time.
- How keys will be rotated as a function of time (expiration) and usage (exhaustion).
- Whether the context uses symmetric or asymmetric keys (or both).

Sometimes, key management may be deferred to the underlying networking environment. When a BPv7 network is deployed as an overlay in an existing network deployment, preexisting key management techniques may be reusable and, thus, not require special consideration for utilized security contexts.

11.4.4 Target Associations

Security contexts determine how bundle information is packaged for input to cipher suites, ways in which cipher suites are used to produce security results, and how those security results are carried back in the bundle. These interactions can be categorized as a function of how they identify target blocks and how many sets of security results they generate.

Security contexts can be developed that identify a single target block (*single-target* contexts) or have the ability to list multiple target blocks (*multiple-target* contexts). Similarly, contexts can generate a single set of results for their target block(s) (*single-result* contexts) or multiple sets of results for each target block(s) (*multiple-result* contexts). The various options for security context associations are listed in Table 11.1.

BPSec Security Blocks are Single Target

The BPSec security blocks (BIB and BCB) are structured to support a single target block per security operation. Therefore, the security services *bib-integrity* and *bcb-confidentiality* are compatible with the single-target associations STSR and STMR.

Each of these associations are discussed in greater detail in the remainder of this section.

Table 11.1 Security context target associations.

Context association	Target type	Result type
Single-target single-result (STSR)	Single	Single
Single-target multiple-result (STMR)	Single	Multiple
Multiple-target single-result (MTSR)	Multiple	Single
Multiple-target multiple-result (MTMR)	Multiple	Multiple

Source: Security contexts can input single or multiple target blocks and produce one or more sets of security results.

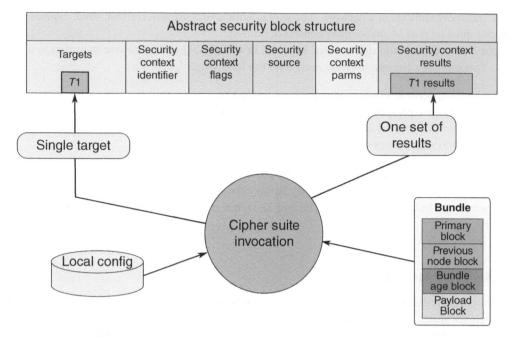

Figure 11.2   Single-Target, Single-Result Security Contexts. STSR security contexts associate a single set of cipher suite results with a single target block in a bundle.

11.4.4.1   Single-Target Single-Result (STSR) Contexts

The Single-Target Single-Result (STSR) association relates a target block with a single set of security results. This means that security operations using this association can be mapped to a single target block in a bundle and invoke a cipher suite operation once to generate the set of results to association with that target.

The operation of an STSR association is illustrated in Figure 11.2. In this figure, the abstract security block comprising a BPSec security block's block-type-specific data field is populated by a single invocation of a cipher suite. The cipher suite might input multiple individual data items, such as from the bundle being processed and from local configuration, but there is a single set of outputs from the invocation associated with a single target block identifier ($T1$).

11.4.4.1.1   Confidentiality Considerations   When a STSR context is used to provide a confidentiality security service over its target block, then the single set of security results for that target will include the produced ciphertext and an integrity mechanism associated with the authenticated encryption operation.

> *Any Security Context used by a BCB MUST utilize a confidentiality cipher that provides authenticated encryption with associated data (AEAD).*
>
> *BPSec, RFC 9172 §3.8,* [1]

The plaintext data in the block-type-specific data field of the security target block is encrypted using the STSR security context and replaced with the generated ciphertext. When the ciphertext is calculated, Additional Authenticated Data (AAD) is also generated as the BPSec requires the use of an Authenticated Encryption with Associated Data (AEAD) confidentiality cipher.

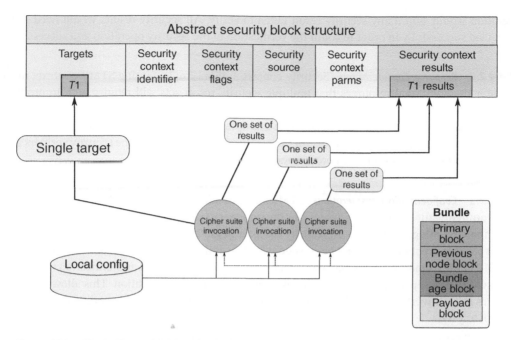

Figure 11.3 **Single-Target, Multiple-Result Security Contexts.** STMR security contexts associate multiple sets of cipher suite results with a single target block in a bundle.

11.4.4.1.2 Integrity Considerations A STSR context generates a single integrity mechanism over the plaintext associated with the target block. If a security context concatenates other information to this plaintext, the context is still considered an STSR context if the assembled plaintext is, itself, still associated with one target block.

11.4.4.2 Single-Target Multiple-Result (STMR) Contexts

The single-target multiple-result (STMR) security context provides a one-to-many mapping between a target block and invocations of cipher suites. STMR security operations invoke cipher suites multiple times to generate multiple sets of security results to be associated with the single target block of the operation.

The operation of an STMR association is illustrated in Figure 11.3. In this figure, the abstract security block comprising a BPSec security block's block-type-specific data field is populated by multiple invocations of a cipher suite. The cipher suite might input multiple individual data items, such as from the bundle being processed and from local configuration, generating (in this case) three separate sets of outputs from three separate invocations associated with a single target block identifier (*T*1).

11.4.4.2.1 Verification Considerations Since an STMR context generates more than one set of security results, the context must explain how these multiple sets of results are processed at security verifiers and security acceptors. Successful verification may mean that the verification of any one of the security result sets counts as the verification of the security operation. In this case, processing of the security operation would stop when the first security result is successfully verified. A failure during security processing in this case would not mean failure to verify the security operation.

The security context may instead require the verification of all of the security result sets as the criteria for successful verification of the security operation. In this case, processing of the security operation would stop if any security result failed to verify. All security results would need to be processed successfully in order to count the verification of the security operation as a success.

11.4.4.2.2 Resource Considerations Security context authors implementing STMR contexts must consider the additional resources that multiple security result sets will require both to produce and process. When choosing an STMR context, network operators must ensure that the security source has the resources necessary to generate multiple security results for each target block, that the network has the bandwidth to transmit the resulting bundle, and that the security verifiers and acceptors receiving the bundle can process these results.

11.4.4.2.3 Confidentiality Considerations STMR contexts allow multiple sets of ciphertext and associated authentication data to be produces for the same set of plaintext. This capability can be useful when there is ambiguity as to which security acceptor might be encountered by the security operation or what encryption key would be considered active at the time a bundle reaches that acceptor.

STMR contexts may also generate both a regular pair of ciphertext and authentication tag as well as an additional, separate integrity mechanism over that pair of information. This allows security verifiers in a network to verify the integrity of encryption materials within needing to attempt to decrypt any data.

11.4.4.2.4 Integrity Considerations Similar to their use for confidentiality, STMR integrity contexts can generate multiple integrity signatures over the target block (and any other associated bundle information). This allows security verifiers and security acceptors to verify the STMR security operation despite changes to topology or key information.

11.4.4.3 Multiple-Target Contexts

The definition of multiple-target security contexts are not directly supported by BPSec, as the Block Integrity Block (BIB) and Block Confidentiality Block (BCB) are structured to associate a single target block with one or more sets of security results for that block.

Figure 11.4 illustrates the difficulty of adjusting a BPSec security block to a multiple-target association. The BPSec abstract security block target list maps to a single set of security results for that target. To support a multiple-target capability, a BIB or BCB would need to ignore the target block identifier and, instead, rely on some security results (or parameters) to identify multiple targets. This approach is not compliant with RFC 9172 and would result in false or hidden dependencies between the security block and its targets, which in turn can cause undesirable security processing behaviors at RFC 9172 compliant BPAs.

However, there are two ways in which multiple-target contexts can be designed and incorporated into a BPv7 network.

1. New security blocks (termed Other Security Blocks) can be defined for use with BPSec blocks that are structured to identify multiple target blocks.
2. Associate the security block using a multiple-target context with the bundle itself by making its target block the bundle primary block, and using the context to store all security results associated with that bundle.

Each of these options is discussed in more detail in the following text. Ultimately, each of these multiple-target security context concepts have some processing issues, which is why there is no default support for these types of contexts in the original BPSec specification.

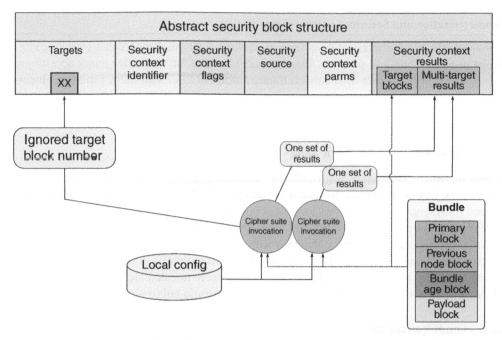

Figure 11.4   BPSec blocks CANNOT be used for multiple-target associations. Attempting to simulate a multiple-target association by ignoring the single-target identifier of a BPSec security block leads to undesirable or erroneous processing.

11.4.4.3.1 New Block Considerations Generating other security blocks to hold multiple targets provides a clean separation of single-target versus multiple-target security operations. However, standardizing these OSBs requires careful consideration of how any multiple-target security block would interact with single-target security blocks. It is recommended that the definition of new OSBs be done with great care and that BPSec security blocks and OSBs not be used together in operational security policy if possible.

11.4.4.3.2 Single Bundle Considerations Associating a multiple-target security operation with the primary block allows for the association of very complex security mechanisms for the bundle. However, this also presumes that these mechanisms are applied once at a security source, as subsequent downstream BPAs would not be able to use BIB or BCB blocks to apply the same security service to the primary block because of the BPSec requirement that security operations within a bundle stay unique.

11.5   Syntactic Considerations

Documenting behavioral considerations help determine what contexts need to be deployed in a given network. Syntactic considerations inform developers on how to correctly provide information to, and apply information from, the algorithms defined in the context. These considerations must be clearly addressed to ensure that different implementations of the same context by different vendors and/or on different platforms remain interoperable. BPSec notes the following items as important to the syntactic processing needed to implement a security context at a BPA.

> **Parameterization and Security Context Fragmentation**
>
> By making use of parameterization whenever possible, a single standardized security context can be used as broadly as possible. This prevents the need for successive minor adjustments to existing security context which avoids fragmentation of BPv7 into many similar but incompatible implementations.

11.5.1 Parameter and Result Encodings

The way in which parameters and results are encoded must be defined by a security context. Because these values are specific to a security context, the identifiers, data types, and appropriate serialized representation is necessary for proper operation and interoperability to ensure that this information can be correctly interpreted by downstream BPAs.

Having a consistent encoding for elements within a security block is additionally important if these parameters and/or results will also be included in an authentication or integrity mechanism covering a security block. If different implementations were to calculate semantically equivalent but bitwise different representations, these mechanisms would fail to validate.

11.5.2 Canonicalization

Similar to the proper encoding of parameters and results, security contexts might choose to define how other elements of a bundle would be encoded. While BPv7 defines standard CBOR encodings for elements within a bundle, there are several reasons why a security context might wish to alter that encoding, thus creating a new canonical form.

- Certain fields within a block that might change during transit might need to be masked out of the canonicalization. In which case, a canonical form of input data might omit the field or otherwise set it to a constant value, such as all 0's.
- Similarly, bits within flag fields may also be excluded from a canonicalization, in which case ignored bits might be set to a specific value, such as 0. This is particularly the case with reserved and unassigned portions of a bitfield.
- Particularly when interfacing with hardware-accelerated protocol handlers, certain values might need to be encoded using fixed-length fields to assist with processing. While this is less important for application data (which is likely opaque to the security context) protocol handlers may auto-expand block flag fields and other parts of the BPv7 or BPSec standard.

11.5.3 Encryption Ciphertext Packing

> **Canonical Form**
>
> This is a standard, universal data serialization definition for fields or objects and is critical for security operations.

The application of a BPSec confidentiality service such as *bcb-confidentiality* not only results in the creation of a BCB, but also changes portions of the target block's block-type-specific data field.[2]

Certain cipher suites generate ciphertext that is the same size as the associated plaintext. Other cipher suites might generate ciphertext that is of a different size than the plaintext. If a cipher

---

2 In particular, this field is expected to have its plaintext replaced with ciphertext on encrypted, and the reverse on decryption.

suite were to perform a data compression on plaintext prior to encryption, the resultant ciphertext might be smaller than the original plaintext. Alternatively, if a cipher suite produces larger cipher-text (or appends additional information to the ciphertext such as an authentication tag) then the ciphertext would have a different size than the plaintext.

BPSec requires that security context definitions explain how to handle differences in the size of plaintext and ciphertext as it relates to changing the content of a target block in a bundle. There are several possible ways in which these differences could be handled, with a few suggestions as follows.

- Ciphertext smaller than plaintext can be padded and used to replace plaintext without changing the target block length.
- Ciphertext larger than the plaintext can be truncated with overflow placed in a special "overflow" security result in the security block carrying the security operation.
- The complete ciphertext may replace the plaintext and the length of the target block be updated.
- The target block data plaintext may be replaced with a constant, meaningless values, such as all 0 padding, and all ciphertext placed as a security result in the security block carrying the security operation.

11.5.4 Handling CRC Fields

BPv7 includes a Cyclic Redundancy Check (CRC) field for blocks in a bundle. The CRC field provides a simple way to detect corruption of a block without requiring BPSec or the processing associated with a BIB and an associated security context. However, when a BPSec security oper-ation is applied to a block that also has a CRC the interaction between these two blocks must be handled with care.

When a BCB encrypts the contents of a target block, and that target block includes a CRC in the target block header, the security context MUST determine how this is handled. The CRC field in the target block will no longer correctly reflect the contents of a target block, because the target block's block-type-specific data field has been replaced with ciphertext. This can be problematic as downstream BPAs that check CRC values might see the block as corrupted and remove the block from the bundle.

While a BIB does not alter a target block, the canonicalization associated with a BIB might be more resilient to errors in certain parts of a block that, otherwise, could cause a CRC check to fail. Consider, if a reserved or unassigned bit flips in a target block header and the canonicalization algorithm defined for the security context used by the BIB masks out those bits, then the integrity of the target block can still be verified. However, in the same circumstances, a CRC calculation over the entire block would fail.

Security contexts can handle this in a variety of ways, such as the following.

- The CRC field may be removed from any target block that has a security operation applied to it.[3]
- The CRC field may be updated with the new contents of the target block (in the case of encryp-tion).
- The CRC field may be stored as a special context result, removed from the block when a security operation is applied, and restored after the security operation is accepted.
- The CRC field may be removed and recalculated later after the security operation is accepted.

---

3 This is because both *bcb-confidentiality* and *bib integrity* services require integrity checks.

11.6   Cryptographic Binding

In the simplest case, a BPSec security operation is applied to a single target block within a BPv7 bundle. However, there are circumstances where more information can be included by the security context. When a security context includes other block information in the generation of cryptographic materials, and requires that other information to be present later when the security operation is verified or accepted, then that information can be said to be cryptographically bound to the security operation.

Definition 11.1   *(Cryptographic Binding)*   The association of two or more data sets using a single cryptographic technique with those data sets as inputs. When two or more data sets are cryptographically bound, then a change in one data set may affect cryptographic processes across other unchanged data sets.

Some examples of such binding includes the following.

- If the primary block is used in addition to the block-type-specific data field of the target block in the calculation of an integrity signature, then the signature will also verify if the primary block of the bundle has not changed. In this way, the integrity mechanism says both that the target block data has not been changed, and that the target block exists in the expected bundle.
- If the security context parameters of a BCB are included as AAD as part of applying a *bcb-confidentiality* security service, then decryption can only happen when these parameters are unchanged at a downstream BPA in the network. This ensures that a security block is not manipulated prior to attempting a decryption of a security operation within that block.
- If the content of other extension blocks are pre-pended to an integrity signature over the payload block of a bundle, then the payload will only be seen as having correct integrity if those other blocks are, also, present in the bundle at a security verifier or acceptor.

By cryptographically binding information, security processing can catch issues associated with extension blocks being dropped from a bundle, or placed in different bundles.

There are three types of information that security context designers should consider when determining how to support cryptographic binding: the identification of data being bound, the way in which that data is presented to cipher suites, and what types of issues these bindings can introduce in a network.

11.6.1   Candidate Data Sets

Additional data sets can be added to the block-type-specific data of the target block. Potential data sets include the following.

11.6.1.1   Other Blocks' Block-Type-Specific Data

The block-type-specific data from blocks other than the security target block can be used to bind additional blocks to the security operation. This technique requires the block-type-specific data from another block present in the bundle to be added to the data set used as input for security result generation. Using the block-type-specific data as a parameter for security result generation requires that block to be present and unchanged in the bundle when security processing occurs.

11.6.1.2 Processing Flags

The processing flags associated with a block or the bundle itself may be included in the data set used for security result generation to ensure that these processing flags are not modified during transit. It may be necessary to ensure that these flags are not modified as they determine how the associated block is handled by BPAs encountered in the network. To ensure consistent block policy, it is recommended that the processing flags are cryptographically bound to the security operation.

11.6.1.3 Other Bundle Elements

Other bundle elements that may be useful to include in the data used to generate security results by a particular security context may include data from the primary block, such as the bundle source and destination, block counts, or other block data fields.

11.6.2 Identifying Data Sets

A security context designer must provide a deterministic method of identifying the data used for security result generation when using cryptographic binding. The order in which these data sets are concatenated matters.

One such option for identifying these data sets is the use of flags. A simple bitfield could be used to identify the presence of each potential data source for the set.

The security context designer could instead require the user or the application to provide the data set as an input at the processing BPA instead. Deferring the assembly of the data set to the user or the application may be necessary in situations where resource constraints on the node itself prevent it from assembling the data set for the security context.

11.6.3 Data Representation

When data sets are bound (with either integrity or confidentiality), then a binding security context must accept as input one or more sets of data and apply to them the shared security service. The presentation of data sets to the security context can be implemented in multiple ways, to include the following.

11.6.3.1 Monolithic Data Input

A monolithic data input can be constructed by catenating the canonical form of block data in a particular order. A monolithic data input has the benefit of a smaller data footprint, as only a single integrity result is generated, but also requires that bound blocks be concatenated in the correct order at every security verifier and acceptor. To help with this ordering, security contexts might either mandate in their specification or customize through parameterization the order in which bound blocks should be concatenated. This approach is illustrated in Figure 11.5.

11.6.3.2 Independent Data Inputs

Independent data inputs can be used to generate multiple security results per input, such that each element in the data set must be processed independently as security verifiers and acceptors. This approach requires more space in the security block as additional results must be calculated, but has the benefit of not requiring a specific ordering to the elements of the data set. This approach is illustrated in Figure 11.6.

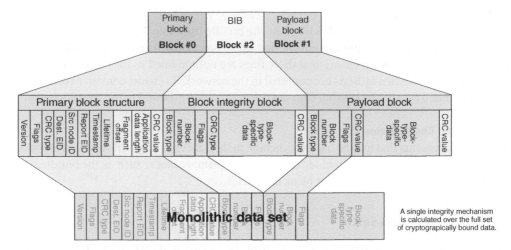

Figure 11.5 **Monolithic data sets construct a single security result.** Multiple elements of a bundle can be bound together by first concatenating them into a single, large plaintext input over which a security result is generated.

11.6.3.3 Scenarios

There are multiple ways in which operations would benefit from the cryptographic binding of different data sets associated with a security operation.

There are four types of information relationships that might benefit from the cryptographic binding capabilities of a security context: bundle binding, self-block content binding, other-block content binding, and manifest bindings.

1. **Bundle Binding**: A block may be bound to a bundle by including one or more required blocks in a shared integrity mechanism. In this way, a required block would need to be present with the bound block in order for the integrity signature to be verified. Including the primary block of a bundle in the integrity mechanism accomplishes this task.
2. **Self-Block Content Binding**: When a security operation is applied to a target block, the default input is the block-type-specific data field of the block. However, the target block also includes other data, such as block processing flags. In cases where more than just the block-type-specific data field must be checked, these other flags can be included. In this way the block-type-specific data and other data are bound to each other as part of considering the block as a whole.
3. **Other-Block Content Binding**: There is a natural dependency between a security operation as carried in a security block and the target block of that operation – both blocks must be present in the bundle. However this type of binding is simply an existence binding – the encapsulating security block must be present, but it could have been changed since its creation at a security source. When a target block is bound to its security block, the contents of the security block are included in the integrity calculation. This type of binding will confirm that portions of the security block relevant to the operation (such as security context parameters) are unchanged since the generation of the security results at the Security Source.
4. **Manifest Binding**: Just as bundle binding can be used to ensure that blocks only exist in the proper bundle, manifest binding can be used to identify cases were a block has been removed from the bundle. When using this binding, an integrity mechanism can be calculated over a required block, such as the bundle primary block and other blocks can be included in the integrity mechanism of the primary block. Then, at other BPAs in the network, if a block that was included in the original integrity mechanism is not present in the bundle, the primary block would fail to verify.

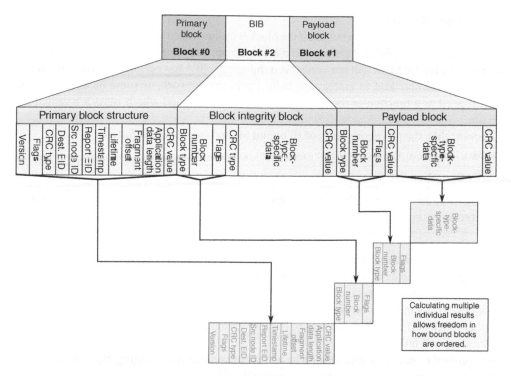

Figure 11.6 Independent data sets build multiple security results. Multiple elements of a bundle can have their security results independently represented to avoid ambiguity in constructing a single monolithic plaintext.

11.6.3.4 Processing Steps

The implementation of cryptographic binding may differ among security contexts. However, the use of a security context which provides cryptographic binding can be generalized to a common set of steps, as follows.

1. Actions at a Security Source.
   1.1 Identify all blocks to which the current block must be bound.
   1.2 Instantiate a security operation over that block using a binding-aware security context.
   1.3 Parameterize the security context with the security target block and all blocks to which it should be bound.
   1.4 Capture the security results from the security context.
   1.5 Place the security operation in a new or existing security block.
2. Actions at a Security Verifier or Security Acceptor
   2.1 Determine the security target and all bound blocks from the security context parameters.
   2.2 Provide the security target and bound blocks as input to the security context.
   2.3 React to the success or failure of integrity verification from the security context.

11.6.4 Common Error Conditions

Some potential issues with binding include the following.

11.6.4.1 Dropped Blocks

A BPA may remove blocks from a bundle if the block is unknown to the BPA and the block's processing flags are configured for that purpose. If a block dropped for this reason was bound to another block, then the bound block will not verify even though the BPA has performed in accordance with BPv7 policy. This may lead to ambiguous behavior in the network, as some blocks may or may not be dropped as a function of the path taken by each bundle and which BPAs are visited along that path.

To prevent this, the processing flags of all bound blocks should be carefully set.

Security Operation Event

This error may correspond to the Security Operation Corrupted at Verifier or Security Operation Corrupted at Acceptor security operation events.

11.6.4.2 Poor Canonicalization

Blocks might have fields other than their block-type-specific data field changed as they journey through the network. For instance, a target block might have its CRC value removed when the target block has its integrity verified using a BIB block. If the target block was bound beforehand to another block and the CRC value was part of the canonicalization of that target block, then the block binding will fail.

To prevent this, block canonicalization by any security context supporting block binding should not include elements of any block that may be modified by a BPA.

Security Operation Event

This error may correspond to the Security Operation Misconfigured at Verifier or Security Operation Misconfigured at Acceptor security operation events.

11.6.4.3 Block Ordering

Blocks may be received by the BPA in a different order than anticipated, causing the data set to be calculated incorrectly. The data set for cryptographic binding must be in the correct order to generate the correct security results. The data set may be complete, containing all data necessary for result generation, but will still produce the incorrect results if that data is not in the proper order.

Security Operation Event

This error may correspond to the Security Operation Corrupted at Verifier or Security Operation Corrupted at Acceptor security operation events.

11.6.4.4 Fragmentation

Fragmentation can modify block and/or bundle processing flags, which will impact the data set used for generating cryptographic material if these flags are included. A security context designer must carefully consider allowing bundle fragmentation when using cryptographic binding.

If the payload block is included in the cryptographic binding at the security source, but the bundle is fragmented during transmission before it reaches its security acceptor, none of the security operations bound to the payload block can be processed successfully until reassembly occurs.

Additionally, when bundle fragmentation occurs, some blocks may be moved to one fragment while the other blocks are part of a separate fragment. If two blocks are cryptographically bound, they may still be separated to different bundle fragments, meaning that none of the security operations requiring the presence of both blocks can be processed successfully until reassembly occurs.

Security Operation Event

This error may correspond to the Security Operation Misconfigured or Corrupted at Verifier security operation events or the Security Operation Misconfigured or Corrupted at Acceptor security operation events.

11.7 Summary

Security contexts are a powerful component of BPSec, allowing for customization of the configuration and policy for security services. Design, behavioral, and syntactic considerations for building security contexts were discussed, and common networking scenarios were provided for each security context type introduced.

When selecting a security context, it is important to keep in mind the common errors that may occur from misconfigurations in the security context design or implementation.

References

1 Birrane, E. J. and McKeever, K. [2022]. Bundle Protocol Security (BPSec), RFC 9172. **URL:** https://www.rfc-editor.org/info/rfc9172.

2 Bellare, M. and Namprempre, C. [2000]. Authenticated encryption: relations among notions and analysis of the generic composition paradigm, *International Conference on the Theory and Application of Cryptology and Information Security*, Springer, pp. 531–545.

12

Security Policy Overview

The potentially challenged nature of BPv7 networks places unique constraints on the establishment and upkeep of security configuration and policy information. Importantly, policy expressions cannot always be negotiated in real-time between secure endpoints, as there might not exist connectivity between those endpoints.

BPSec, in particular, requires an expressive security policy to handle the processing of security operations in a bundle. This policy being made more complex by the fact that security operations can be applied block-by-block and not bundle-by-bundle, that these operations can use different security contexts, and that Bundle Protocol Agents (BPAs) have multiple roles of sources, verifiers, and acceptors.

This chapter outlines a policy model that can be used as a basis for developing security policy expressions and software implementations for BPv7 networks. This includes a discussion of the ways in which policy can be communicated in a network and common events and actions that should be considered.

After reading this chapter you will be able to:

- Identify why DTN security policy establishment is handled differently from other networks.
- List the success and error events in the security operation lifecycle.
- Categorize processing actions and explain the difference between required, optional, and prohibited actions.
- Create security policy configurations for various scenarios.

12.1 Overview

The diversity of BPv7 network constraints, combined with the diversity of BPSec operations and security contexts, provides a powerful, but complex solution space for security policy configuration. The implementation of a policy engine for BPSec must be able to both exist and operate local to an individual BPA and also stay updated as a function of the capabilities of the network itself.

Definition 12.1 *(BPSec Policy Engine)* Software, resident on a Bundle Protocol Agent (BPA), that can be queried to determine processing related to a given security operation either in a bundle or to be added to a bundle. This includes the security role that the BPA takes relative to the security operation, and any special events and associated actions that must be processed by the BPA acting in that role.

Securing Delay-Tolerant Networks with BPSec, First Edition. Edward J. Birrane III, Sarah Heiner, and Ken McKeever.
© 2023 John Wiley & Sons, Inc. Published 2023 by John Wiley & Sons, Inc.
Companion website: www.wiley.com/go/birrane/securingdelay-tolerantnetworks

Each BPA that processes security blocks must implement at least one[1] policy engine. While a single policy engine provides a single point of reference for policy lookups, this engine might be configured by a variety of policy sources.

This chapter discusses design elements and considerations of a policy engine that accepts various types of policy information from various information sources to construct and adhere to a policy model for a local BPA. This policy model also identifies actions to be taken as part of the successful and unsuccessful processing of security operations. Finally, an example syntax for a policy engine is provided.

12.2 Policy Information Sources

Traditionally, security policy information comes from operators that can provide authoritative information that describes the required behavior of a policy engine. The delayed and disrupted nature of some BPv7 networks makes this type of operator-in-the-loop operation difficult or impossible to serve as the sole source of policy information.

A scalable BPSec policy engine is one that provides a hybrid approach to configuration, allowing the use of multiple information sources. In-band information, found in the bundle contents, and out-of-band information, generated locally at a BPA, can both be used to establish security policy [1].

Definition 12.2 *(In-Band Policy)* Information communicated to a policy engine through the network being managed. This includes information captured in the payload or extension blocks of a bundle,

Definition 12.3 *(Out-of-Band Policy)* Information communications outside of the network being managed, such as through local autonomy on a BPA or as communicated to the local BPA through other physical interfaces into the system.

The difference between in-band and out-of-band communications is illustrated in Figure 12.1. Information may come to a local BPA using traditional in-band means as the payload of messages from authoritative sources (a) or from other sources of network traffic (b). Out-of-band information reaches the BPA through a mechanism other than the network itself, such as from local autonomy applications running on the local node itself (c).

12.3 Policy Information Types

There are many types of policy sources that can be used to configure the policy engines at individual BPAs. While the specifics of any single policy source might be tied to the implementation and behavior of a specific network, these sources can be categorized by the way in which the policy information is generated by the source.

Definition 12.4 *(Security Policy Source)* Software, either resident on the local BPA or remotely over a network that generates policy statements that can be configured on a local policy engine to affect the behavior of that engine.

---

1 While implementing multiple policy engines is possible, the complexity of preventing overlap and policy collisions makes multiple engines inadvisable.

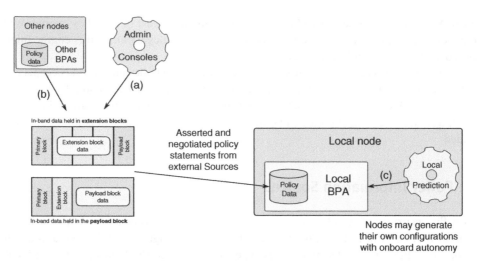

Figure 12.1 Policy information comes from multiple sources. Ready access to administrative consoles in a BPv7 network is limited in certain network architectures. Alternative sources of information must be considered for challenged networks.

As discussed in Section 4.6, there are at least three ways to approach policy configurations: negotiating, asserting, and predicting. Understanding how these sources are used is important to understanding how sources might be deployed in a network to inform policy decisions – and what kinds of sources are required for certain security outcomes.

12.3.1 Negotiating Sources

Negotiation sources configure security policy at a BPA as a function of information exchange with other BPAs in the network. This exchange may be to receive some policy resident on another BPA or to receive some information that allows for the local computation of security policy at the local BPA.

Not every link, or every path between secure endpoints in a BPv7 network is challenged. Even in extreme cases, there are certain subpaths in the network that use reliable links. Consider the case of Interplanetary Internetworking, where a planetary lander communicates through a series of relay orbiters to ground stations on Earth and, ultimately back to some operations center. While the end-to-end path might be very challenged, communications between ground stations and operations centers is likely no more challenged than other terrestrial Internet traffic.

In these cases, just-in-time negotiation of policy can occur as a function of handshaking protocols, link measurement, and other algorithms than might result in the configuration of security services at a BPA. As an example, security policy might specify the use of an asymmetric key as verified by a third party certificate authority for portions of the network that exist on the terrestrial Internet. In this case, the policy determination that such an authority is present, and that such a key approach is viable, might be discovered opportunistically as part of the start of the secure message exchange.

12.3.2 Asserting Sources

Asserting sources configure security policy at a BPA as a function of preplaced information from an authoritative source that can be referenced by the local BPA without requiring any additional communication. Asserting sources are the default source of many types of information in challenged networks because, once configured, no other information exchange is needed for the local policy engine to operate.

The difference between an asserting source and a negotiating source is the preplaced nature of the policy information. Both can be seen as authoritative sources of information, but a negotiating source is used only at the time the information is needed and in the context of a specific security activity.

Consider a challenged network where a BPA might only have connectivity with a network operations center for brief periods of time. During periods of connectivity the operations center may update the security policy of a BPA with enough information to last until the next contact time, which could be days, weeks, or months in the future. In this example, because the policy directives would be received from a presumed authoritative source (an operations center) and not be part of a round-trip exchange in the context of an ongoing security activity, the operations center would be considered an asserting source for policy.

12.3.3 Predicting Sources

Predicting sources configure security policy locally on a BPA as a function of the time and state of the network as observed by the BPA. Predicting sources exemplify the increasing trend of autonomous networked operations for challenged environments or, otherwise, when disconnected from a messaging endpoint.

Having BPAs change their security posture in a non-deterministic way is both critical to learning how to build reactive, self-monitoring systems and a security vulnerability if not done correctly. Predicting sources of policy can be helpful when they identify times when additional (and perhaps more restrictive) policy should be applied to a BPA. Consider a case when a BPA believes that traffic of a certain sensitivity might be coming through the BPA as a function of some higher-order autonomy.[2] In such a situation, a BPA might change its policy to add encryption to messages that might not already be encrypted, or to verify signatures on traffic that would otherwise not be verifies.

12.4 Security Operation Events

A security policy model relates policy information types, from different information sources, in the context of applying policy to security within a BPv7 network. The atomic unit of security within the BPSec is the security operation, as it is the mechanism by which bundle blocks are secured through the use of security contexts. Therefore, the start of a policy model must describe the use of security operations across the security operation lifecycle.

12.4.1 The Security Operation Lifecycle

Extension blocks (to include BPSec security blocks) have a lifecycle that is related to, but independent of, the lifecycle of the bundle that contains them. Blocks cannot exist in the network before a bundle is created, or after a bundle has been deleted, but can be added or removed over the lifetime of the bundle.

Similarly, security operations have a lifecycle that is related to, but independent of, the lifecycle of the security block that contains them. Security operations do not exist in the bundle before a

---

2 Perhaps based on GPS position, or local time, or message traffic analysis.

security block is created, or after a security block has been destroyed, but they can be added to, or removed from, security blocks over the lifetime of those blocks.

Definition 12.5 *(Security Operation Event)* Any event on a BPA that involves the creation, updating, reading, or removal of a security operation in a bundle.

Any individual actions relating to a security operation can be termed a security operation event. These events can occur at any BPA in a BPv7 network as a function of security policy configuration. Together, the set of all possible security operation events forms the lifecycle of a security operation.

Definition 12.6 *(Security Operation Lifecycle)* The set of Security Operation Events that may occur in the creation, processing, and termination of a BPSec security service as performed at BPAs in the roles of security source, security verifier, and security acceptor for a single security operation.

The security operation lifecycle, illustrated in Figure 12.2, is independent of the security service or security context associated with a given security operation. The lifecycle is finite, composed of thirteen events that capture all possible security occurrences during block processing. The lifecycle is also consistent, meaning that the events composing the lifecycle will not change as a function of the configuration or state of a BPv7 network.

In order to build a complete security operation lifecycle, events indicating both success and failure during security processing must be captured. The events capturing potential security failures include identification of a missing, misconfigured, or corrupted security operation.

A brief overview of each of the states of this lifecycle is provided as follows.

12.4.1.1 Security Source Events

The lifecycle is initiated with the **Source for Security Operation (1)** event, which designates the current BPA as a security source, indicating the presence of a security policy rule which requires the security operation's addition to the bundle.

Successful addition of the security operation to the bundle results in the transition to the **Security Operation Added at Source (3)** event. A security operation is represented in a bundle as a security block. A security operation is considered to be added to the bundle when it is represented as a security block populated with the necessary security results and security context information.

If the security operation cannot be added to the bundle at the security source, it is considered misconfigured. The **Security Operation Misconfigured at Source (2)** event causes the security operation to be removed, marking the end of the security operation's lifecycle. The security operation may be misconfigured at the source if a security result could not be calculated, indicating an error in the generation of the cryptographic material required for that security operation, another security operation exists in the bundle that prohibits the addition of another, or if a security context or configuration parameter was incorrect.

12.4.1.2 Security Verifier Events

A bundle may encounter zero to many security verifier BPAs along its path. If the bundle is transmitted to a BPA identified as a security verifier for the security operation, a transition occurs to the **Verifier for Security Operation (4)** event. There are three security failures that can occur at a security verifier.

If the required security operation cannot be located in the bundle, indicating that the security block representing that operation does not exist, the **Security Operation Missing at Verifier (5)** event occurs.

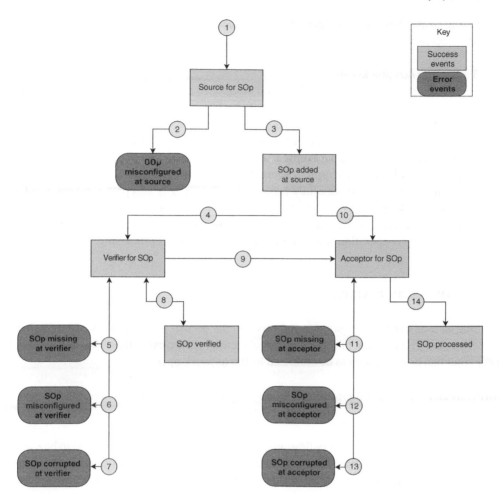

Figure 12.2   The security operation lifecycle. The security operation lifecycle is composed of security operation events. These events are identified by the BPA and processed according to local security policy.

The security operation may be identified as misconfigured (**Security Operation Misconfigured at Verifier [6]**) for a number of reasons, including the use of an incorrect security context or parameter to generate its cryptographic material. This security event occurs when necessary parameters for the security operation verification are missing, either from the policy configuration at the BPA or the security block itself, or where there is a conflict between parameters from the two sources. A security operation may also be identified as misconfigured if the security context it identified is not as expected.

If the **Security Operation Corrupted at Verifier (7)** event occurs, verification of the security operation represented in the bundle was unsuccessful. As the security operation is not identified as misconfigured, it is otherwise valid (i.e. uses the expected security context and parameters) but could not be verified. An example of a corrupted security operation in this case is a computed integrity signature that does not match the signature the security block holds as a security result.

The **Security Operation Verified (8)** event is used to indicate successful verification. From this point, the bundle is ready to be forwarded and may encounter other security verifiers, transitioning to the **Verifier for Security Operation (8)** event again, or arrive at its security acceptor.

12.4.1.3 Security Acceptor Events

When the bundle reaches the BPA that has been designated as the endpoint for the security operation, its security acceptor, the **Acceptor for Security Operation (10)** event is triggered.

Failures similar to those at a security verifier BPA may also occur during the final processing of the security operation at its security acceptor. The security operation may be identified as missing (**Security Operation Missing at Acceptor (11)**, misconfigured (**Security Operation Misconfigured at Acceptor [12]**), or corrupted (**Security Operation Corrupted at Acceptor (13)**).

If the security operation is successfully processed at the security acceptor, the **Security Operation Processed (14)** event is encountered, indicating the successful completion of the security operation's lifecycle. As the security acceptor serves as the endpoint for the security operation, that removal of the security operation from the bundle is included in its processing.

12.5 Processing Actions

Processing actions can be configured as responses to the occurrence of security operation events. The execution of processing actions, as a matter of security policy, can impact the bundle itself, the bundle contents, or create data to be evaluated by the network operator.

Definition 12.7 *(Processing Action)* Processing that occurs on a BPA as a result of some security operation event that has occurred on that BPA.

When the action must be run every time an event occurs, the action is considered to be *required*. If an action only runs in response to an event when required by policy, it is considered to be *optional*. Processing actions that must never be performed in response to an event are considered to be *prohibited*.

A processing action can be required for a certain security operation event, optional for another, and prohibited for a third event.

The set of processing actions that must be executed in response to an occurrence of a security operation event may consist of both required and optional processing actions. Required processing actions are performed by the BPA each and every time the security operation event is encountered, regardless of other bundle characteristics. Optional processing actions are configured as a matter of security policy and are associated with the occurrences of specific security operations and their corresponding events. The BPA must consult local security policy to determine if optional processing actions are enabled for a particular security operation.

12.5.1 Processing Requirements

As a function of security policy, the processing actions configured for a security operation event must be executed when that event occurs. This relationship is represented below, where P represents the security policy function making the association between the event and processing actions, E is the security operation event for which the processing action(s) are identified, and A is the set of processing actions that must be executed.

$$P(E) \to A$$

Note that the set of processing actions A for an event can be described as the union of the set of required processing actions R and set of optional processing actions O associated with that event. There may be zero to many actions in each of these sets.

$$A = R \cup O$$

12.5.1.1 Required Processing Actions

Required processing actions for a security operation event must be executed each time an occurrence of that event is identified by the BPA. Required processing actions are associated with security operation events regardless of security policy configuration.

For instance, the **Source for Security Operation** event is associated with a required processing action. The event indicates that the current BPA has been identified as the security source for a security operation. That BPA executes the required processing action by adding a security block to the bundle, representing the security operation in the bundle.

In addition, consider the **Security Operation Processed** security operation event, indicating that the security operation in question has been successfully verified and processed by its security acceptor. The BPA must then remove the security operation from the bundle, as this is a required processing action for that event. A security operation which has been processed is not to be retained in the bundle as it no longer provides the security service which it represents.

12.5.1.2 Optional Processing Actions

Optional processing actions are configurable as a matter of security policy for security operation events. Any enabled, optional processing actions are executed when an occurrence of that security operation event is identified. Multiple optional processing actions can be enabled for a single security operation event. For this reason, a predetermined order of application must be followed for processing actions.

Consider a scenario in which the optional processing actions enabled for a particular security operation event include the actions to retain the bundle at the current BPA, failing bundle forwarding, and removing the security operation from the bundle. The order in which these processing actions are executed is important as there are two versions of the bundle which may be stored at the BPA depending on which action is carried out first.

It is recommended that any processing actions that impact the bundle contents (removing a security operation, for example) are executed after any data generation actions are taken. In this example, the bundle should first be stored at the BPA, creating a record of the security operation which caused the event to occur. After generating this data, the security operation may be discarded to ensure it has no way of persisting in the network.

12.5.1.3 Prohibited Processing Actions

Prohibited processing actions cannot be executed in response to the occurrence of the security operation events they are associated with. Processing actions are prohibited where they would interfere with proper security processing.

Consider the security operation events that may occur at a security acceptor, such as the **Security Operation Processed** event. A processing action such as the retention of the security operation has to be prohibited for this event in order for security processing to proceed as expected. The processing of a security operation by its security acceptor, which serves as the endpoint for that security operation, must conclude with the removal of that security operation's representation from the bundle. This removal indicates the end of the security operation lifecycle. If the security operation was permitted to be retained in response to this event, it must do so outside of the processing requirements defined by BPSec.

12.5.2 Processing Action Categories

Processing actions can be categorized as one of the following:

- **Data Generation**: These actions result in the production of diagnostic data which can be used by a network operator to identify the occurrence and cause of a security event. This can include generated new data, such as a report, or storing existing data, such as a whole bundle. Data is typically generated to enable forensic analysis of bundle traffic and its associated security operations.
- **Block Manipulation**: These actions affect elements of a particular block in the bundle. Examples of these actions include modifying or temporarily overriding the Block Processing Control Flags (BPCFs) for a block.
- **Bundle Manipulation**: These actions affect the bundle being processed. Bundle manipulation actions may modify bundle transmission or the bundle contents, such as by adding or removing blocks from that bundle, or removing the bundle from the network.

While the definition and implementation of these actions must occur in the context of a specific policy engine, it is recommended that data generation policy actions are executed first, as they preserve information for future analysis. Similarly, block manipulation actions should occur prior to bundle manipulation actions, as block changes might affect subsequent bundle processing. Generally, actions should be implemented from least to most impactful.

A complete list of processing actions, as well as their associated categories, is provided in the next three sections.

12.5.2.1 Data Generation Actions

All data generation processing actions provide information to either the current BPA or some other report-to BPA configured by a network operator. While the full set of generated reports is, itself, a function of a particular BPv7 network deployment and policy engine, the following data elements should always be reportable by a BPA.

1. Request storage of the bundle at the current BPA
2. Report occurrence of the security operation event with reason code

---

Operational Security Concerns

Data generation processing actions must be used with caution, as they create data that may outlive the bundle itself. This information – the bundle contents or BPSec reason code – may be a useful tool for a network operator during forensic analysis, but may also provide sensitive information to an unauthorized user if not handled properly.

Each of these data elements is discussed in more detail below.

12.5.2.1.1 *Request Bundle Storage* Execution of this processing action causes the BPA to issue a request for storage of the bundle in its current state. The bundle state is preserved to enable offline forensic analysis – similar to placing a bundle in quarantine. A network analyst may use this action, for example, to examine corrupted security operations of critical bundle blocks without allowing corrupted bundles to persist in the network.

The request for bundle storage is an optional processing action which should be considered in reaction to the following security operation lifecycle events.

- Security Operation Misconfigured at Source
- Security Operation Misconfigured at Verifier

- Security Operation Missing at Verifier
- Security Operation Corrupted at Verifier
- Security Operation Misconfigured at Acceptor
- Security Operation Missing at Acceptor
- Security Operation Corrupted at Acceptor

12.5.2.1.2 Report Reason Code The report reason code processing action generates a bundle status report using the reason code provided by the network operator. The following security reason codes are defined by BPSec in Section 7.1 [2] and can be included in administrative status reports:

Missing Security Operation: Issued from a security verifier or security acceptor. This reason code is used to indicate the absence of a required security operation from a bundle.

Unknown Security Operation: Issued from a security verifier or security acceptor. This reason code is used to indicate the presence of a security operation that cannot be understood and processed by the current BPA. This security reason code is not issued from a security source because a security operation only exists once it has been successfully created by a security source.

Unexpected Security Operation: This reason code is used to indicate the presence of a security operation for which the BPA is not a security verifier or security acceptor. In other words, the receiving BPA does not have the responsibility for processing that particular security operation. This reason code should not be considered to be an error condition as not every BPA is the security verifier or security acceptor for every security operation represented in every bundle it receives. This reason code may be used by a network operator to identify potential security policy misconfigurations.

Failed Security Operation: This reason code is used to indicate a security operation that has failed security processing. At a security source, this reason code is issued if a security block cannot be added to the bundle when required. At a security verifier, this reason code is issued if the security target block for the operation fails to verify. At a security acceptor, the reason code is issued if the security operation cannot be fully processed and removed from the bundle.

Conflicting Security Operations: This reason code is used to indicate the presence of two or more security operations in a single bundle that are not conformant to BPSec processing rules. When such a conflict occurs security processing is not permitted to proceed.

It is expected that different network implementations and standards will define additional reason codes to assist with debugging and data generation actions. The generation of any report with a reason code is an optional processing action that should be considered for the following security operation lifecycle events.

- Security Operation Misconfigured at Source
- Security Operation Misconfigured at Verifier
- Security Operation Missing at Verifier
- Security Operation Corrupted at Verifier
- Security Operation Misconfigured at Acceptor
- Security Operation Missing at Acceptor
- Security Operation Corrupted at Acceptor

12.5.2.2 Block Manipulation Actions

Block manipulation processing actions impact a single block in a bundle. These actions result in the modification (temporary or permanent) of blocks related to security operations, such as the security block and the target block.

Table 12.1 Actionable block processing control flags.

Bit	Flag field
Bit 0 (0×01)	Block must be replicated in every fragment
Bit 1 (0×02)	Transmit status report if block can't be processed
Bit 2 (0×04)	Delete bundle if block can't be processed
Bit 4 (0×10)	Discard block if it can't be processed

See BPv7, RFC 9171 §4.2.4, [3] for more information on block processing control flags.

Two common block manipulations that should be considered for any policy engine involve how the BPCFs of a given block should be handled. These flags are set by a block source, but also can represent manipulation or corruption by attackers in the network.

These flags, defined by the BPv7 specification, are summarized in Table 12.1.

Local policy might choose to override (or overwrite) these flags in response to security processing events, such as through the following recommended actions.

1. Override the BPCFs of the target block
2. Override BPCFs of the security block

12.5.2.2.1 Override Target Block Processing Control Flags Overriding the BPCFs of the target block changes the way that the security target block is processed by the BPA. This action temporarily overrides the target's BPCFs.

This processing action is optional at the following security operation events:

- Security Operation Misconfigured at Verifier
- Security Operation Missing at Verifier
- Security Operation Corrupted at Verifier
- Security Operation Misconfigured at Acceptor
- Security Operation Missing at Acceptor
- Security Operation Corrupted at Acceptor

12.5.2.2.2 Override Security Block Processing Control Flags The override security block's BPCFs processing action is used to temporarily modify the flags of the security block representing the working security operation in the bundle. This processing action can also be used to set the security BPCFs when that block is first added to the bundle, in response to the Security Operation Added at Source event.

This processing action is optional at the following security operation events:

- Security Operation Added at Source
- Security Operation Misconfigured at Verifier
- Security Operation Corrupted at Verifier
- Security Operation Misconfigured at Acceptor
- Security Operation Corrupted at Acceptor

12.5.2.3 Bundle Manipulation Actions

Processing actions that can be categorized as bundle manipulation actions affect the bundle as a whole. These actions may change the way in which the bundle is transmitted or the information that the bundle holds. Blocks may be added or removed from the bundle as a result of the execution of a bundle manipulation processing action.

1. Retain the security operation in the bundle
2. Remove the security operation from the bundle
3. Remove the security target block from the bundle
4. Remove all security operations for the security target block from the bundle
5. End bundle transmission at the current BPA

12.5.2.3.1 *Retain Security Operation* The BPA preserves the security block representing the associated security operation in the bundle to carry out this processing action.

Retention of a security operation is a required processing action at the following security operation events:

- Security Operation Added at Source
- Verifier for Security Operation
- Security Operation Verified

At a security source, where the security operation has just been added and successfully represented in the bundle, retention of that security operation is required. This ensures that the security operation is present in the bundle when it is dispatched for transmission from the security source.

The other security operation events for which this processing action is required are both security verifier events. When the BPA assumes the role of security verifier, and when the security operation is successfully verified, the retention of the security operation in the bundle is required. A bundle may encounter multiple security verifiers in its path, indicating the need for the security operation to remain a part of the bundle in order to be verified at these different hops.

Security operation retention is not an optional processing action for any security operation events, meaning that security policy cannot be used to configure this processing action.

12.5.2.3.2 *Remove Security Operation* The removal of a security operation from the bundle may result in the removal of the security block representing that operation, as illustrated in Figure 12.3.

However, if the security block has multiple target blocks, indicating that the security block represents multiple security operations, only the affected security operation is removed from the security block, as illustrated in Figure 12.4.

Removal of a security operation is a required processing action at the following security operation events:

- Security Operation Misconfigured at Acceptor
- Security Operation Corrupted at Acceptor
- Security Operation Processed

Removal of the security operation is required for all events associated with the security acceptor role, except for the Security Operation Missing at Acceptor event, for which this processing action is not possible. At the security acceptor, removal of the security operation is required whether processing is successful or not.

Removal of a security operation is an optional processing action at the following security operation events:

- Security Operation Misconfigured at Source
- Security Operation Misconfigured at Verifier
- Security Operation Corrupted at Verifier

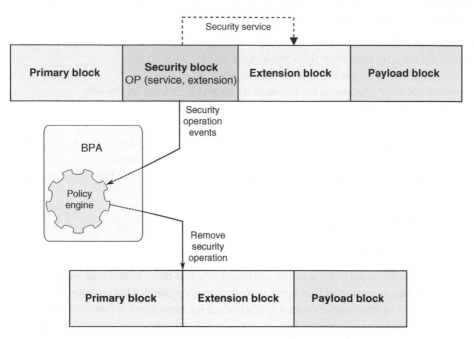

Figure 12.3 Remove lone security operation. If a security block has a single security operation remaining, then this processing action will cause the security block to be removed from the bundle.

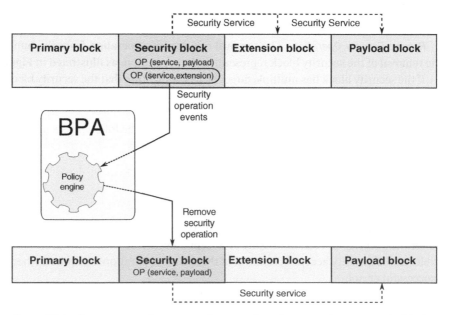

Figure 12.4 Remove one of many security operations. In cases where a security block aggregates multiple security operations, this processing action only removes the single security operation from the bundle.

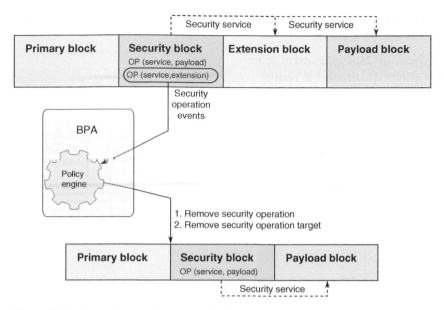

Figure 12.5   Removing security target blocks. When removing a target block from the bundle, any security operation over that target block must be removed from its security block.

The security operation may be configured to be removed from the bundle at the security source or verifier as a matter of security policy. The security operation may also be removed from the bundle if it is found to be corrupted at a security verifier, preventing the security operation from being processed at other verifier or acceptor BPAs.

12.5.2.3.3   Remove Security Operation Target   The security operation's target block is the block in the bundle to which the security service was applied. That security target block is removed from the bundle when this processing action is executed, as illustrated in Figure 12.5.

Removal of a security operation's target block is an optional processing action at the following security operation events:

- Security Operation Misconfigured at Source
- Security Operation Misconfigured at Verifier
- Security Operation Missing at Verifier
- Security Operation Corrupted at Verifier
- Security Operation Misconfigured at Acceptor
- Security Operation Missing at Acceptor
- Security Operation Corrupted at Acceptor

Security target block removal is a configurable processing action to be used in response to identification of a misconfigured, missing, or corrupted security operation.

12.5.2.3.4   Remove All Security Target Operations   Execution of this processing action results in the removal of all security operations associated with the target block of the current security operation. BPSec requires that only a single security service be applied to a single target block, thus this processing action may result in the removal of up to two security operations: one applying a *bib-integrity* service to the security target block and one applying *bcb-confidentiality* to the security target block, as illustrated in Figure 12.6.

Removal of all of a target block's security operations is an optional processing action at the following security operation events:

- Security Operation Misconfigured at Source
- Security Operation Corrupted at Verifier
- Security Operation Corrupted at Acceptor

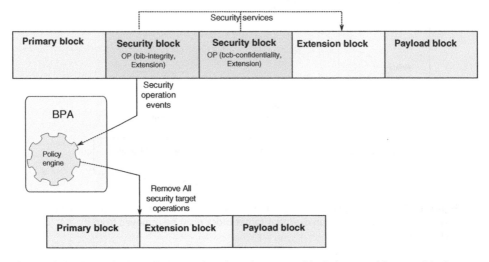

Figure 12.6 Removing security operations based on target block. Issues with target blocks may cause a policy engine to drop operations over a given block. Security operation uniqueness requirements imposed by BPSec mean that this will only ever result in the removal up to two operations: *bib-integrity* and *bcb-confidentiality*.

12.5.2.3.5 Fail Bundle Forwarding The fail bundle forwarding processing action results in the termination of bundle transmission at the current BPA.

The termination of bundle transmission is an optional processing action at the following security operation events:

- Security Operation Misconfigured at Source
- Security Operation Misconfigured at Verifier
- Security Operation Missing at Verifier
- Security Operation Corrupted at Verifier
- Security Operation Misconfigured at Acceptor
- Security Operation Missing at Acceptor
- Security Operation Corrupted at Acceptor

12.6 Matching Policy to Security Blocks

Security policy identifies a required security operation for a bundle. Recall that a security operation identifies both the security service and target block of that service. In order to match policy to a security block, the role of the processing BPA must be considered.

At a security source, the target block type of the required security operation is used to identify the block(s) in the bundle for which the security operation applies. If the bundle contains one or more blocks of the identified type, the security operation can be applied. The security block matching the policy statement represents the security operation that statement describes, and is added to the bundle by the security source.

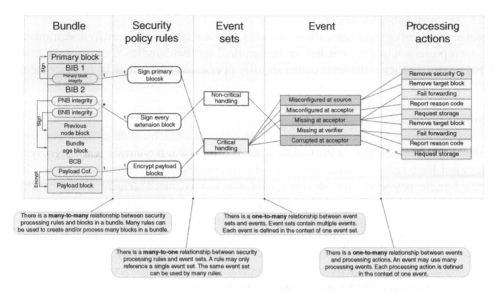

Figure 12.7   A sample mapping of security policy structures and their interactions. [4]

At BPAs that are receiving a bundle with security blocks present to be processed, either a security verifier or security acceptor, identifying the security block that matches the policy statement starts with the security service described by the security operation. Using the security service, such as *bib-integrity* or *bcb-confidentiality*, the BPA is able to determine the security block(s) in the bundle that may match the security policy rule to be applied. From the subset of security blocks identified, the BPA then uses the security target block type to determine which security block matches the policy statement.

These relationships amongst bundle elements, policy rules, events, and actions are illustrated in Figure 12.7.

12.6.1   Types of Policy Statements

Different types of security policy statements are necessary in order to provide the full capability set for the security policy engine described in this chapter. These security policy statement types are:

- Required policy statements.
- Optional policy statements.
- Constraining policy statements.

12.6.1.1   Required Policy Statements

The common use for security policy statements is reflected by the **required** policy statement type. A required security policy statement describes a security operation that is required to be represented in the bundle.

At a security source, a required policy statement describes a security operation that *must* be added to the bundle before it is forwarded by that BPA.

At a security verifier or security acceptor, a required policy statement describes a security operation that *must* be represented in the bundle when it is received. If that security operation is not in the bundle, one of the **Security Operation Missing** events occur. The security verifier or acceptor is required to process the security operation that was received.

12.6.1.2 Optional Policy Statements

Optional security policy statements allow a network operator to describe security operations which should be represented in a bundle, but are not required for that bundle to persist in the network. Optional security policy statements define security operations which *may* be represented in the bundle.

An optional security policy statement is not used at a security source. A security operation is to be required only at a security source.

At a security verifier or security acceptor, an optional policy statement describes a security operation that *may* be represented in the bundle when it is received. If a bundle is received by a BPA that does not have access to a security policy statement describing a security operation it contains, that bundle may be rejected with the reason of an unrecognized security operation. When an optional security policy statement is present at a security verifier or acceptor, the BPA will process the described security operation if it is present in the bundle.

If the security operation described by the optional security policy statement is not represented in the bundle, it does not indicate the occurrence of a **Security Operation Missing** event as this event is only associated with required security operations. An optional security policy statement cannot cause a **Security Operation Missing** event to happen.

Optional Policy Statements

Optional policy statements can be used to perform courtesy verification of security operations when present in a bundle and supported by the BPA. In these cases, bundles either without security services or that use security contexts not supported by the BPA, are not considered to be in error.

12.6.1.3 Constraining Policy Statements

Constraining security policy statements describe a security operation that is prohibited. The security operation associated with the statement must not be added to any bundle and if an instance of that security operation is identified in a received bundle, the BPA must treat that bundle as corrupted. Constraining security policy statements can be used to identify portions of the network that are not trusted. For instance, a constraining policy statement may describe all security operations originating from a single BPA. In that case, the BPA is identified as untrusted and any security operations it adds to a bundle will be identified as corrupted.

At a security source, a constraining policy statement describes a security operation that *must not* be added to the bundle by that BPA. A bundle should never originate from that BPA containing the associated security operation.

At a security verifier or acceptor, a constraining security policy statement describes a security operation that *must not* be represented in the received bundle. If that security operation is found in the received bundle, it is identified by the BPA as corrupted.

12.6.2 Associating Events and Actions

The relationship between a security operation event and its associated processing actions must be established. An event is coupled to both the required and optional processing actions to be performed when an occurrence of that event is identified by the BPA.

As seen in Figure 12.8, security operation events are mapped to zero or more processing actions as a matter of local security policy configuration. In this figure, solid arrows represent a mapping to a *required* processing action while dashed arrows are used to show the mapping to an *optional*

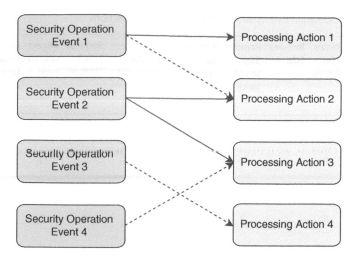

Figure 12.8   Security operation events map to multiple processing actions. Security policy may define required (solid lines) and optional (dashed lines) relationships between the occurrence of a security operation event and a predefined action on a BPA.

processing action. Note that a security operation event can be associated with both required and optional processing actions, as is the case with Security Operation Event 1. Security Operation Event 2 provides an example of an event that is associated with required processing actions only. Security Operation Events 3 and 4 map only to optional processing actions. A processing action may be required for some security operation events and optional for others, as displayed by processing actions 2 and 3.

12.7   A Sample Policy Engine

This chapter has laid out the logical components necessary for a BPSec policy model. These include policy statements which capture the security operation event(s) for which security policy is configured for and the required, optional, and prohibited processing actions that can be associated with these events.

There are several methods for implementation of this security policy engine. This section discusses one possible engine design, modeled after the BPSec policy implementation in the NASA Interplanetary Overlay Network (ION) open-source software distribution [5, 6].

12.7.1   System Policy Engine Overview

BPSec policy is defined in the context of security policy rules. Each security policy rule defines: the bundle(s) that the rule applies to, a required security operation for those bundles, the blocks in the bundle for which the rule applies, the role the BPA applying the rule must take on, and the security operation events that the BPA must execute processing actions for as a matter of security policy.

Security policy rules provide three sets of criteria: filter, specification, and event. These criteria sets are defined in detail in the sections below.

12.7.1.1   Filter Criteria
Filter criteria provides the identifying features of the bundle(s) to be matched to the security policy rule. Filter criteria can include bundle characteristics, such as the bundle source and destination, as well block characteristics such as the security target block type.

Table 12.2 Security Operation Events and Associated Processing Actions.

	Retain SOp BPCFs	Remove SOp	Remove SOp target block	Remove all security target operations	Fail bundle forwarding	Request bundle storage	Report reason code	Override security target block BPCFs	Override security block's BPCFs
Source for security operation	P	P	P	P	P	P	P	P	P
Security operationmiscon-figured at source	P	O	O	O	O	O	O	P	P
Security operation added at source	R	P	P	P	P	P	P	P	O
Verifier for security operation	R	P	P	P	P	P	P	P	P
Security operationmiscon-figured at verifier	P	O	O	P	O	O	O	O	O
Security operation missing at verifier	P	P	O	P	O	O	O	O	P
Security operation corrupted at verifier	P	O	O	O	O	O	O	O	O
Security operation verified	R	P	P	P	P	P	P	P	P
Acceptor for security operation	P	P	P	P	P	P	P	P	P
Security operationmiscon-figured at acceptor	P	R	O	P	O	O	O	O	O
Security operation missing at acceptor	P	P	O	P	O	O	O	O	P
Security operation corrupted at acceptor	P	R	O	O	O	O	O	O	O
Security operation processed	P	R	P	P	P	P	P	P	P

A security operation (SOp) event can be mapped to each processing action with that action being designated as a Required processing action (R), Optional processing action (O), or Prohibited processing action (P) for that event.

Security Policy Rule ID: A value to uniquely identify the security policy rule. This value may be assigned by a user or auto-generated by the security policy engine.

Security Processing Role: The role that the BPA must play when applying the security policy rule to a bundle. Security processing roles are: security source, security verifier, or security acceptor.

Bundle Source: The endpoint from which the bundle originates.

Bundle Destination: The endpoint at which bundle transmission must end as the BPA is responsible for receiving the traffic.

Security Source: The endpoint identifier of the BPA responsible for adding the specified security operation to the bundle. Note that the security source may be different from the bundle source.

Security Target Block Type: The block type of the security operation's target block. Note that block type, not unique block number, is used here, meaning that it is possible for a single rule to specify multiple security target blocks and therefore multiple security operations.

Security Context: The security context to use when applying the security operation to the bundle.

Focus: ION Policy Rule Specificity

ION bpsecadmin console output

The Interplanetary Overlay Network (ION) contains a reference implementation of BPSec policy in versions 4.0.2 and later that can be configured using a command-line utility. This reference implementation deconflicts security policy rules using specificity scores, assigned based on the detail provided in the bundle source, bundle destination, and security source fields of the rule and uses the age of the rule as the final deciding factor when two rules have the same specificity. Wildcard characters are permitted in the three fields used to calculate the rule specificity. A field receives a higher weight if it is more specific. As illustrated in this example, the bundle source endpoint identifier field "ipn:2.1" would be weighted more than ""ipn:2.*"" which is still more specific than "*." Security policy rules also receive a higher specificity score when they have more fields defined. In the sample output above, rule 4 receives the highest weight, 10, as it provides EIDs for all three fields, bundle source, bundle destination, and security source, used to calculate specificity. In cases where the specificity score from these three fields is the same, ION selects the security policy rule that was created more recently as the most specific.

12.7.1.2 Specification Criteria

Specification criteria provide a description of the required security operation the security policy rule represents. This set of information includes the security service, security context, and any parameters necessary to establish that context.

Security Service: The security service to be provided by the security operation.

Security Context: The security context to use when applying the security operation to the bundle.

Security Context Parameters: Any additional information necessary to supplement the identified security context. Sample security context parameters include an Initialization Vector (IV), key name, or salt.

12.7.1.3 Event Criteria

The event criteria define the security operation events for which security policy is configured in the form of optional processing actions.

Security Operation Event: The security operation event for which optional processing actions are enabled.

Processing Actions: The optional processing action(s) to enable for the specified security operation event.

Processing Action Parameters: Any additional information necessary to supplement the enabled processing actions. Sample processing action parameters include a security reason code or BPCF masks.

12.7.2 Policy Configuration Examples

The BPSec policy architecture is designed to be easily configurable for both simple and complex security use cases. As security policy is configured as a response to the occurrence of security events, each sample security policy configuration outlines the building of both event sets and the security policy rules that reference them.

12.7.2.1 Minimizing Illegitimate Traffic

Illegitimate traffic can be minimized in the network by use of BPAs acting as security verifiers for the required security operation(s) for that traffic. A network operator can use security policy to identify BPAs as security verifiers, which must verify the security operation(s) required to be present in a bundle, as well as configure the actions to take if this verification fails and the bundle contents can no longer be trusted.

The network operator first creates an event set to hold the configured, optional processing actions to be executed in response to security failure events at the verifying BPAs. The event set should contain all three possible security operation failure events at the verifier: **Security Operation Missing at Verifier**, **Security Operation Misconfigured at Verifier**, and **Security Operation Corrupted at Verifier**. The optional processing actions configured for each event should remove the blocks in the bundle that have failed security verification and therefore represent information that cannot be trusted. For both the **Security Operation Misconfigured at Verifier** and **Security Operation Corrupted at Verifier** events, the optional processing actions to remove the security operation and remove the security operation's target block from the bundle should be enabled. At the **Security Operation Missing at Verifier** event, only the removal of the security operation's target action is needed as it is implied that the security operation itself is not represented in the bundle as it has been identified as missing.

The network operator then uses a security policy rule at each BPA that is configured to serve as a security verifier for the security operation. These security policy rules should reference the event set created in the step above to discard the blocks in the bundle which fail to verify.

12.7.2.2 Analysis of Security Failures

A network operator may be interested in understanding the cause of security failures in the network. One way of determining this cause is performing forensic analysis on the bundle which caused the security failure to occur. The processing actions prompting bundle storage and the halt of bundle transmission can be used to create the information necessary to perform this forensic analysis.

The network operator first creates an event set capturing all security failure events: **Security Operation Misconfigured at Source, Security Operation Missing at Verifier, Security Operation Misconfigured at Verifier, Security Operation Corrupted at Verifier, Security Operation Missing at Acceptor, Security Operation Misconfigured at Acceptor**, and **Security Operation Corrupted at Acceptor**. For each of these security operation events, two optional processing actions are configured: request bundle storage and fail bundle forwarding.

Each security policy rule created by the network operator should reference this event set to handle security failures. When a failure occurs in the network, the bundle is stored at the current BPA in the state which caused the security failure to occur, and is not transmitted further. A network operator can then examine the bundle contents to determine the root cause of the failure, determining if a target block was removed from the bundle, a misconfiguration occurred with the necessary security context parameters, or the current BPA was unable to identify the security operation.

12.8 Summary

BPSec policy provides a hybrid approach to the establishment of the context necessary for processing security events. This hybrid approach uses both in-band information, provided by the security block itself, and out-of-band information, maintained by the BPA, to require security operations in the bundle and process both security success and failure events.

An effective security policy must enable consistent handling of the occurrence of security events. BPSec policy identifies security events as those presented by the security operation lifecycle and supports the configuration of processing actions as the reactions to these events.

References

1 Birrane, E. and Heiner, S. [2020]. A novel approach to transport-layer security for spacecraftconstellations.

2 Birrane, E. J. and McKeever, K. [2022]. Bundle Protocol Security(BPSec), RFC 9172. **URL:** https://www.rfc-editor.org/info/rfc9172.

3 Burleigh, S., Fall, K. and Birrane, E. J. [2022]. Bundle Protocol Version 7, RFC 9171. **URL:** https://www.rfc-editor.org/info/rfc9171.

4 Birrane, E. and Heiner, S. [2021]. Towards an interoperable security policy for space-basedinternetworks.

5 Burleigh, S. [2007]. Interplanetary overlay network: an implementation of the DTN bundleprotocol.

6 Burleigh, S. C., Scott, K. L. and Lyons, B. E. [2019]. NASA ION (Interplanetary Overlay Network) developer course materials, *Technical report*.

13

Achieving Security Outcomes

Security extensions, such as those provided by BPSec, are one portion of a larger security ecosystem. The construction of a trusted network supporting secured information exchange requires multiple portions of the BPv7 ecosystem to work together to achieve desired security outcomes.

This chapter defines the concept of a security outcome as a system-level expression of desired security behavior coupled with the policy, configuration, block definitions, and security context capabilities that must work together to achieve that outcome.

After reading this chapter you will be able to:

- Explain how policy, configuration, extension blocks, and security contexts must work together to provide secure communications.
- Select amongst options for security context configuration.
- Design new extension blocks to carry special information needed to enable security outcomes.
- Identify critical relationships amongst blocks in a bundle.

13.1 Security Outcomes

The overall purpose of a security policy is to achieve some set of security outcomes for message exchange in a network. These outcomes, as expressions of operational capabilities, can be very similar even across networks exhibiting very different behaviors and capabilities. In that context, security outcomes provide the approach necessary for trusted bundle exchange in any of the networking environments in which BPv7 might be deployed.

Definition 13.1 *(Security Outcome)* The achievement of a particular security service within a network. Outcomes might be associated with specific blocks in a bundle, bundles in a network, or configurations of a policy engine.

The remainder of this section describes the ways in which BPSec concepts work together to implement security outcomes in a network, and provides a common template for the explanation of how to achieve common outcomes that comprise the remainder of this chapter.

Securing Delay-Tolerant Networks with BPSec, First Edition. Edward J. Birrane III, Sarah Heiner, and Ken McKeever.
© 2023 John Wiley & Sons, Inc. Published 2023 by John Wiley & Sons, Inc.
Companion website: www.wiley.com/go/birrane/securingdelay-tolerantnetworks

13.1.1 Outcome Components

BPSec policy engines apply security operations differently than other security protocols because of the higher fidelity of BPSec security services and the flexibility provided by security contexts. Therefore, achieving a security outcome with BPSec requires additional information to fully configure security processing.

Any security outcome is achieved through the proper specification of three types of information: the policy statements that apply to a specific situation, the capabilities of the applied security context, and the data resident on the local node to configure the context.

- **Policy Statements**: The expression of a policy, as enforced by a Bundle Protocol Agent (BPA's) policy engine, is the genesis of any security outcome applied at a node. The correct statements, based on information from various policy sources, must be inspected by a BPA to determine what actions should be taken regarding security processing.
- **Context**: The capabilities of a security context define how information is provided to cipher suites, and what information might be carried in a bundle itself. Certain outcomes require certain capabilities of a security context.
- **Configuration**: The information expected to reside on a node, either as part of a policy or context configuration, must be reasonable given the likely operational capabilities of a network. For example, key negotiation cannot be the configured approach for a security outcome in a network that cannot support contemporaneous round-trip data exchange.

13.1.2 Outcome Descriptions

Every security outcome can be described as a function of the desired resultant behavior of the network and the methodology used to achieve that behavior.[1] As such, each outcome can be described with an overview, methodology, and discussion of potential issues.

1. **Overview**: An outcome overview describes the overall security objective being achieved by the security outcome, with discussion on any unique terms or concepts, and ways in which differing network capabilities may affect how the outcome is achieved.
2. **Methodology**: The methodology describes the set of policy statements, context capabilities, and local configurations needed to both express the need to achieve the outcome, and to implement the outcome at a BPA.
3. **Potential Issues**: To the extent that outcomes are based on correct methodology, each outcome includes a description of potential issues and, where appropriate, thoughts on how to mitigate those issues in practice.

13.2 Verifying BIB-Integrity

There are multiple scenarios that may be served by verifying the integrity of blocks prior to the bundle destination. While the goals in these scenarios are different, similar security policy configurations requiring the use of in-network integrity verification can be used to meet these goals.

To ensure the integrity of certain blocks in a bundle, BPAs other than a security operation's security acceptor may be configured to serve as a security verifier. Such BPAs verify the integrity mechanism(s) held by one or more operations providing *bib-integrity* services (and, thus, resident in Block Integrity Block [BIBs]).

---

1 As a combination of the policy, context, and configuration that implement the behavior.

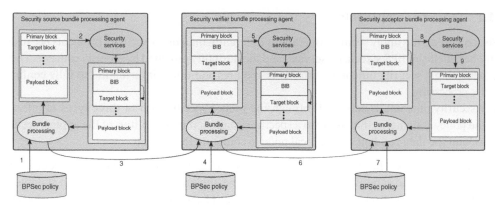

Figure 13.1 In-network integrity verification.

Definition 13.2 *(In-Network Integrity Verification)* Evaluating, using a security verifier, that certain contents of a block have not changed since the application of an integrity mechanism at a security source.

13.2.1 Overview

In-network integrity verification can minimize the amount of illegitimate traffic transmitted in a network. Allowing altered traffic to stay in the network consumes networking resources. Within the Delay Tolerant Networking (DTN) architecture, which relies upon time-variant routes and store-and-forward features, these resources include computational resources and on-board storage in addition to transmit energy and link utilization.

At each bundle hop, security verifiers will check the integrity of specified target blocks and those that fail verification can be handled in an appropriate way, such as removing the information that failed verification. Removing unverifiable traffic is the primary way of reducing potentially wasteful traffic in the network.

The mechanism of in-network integrity verification can be used for specialized scenarios such as primary block integrity. Recall that the primary block in each bundle provides critical routing information such as the bundle's source and destination.

By ensuring primary block integrity, a user can be sure that the bundle is being forwarded to its original, intended destination. To provide this service, the bundle source is configured to serve as the security source for a security operation providing *bib-integrity* over the primary block. Each BPA encountered in the bundle's path should be configured to serve as a security verifier to check the *bib-integrity* operation and make sure that the primary block has not been modified since the bundle was first transmitted from its source.

Security verifiers may also be configured to require the *bib-integrity* service over all encountered primary blocks – in this case a missing security operation can be considered a failed security operation.

13.2.2 Methodology

The following set of bundle transmission steps, illustrated in Figure 13.1, represents a standard approach to in-network integrity verification that can be adapted to serve any of the scenarios described in the previous section.

1. A bundle source, identified as a security source by local security policy, identifies a block that is the target of a required *bib-integrity* security operation.
2. The security source creates a BIB targeting the identified block, resulting in the generation of an integrity mechanism over the security target block's contents. The BIB is added to the bundle before it is forwarded by this BPA.
3. The bundle is forwarded, encountering waypoint nodes in its path that are configured to serve as security verifiers for the *bib-integrity* operation represented in the bundle.
4. At each security verifier, security policy is present, which indicates that the BPA must process the BIB to verify the integrity mechanism of its security target block.
5. The security verifier processes the BIB to check the integrity signature over the security target block's contents.
6. The bundle is forwarded by the security verifier upon successful *bib-integrity* verification.
7. The bundle arrives at a security acceptor for the *bib-integrity* operation (which might be the bundle destination).
8. The BIB is processed so that the integrity of the security target block can be verified.
9. When the security operation has been successfully processed, it is removed from the BIB. If it was the last operation in the BIB, then the BIB is removed from the bundle.

13.2.3 Potential Issues

The combination of BIBs to apply integrity mechanisms over target blocks, and the proliferation of security verifiers throughout the network, is a security outcome for which the BPSec was created. However, there are cases where this can cause unintended side effects.

The following potential issues should be considered in the creation of a security verifier policy and the definition and usage of security contexts that apply that policy.

- **Computational Load**: Independently checking the integrity of every block in every bundle at every BPA in a network may place a large computational load on a BPA, particularly when BPAs are implemented on resource-constrained nodes. In-network verification should be focused on very well-resourced nodes acting as gateways in a network or, if possible, the verification of different types of blocks might be distributed across multiple nodes along common network paths and gateways.
- **Corrupted BIBs**: A false negative may occur when the BIB holding an integrity security result is corrupted, rather than the security target block. For integrity mechanisms that only measure whether a locally calculated value matches a previously calculated value, a corrupted BIB will fail to verify, even is that target block remains unchanged.
- **Policy Distribution**: The more security verifiers that are distributed throughout the network, the more the policy of each such verifier must be kept up to date, particularly as it is related to key distribution and usage. If a security verifier has an outdated policy configuration, it may use an older (or newer) key to check a signed integrity mechanism, which would cause the verification to fail. While this may be the intended behavior around correct key rotations, it may not be the intended behavior when a planned key change has failed to propagate to all BPAs in the network.

13.3 Verifying BCB-Confidentiality

Checking the integrity of block information at security verifiers is an important way to remove bad data early in a store-and-forward network. However, verifying encrypted information at waypoint

BPAs is harder, because the *bcb-confidentiality* service, by definition, replaces user plaintext with ciphertext. Decrypting information is a prohibited action at a security verifier because doing so would destroy the confidentiality of the security target block and lost the authentication properties of having encrypting be signed by the security source.

Definition 13.3 *(In-Network Confidentiality Verification)* Evaluating, at a security verifier, that the ciphertext and associated additional authenticated data has not changed since the application of the confidentiality mechanism at a security source. Notably, this verification is done without performing decryption of ciphertext.

13.3.1 Overview

To provide for in-network confidentiality verification, a security context that provides an ancillary integrity mechanism over ciphertext must be used. Such a security context would not only generate ciphertext and an associated authentication tag, but also a mechanism for checking the integrity of these values without performing decryption.

13.3.1.1 Security Context Options

There are multiple ways in which a security context could be defined to produce integrity mechanisms that could be verified without performing decryption. Three possible ways are listed in the following text and illustrated in Figure 13.2.

1. **Expose the Cipher Suite Verification Interface**: As part of the decryption process that occurs at a security acceptor, the ciphertext and other associated data are input to a cipher suite that first checks the integrity of the input and then performs decryption. Often, the verification process is part of the decryption application programming interface and, thus, can only be run as part of attempting a decryption. However, if this interface were separated and made available to the security context implementation of a BPA it could be used for non-decryption verification checks.
2. **Compute an Independent Security Result over Inputs**: As part of creating a Block Confidentiality Block (BCB), an independent integrity mechanism could be used to generate an integrity-only security result over the set of inputs (security target block and other associated data). Verification of the *bcb-confidentiality* service, in this case, could be defined as integrity verification of this security result without, otherwise, examining ciphertext, authentication tags, or other results from the encrypting cipher.
3. **Compute an Independent Security Result over Other Results**: Similar to the aforementioned approach, a new integrity-only security result could be used to calculate an integrity mechanism. However, in this case, the mechanism might only be calculated over the results of the encrypting cipher: the ciphertext placed in the security target block and any generated authentication tag.

Of these, the approach that results in the least additional processing and block size is to expose the verification interface of an existing cipher suite. Based on the cipher suite, if this is not possible, then the next recommended option is to calculate a new security result holding the integrity of other security results; this balances processing power and integrity protection. Ensuring that an authentication tag has not changed is different than verifying that all of the additional authenticated data have not changed, but the authentication tag can be verified faster than individually verifying each of the inputs to that tag.

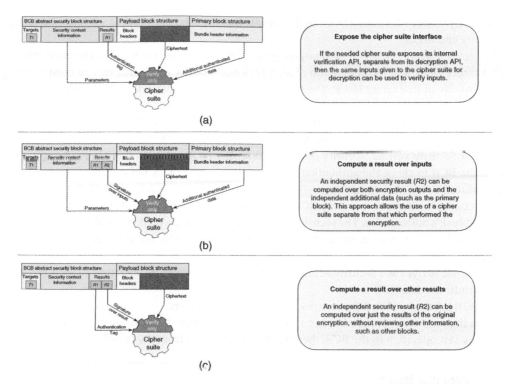

Figure 13.2 Verifying confidentiality. Confidentiality services can be verified without decryption with (a) the support of cipher suites, (b) calculating an independent integrity mechanism over ciphertext and other plaintext, or (c) calculating an integrity mechanism over cipher suite outputs. Each of these mechanisms provides various levels of verification that decryption at a security acceptor remains plausible.

13.3.2 Methodology

The following set of bundle transmission steps, illustrated in Figure 13.3, represents a standard approach to in-network confidentiality verification that can be adapted to serve any of the scenarios described in the previous section.

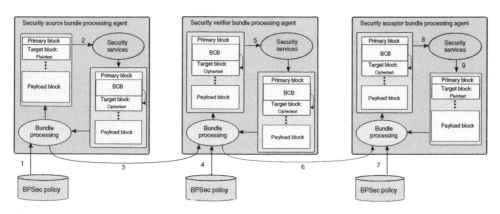

Figure 13.3 In-Network confidentiality verification.

1. A security source identifies a block in the bundle that must be the target of a *bcb-confidentiality* security operation.
2. The security source creates a BCB targeting the identified block, resulting in the encryption of that target block and the generation of an additional security result that can be used to verify authenticity of that target block. The BCB is added to the bundle before it is forwarded by this node.
3. The bundle is forwarded, encountering waypoint nodes in its path that are configured to serve as security verifiers for the *bcb-confidentiality* operation represented in the bundle.
4. At each security verifier, security policy is present, which indicates that the node must process the BCB to check the authenticity of its security target block before bundle transmission continues.
5. The security verifier processes the BCB, using the stored security results to check the security target block's authenticity. The security target block is not decrypted at any time.
6. The bundle is forwarded by the security verifier upon a successful *bcb-confidentiality* authenticity check.
7. The bundle arrives at a security acceptor for the *bcb-confidentiality* operation (which might be the bundle destination).
8. The BCB is processed and the security target block is decrypted.
9. When the security operation has been successfully processed, it is removed from the BCB. If it was the last operation in the BCB, then the BCB is removed from the bundle.

13.3.3 Potential Issues

The same issues present for in-network integrity verification also exist for in-network confidentiality verification. Namely, this security outcome also increases the computational load of a BPA, might generate false positives when the BCB is corrupted, and requires coordinated policy distribution.

There is an additional concern relating to key selection. Several cipher suite algorithms explicitly require that the same key is not used for different types of cryptographic operations. In particular, the same key should not be used for both encryption and the generation of a signature over an integrity mechanism. In cases where a separate security result is generated using a different algorithm than that producing the ciphertext a new key must be used. This requires an additional key to be managed and potentially wrapped and communicated as a security parameter for the security context.

13.4   Whole-Bundle Authentication

A BPSec security operation is the mapping of a security service to a target block within a bundle. One of the powerful features of BPSec is the ability to target individual blocks with individual security services such that multiple such services coexist over the myriad of information present in a single bundle. While block-level granularity is a needed feature of a BPv7 security mechanism, it makes it harder to address the question of how to apply whole-bundle operations, such as bundle authentication.

Definition 13.4   *(Bundle Authentication)*   The determination that a particular bundle has been generated by the asserted bundle source and not modified since transmission from that source.

13.4.1 Overview

Whole-bundle authentication should be used in cases where a bundle is not expected to be changed in any way between the bundle source and the bundle destination. The application of such an authentication mechanism can be done by BPSec when used in coordination with a security context that allows for customized cryptographic binding.

A BIB should be used to hold the whole-bundle authentication result. Since this BIB would likely contain security context identifiers and associated parameters specific to its use for whole-bundle authentication, it would exist as its own security block and not, otherwise, combined with other BIBs that might also exist within the bundle.

There are a number of ways in which whole-bundle authentication could be achieved, and this section makes a specific recommendation on how to achieve this result. There are four important considerations associated with any method to apply authentication to an entire bundle.

1. The target block of the ensuring security operation.
2. The way in which a security result is represented.
3. The scope of the "whole bundle."
4. The needed capabilities of a supporting security context.

These items are discussed in greater detail as follows.

13.4.1.1 Target Block Selection

The BPSec requires that a security operation be applied to a single target block within a bundle. In the case of whole-bundle authentication, there is no single target block, as all blocks in the bundle are included in the security context calculations.

To resolve this, the whole-bundle security result(s) must be associated with a single block that is considered to be representative of the bundle itself. There are three possible choices for selecting such a block, as follows.

- **Primary Block**: The primary block of a BPv7 bundle is immutable and uniquely identifies the bundle in the network.
- **Payload Block**: The payload block of a BPv7 bundle is also immutable in the sense that changes to the payload of a bundle should not happen while the bundle is in the network. Further, payload data is the primary reason why the bundle exists in the network at all.
- **Specialty Block**: A special bundle block may be created for the purpose of serving as the target block for the BIB holding the whole-bundle security result.

Two of these three options are illustrated in Figure 13.4, which shows the use of the primary block or a specialty block as the target for a security block containing information for multiple blocks in a bundle. Of these three options, the recommended practice is the generation of a specialty block. The use of such a block has a few advantages over targeting the primary or payload block for this purpose.

First, using either the primary or the payload block to hold whole-bundle security results prevents any other security results from being associated with those blocks. Each security operation in BPSec must be unique, and two integrity operations targeting the primary block (one for whole-bundle authentication and another for primary block only authentication) would be considered a disallowed duplication.

Second, the presence of such a specialty block would also serve as a proactive signal that the bundle in question has a whole-bundle authentication mechanism on it. This allows for special processing rules associated with BPAs to, perhaps, otherwise limit the way in which a bundle might

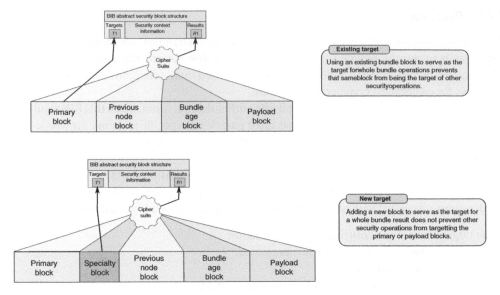

Figure 13.4   Whole-bundle integrity can be implemented as multi-block integrity. The BPSec security service *bib-integrity* requires a target block, which can either be an existing block associated with the bundle (such as the primary block) or a specialty block added just for this purpose.

be processed so as to not change elements of the bundle. Potential bundle changes are discussed in Section 13.4.3.

13.4.1.2   Security Result Definition
Since a bundle is a collection of blocks, it is possible to calculate an authentication mechanism over the bundle in one of two ways:

- **Single Result**: All blocks in the bundle may be catenated and a single result generated over that catenation of blocks. In this case, verification of the single result would verify the authenticity of the bundle.
- **Multiple Result**: Each block may have its own security result generated. In this case, verification of every result would be needed to verify the authenticity of the bundle.

These two approaches are illustrated in Figure 13.5, which show the use of multiple security results carried in a BIB versus a single result computed over catenated blocks in the bundle. Of these two options, the recommended practice is to generate a single security result. Since the desired security outcome in this instance is authentication of the whole bundle, a single result has the advantage of using less space in the bundle than multiple results. This is especially true when the size of the security result is a fixed length, such as the case when generating a signed hash value.

13.4.1.3   Whole-Bundle Scope
Capturing whole-bundle authentication in a security block that is, itself, part of a bundle creates a circular dependency: part of the bundle is defined as a security result calculated over the bundle. There are a few ways in which this dependency could be addressed.

- **Exclusion**: The BIB holding the security result for the whole bundle might, itself, be excluded from the scope of bundle authentication. As something that only exists to capture the whole-bundle result, it could be seen that the BIB is separate from the whole-bundle result.

- **Partial Inclusion**: Since the BIB holding the security result for the whole bundle does exist within the bundle, those portions of the BIB separate from the security result could be included as part of the contents of the bundle and, thus, included in the whole-bundle authentication mechanism.

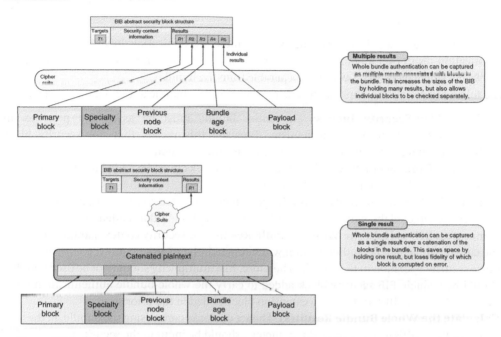

Figure 13.5 Two ways to capture whole-bundle authentication. Whole-bundle authentication can be implemented using multiple results or single result. Multiple results authentication verifies the bundle by verifying each block in the bundle. Single result authentication verifies the bundle by verifying a single catenated representation of the bundle.

Of these options, the recommended practice is to exclude the BIB from the whole-bundle result. Attempting to include only portions of the BIB over the whole-bundle target block would require a nonstandard canonicalization of that BIB to exclude the security result. Any error or change in the BIB would result in a failure to authenticate the whole-bundle, so excluding contents of the BIB from authentication would not alter the intended security outcome.

13.4.1.4 Security Context Capabilities

This security outcome relies on certain capabilities present within a security context. Since this outcomes relies on applying a security service over multiple blocks, the context must not only support the authentication service, but also support some type of method of reasoning about more than one block in the bundle.

To achieve this outcome, a security context must be used that provides for the following capabilities.

- **Authentication**: The algorithms supported in the cipher suites available to the security context must implement some authentication mechanism, such as a signed hash over the contents of the bundle.
- **Block Identification and Ordering**: The ability to include, as a security context parameter, the ordered list of block identifiers that comprise the bundle. The ordering of blocks when calculating a security result (or when checking a result) is important as the same bit stream must be calculated and input to cipher suite algorithms at each verifying node in the network.

- **Security Result Generation**: The ability to ingest the entire contents of other blocks in a bundle and provide them as inputs to some cipher suite algorithm that generates an authentication mechanism over its input.

13.4.2 Methodology

The recommended method of achieving whole-bundle authentication involves calculating an authentication result over all blocks that comprise the bundle, and then associating that security result with a single block that is, itself, inseparable from the bundle.

The following processing steps can be taken to apply whole-bundle authentication to a bundle.

1. **Serialize Non-Security Blocks**: The entire bundle must be assembled prior to any whole-bundle authentication mechanism. Typically this means that non-security blocks in the bundle have been serialized and are ready for transmission.
2. **Apply Any Other Security Blocks**: Other security blocks must be added to the bundle prior to any security block holding whole-bundle result(s).
3. **Add Whole Bundle Authentication Target Block**: An extension block that can be used as the security target of the whole-bundle authentication BIB should be added to the bundle.
4. **Generate a BIB to Hold the Whole-Bundle Result**: The security context parameters used in the case of whole bundle authentication will not match the security context parameters of any other security operation in the bundle. Therefore, BIB multiplicity should not be used and there should be a single BIB security block added to carry the whole-bundle authentication result. The target of this BIB should be the extension block added in the prior step.
5. **Calculate the Whole Bundle Result(s)**: The set of blocks comprising the bundle, in the order specified in the BIB security context parameters, should be input to the security context, which will produce a security result that should be added to BIB.

The processing of this particular BIB should be in accordance with any other BIB processing as security verifiers and acceptors in the network.

13.4.3 Potential Issues

By intuition, the capability to assure that a bundle is both from the asserted source and unmodified since being transmitted from that source should be fundamental to the trusted operation of a network. However, BPv7 differs from other transport mechanisms in that bundles may carry additional information related to the network, a node, or some other processing or behavior not strictly associated with the bundle payload.

This means that certain blocks within a bundle might change as a result of the nominal, secure processing of a bundle in the network. Consider the following processing actions that may result in changes to a bundle between the bundle source and the bundle destination.

- **Added Blocks**: A network might define security sources other than a bundle source, in which case security blocks will be added to a bundle after the bundle has been created and sent from its bundle source. Since a new block would be considered part of the whole bundle, the bundle would fail to authenticate after being processed at a security source in the network.
- **Removed Blocks**: Blocks might be removed from a bundle as they are processed. For instance, if a network defines a security acceptor that is different than a bundle destination, the security acceptor will process and remove one or more security blocks from a bundle. If those security blocks were present at the time of bundle creation, the removal of those blocks would cause the bundle to no longer authenticate after the bundle leaves the security acceptor.

- **Updated Blocks**: A network might require the use of the Previous Hop Notification block that, by definition, is updated (added/removed) at every bundle hop in a network. For example, the Previous Node and Bundle Age extension blocks will change at every hop in a BPv7 network. While the bundle would retain the same block type, if the block contents are updated the whole bundle would fail to authenticate.
- **Fragmentation**: A bundle might be fragmented, in which case bundle authentication cannot occur until final reassembly at the bundle destination.

For these reasons, whole-bundle authentication is not a recommended security outcome in a practical, operational network. Instead, it is recommended that BPAs authenticate subsets of blocks instead.

13.5 Protected Bundle Composition

Altering the contents of a block is not the only way in which the composition of a bundle might be changed. Importantly, blocks may be added or removed from the bundle as a way to alter how the bundle is processed by a BPA and how that processing might change the state of the BPA itself. A security outcome is needed that can protect the bundle composition because changes to the bundle contents create new and different relationships between blocks, alter the processing of the bundle, and potentially alter the status of BPAs processing the bundle.

Definition 13.5 *(Bundle Composition)* The set of individual blocks that comprise the bundle. Changing the composition of a bundle involves adding or removing blocks from the bundle. This may include adding multiple instances of the same block type to a bundle.

Not all bundle composition changes are harmful – in fact, such changes are expected and even needed as part of the proper functioning of a BPA. Several examples of beneficial bundle composition changes are given in Chapter 13.4.3 as reasons why whole-bundle authentication might be impractical.

13.5.1 Overview

The challenging part of a security outcome focused on detecting *harmful* changes to a bundle's composition is defining what blocks in a bundle represent what kinds of information. The addition or removal of certain blocks may be inconsequential to the overall trusted processing of a bundle whereas changes relating to other bundles might be seen as preventing the trusted processing of the bundle in the network.

Any discussion relating to the various combinations of allowed and disallowed block additions and removals requires reasoning about the relationship amongst individual blocks and between blocks and bundles. There might be either a strong or loose association between blocks and the bundle that carries them. In certain cases, migrating a block from one bundle to another might be acceptable. In other cases, this migration might cause an error.

13.5.1.1 Block and Bundle Relationships

The relationship between a block and a bundle (and between two blocks) is a function of the purpose of the bundle, and decisions made at the time of a bundle's creation at the bundle source. Certain blocks, such as the primary and payload blocks, will always have a special role in a bundle. Other blocks may have important roles in some bundles and less so in other bundles.

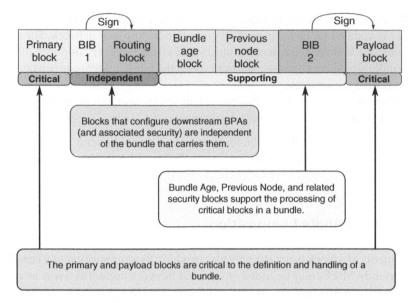

Figure 13.6 Block and bundle relationships. Not every block in a bundle has the same importance to the processing of the bundle. Some blocks are critical to the bundle while other blocks only support that information. Some blocks may be completely independent of the bundle and only included to efficiently carry other kinds of information to a downstream BPA.

To reason about these relationships, some terms must be defined relating to the role that a block plays in a bundle, noting that a block's role might be independent of the block type, or may change between two instances of the same block in a bundle.

There are three roles that a block might have for a particular bundle: critical, supporting, and independent. These relationships are shown in Figure 13.6 and described in more detail next.

Definition 13.6 (*Critical Block*) Any block that must be present for the proper processing of a bundle. Removal of a critical block would cause the bundle to no longer be useful in the network. Critical blocks are part of the bundle at the time of the bundle's creation at the bundle source.

Critical blocks are those identified at a bundle source as being required for the proper syntactic and semantic processing of the bundle through the network. Were a network to require the existence of a *Ticket to Ride* extension block [[1], Chapter 15] then the absence of such a block could cause a downstream BPA to remove the bundle from the network.

Generally, critical blocks are uniquely tied to the bundle in which they appear, such that copying a critical block from one bundle and placing it in another would result in unexpected processing. Consider the Bundle Age block that tracks the time that a bundle has persisted in the network. This block contains information that is only relevant to the current bundle and duplicating that block in some other bundle would give a false reporting of the other bundle's age.

Other blocks in a bundle may exist that act as a support for critical blocks, such as ensuring the integrity or confidentiality of critical blocks. These supporting blocks are not, themselves, critical for bundle processing but are helpful to ensuring the protection of critical blocks in the network.

Definition 13.7 (*Supporting Block*) A block that is bound to a critical block or another supporting block. Supporting blocks may come and go over the lifetime of a bundle in accordance with regular bundle processing. However, removing a supporting block prematurely may make a critical block (or the entire bundle) unintelligible or unable to be processed.

A simple example of a supporting block is a BCB holding a *bcb-confidentiality* operation over a critical block. Such a BCB might be added at a security source other than the bundle source and would not, itself, be considered a critical block. However, if the BCB were to be removed from the bundle by any BPA other than the security acceptor of that service, the critical block would remain unintelligible. Similarly, were a malicious BCB to be added to a bundle claiming to target a critical block, then attempts to process such a BCB would result in a failure to "decrypt" the critical block and loss of that block.

All other blocks in a bundle can be considered independent of the bundle; they may be beneficial but not required for bundle processing.

Definition 13.8   (Independent Block)   A block that is not dependent on the bundle in which it is carried. Independent blocks may be removed from one bundle and placed in another without loss of bundle identity or block utility.

There are three types of independent blocks that might exist within a bundle: blocks that carry network/link state, those that are otherwise shared across multiple bundles, and those blocks that depend only on independent blocks.

- **Network and Link State Blocks**: BPAs may tag certain bundles in a network with information representing the state of the network or a given link as a measurement function unrelated to the bundle used to transport that bundle. Consider a block that carries contact information relating to when a particular network link will be removed from the network. Upon receipt of such a block at a BPA, such as a bundle destination, might copy that block into a new outbound bundle to propagate this information.
- **Blocks Shared Across Bundles**: In certain circumstances, common blocks might be precalculated for placement in one or more bundles. This can happen in very high-rate applications where multiple bundles might share a quality of service, flow label, or other common attribute.
- **Blocks Associated with Independent Blocks**: Any block solely dependent on an independent block is, itself, an independent block. Consider a BCB over an independent block – were the BCB security target block to be moved to a new bundle, then the BCB over that block would also move to the new bundle.

13.5.1.2 Harmful Bundle Manipulation

The block roles of critical, supporting, and independent allow for a better way to express what changes to a bundle's composition would be considered *harmful* versus those changes that are either beneficial or, at a minimum, irrelevant, to the processing of the bundle in the network.

Bundle manipulations that change (directly or indirectly) the processing of a critical block are considered harmful. Some examples of harmful and not-harmful manipulations are illustrated in Figure 13.7 and explained in more detail next.

- **Critical Block Removal**: Any critical block removed from a bundle prior to the bundle destination is harmful as the block is, by definition, critical to the successful processing of the bundle.
- **Critical Block Addition**: Critical blocks must exist in a bundle from the time the bundle is created – any BPA other than the bundle source attempting to add a critical block would imply that the added block was somehow not needed prior to the bundle's arrival at the current BPA. This is, by definition, not the case and so any attempt to add a critical bundle by a BPA other

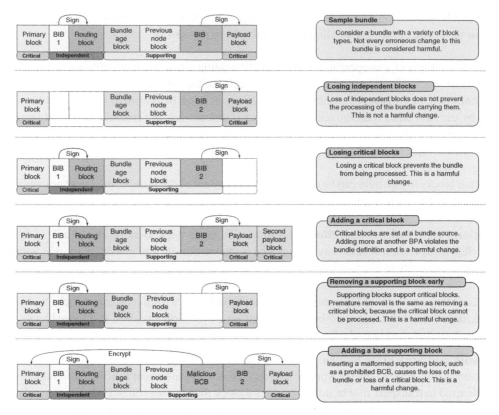

Figure 13.7 Harmful bundle manipulations. Any manipulation of a bundle that results in a failure to process a critical block is considered harmful.

than the bundle source can be seen as an attempt to alter the processing of a bundle in a way not intended by the bundle source.

- **Premature Removal of a Supporting Block**: The removal of a supporting block prior to the expected disposition of that block would, by definition, remove some supported service on the critical block it is supporting. This might leave the critical block open to manipulation or otherwise make the block unintelligible.
- **Malicious Insertion of a Supporting Block**: The addition of a supporting block might be used in an attempt to invalidate the processing of a critical block by providing fictitious or otherwise malicious "support." Adding a fake BIB or BCB over a critical block would render the critical block unusable as security services would incorrectly presume that the critical block has been manipulated.

13.5.1.3 Identifying Critical Blocks

The determination of which blocks in a bundle are critical must be performed at the bundle source for the simple reason that were any other BPA to declare a block as critical would imply that the block was not critical for the portion of the bundle path between the bundle source and that BPA. This leads to two observations relating to the identification of these blocks.

- The bundle source must identify critical blocks in the bundle.
- The bundle source must provide this information for other BPAs in the network.

There are multiple ways in which critical blocks can be identified, both at the bundle source and other BPAs in the network, and this section makes a specific recommendation on how to achieve this result. There are three important considerations associated with any method of identifying critical blocks in a bundle, as follows.

1. Source versus Destination criticality.
2. Strict or Permissive Support Blocks
3. In-Band or Out-of-Band Selections

The remainder of this section discusses each consideration in more detail.

13.5.1.3.1 *Source or Destination Criticality* There are two ways in which a block could be considered critical – either the block is required for the proper processing of the bundle along the network path or the block is only required for the proper processing of the bundle at the bundle destination. In the former case, critical blocks can be interpreted as blocks that must be present and preserved at the source of the bundle. In the latter case, critical blocks are only those that must be present at the bundle destination to aid in processing the bundle.

Figure 13.8 illustrates the difference between "required from source" and "required at destination" when adding a supporting Previous Node block (PNB) to a bundle at the bundle source. Only a single PNB is allowed in a bundle at one time, and upon bundle receipt the existing PNB is processed and removed and a new PNB added prior to bundle transmission.

If the bundle source uses *source criticality* then a block such as the PNB would be included in the set of critical blocks because it existed at the bundle source. The first BPA receiving the bundle

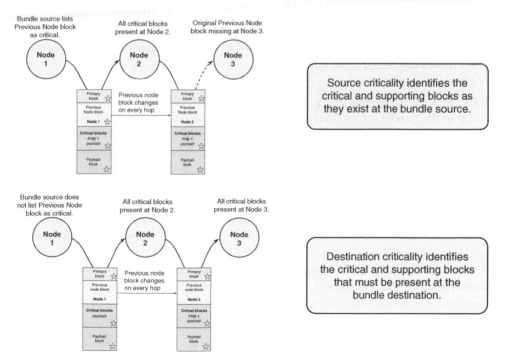

Figure 13.8 **Source and destination criticality.** Identifying critical and supporting blocks is an important part of scoping bundle authentication. Supporting blocks present at a bundle source might not be present at a bundle destination, and this must be taken into consideration when scoping bundle authentication.

would process this PNB, after which it would be removed from the bundle. This implies that the bundle would be considered to have its critical blocks intact only at the first hop in the network.

Alternatively, if the bundle source uses *destination criticality* then only blocks required to exist in the bundle at the bundle destination are included in the set of critical blocks. The PNB would have no expectation of persisting in the bundle beyond the first hop and, thus, would not likely be included in the critical block set.[2]

It is recommended that the choice of criticality determination be the set of blocks that must be present at a bundle destination, as this results in a slightly smaller set of critical blocks, with a greater focus on data delivery.

13.5.1.3.2 *Strict or Permissive Support Blocks*

The addition of support blocks may beneficially aid in the processing of critical blocks or, when abused, might prevent the successful processing of those blocks. A bundle source should decide on a policy of whether downstream BPAs may add supporting blocks to a bundle.

A *strict* supporting block policy would prevent the addition of supporting blocks at any node other than the bundle source. With this policy, all support required for critical blocks are present at the bundle source and expected to be preserved through the network to the bundle destination.

A *permissive* supporting block policy allows other BPAs to add supporting blocks to the bundle as needed as a function of the policy at each of these downstream nodes. In this case, the bundle source does not need to provide all support needed for its critical blocks, and allows support to be delegated to other BPAs as a function of their own capabilities and the state of the network.

It is recommended that a strict supporting block policy be adopted when trying to detect harmful bundle composition changes. The benefit of a strict policy is that, by allowing no other BPAs to add supporting blocks, the set of supporting blocks present at the bundle source can be codified. This provides a useful way to determine when a supporting block was prematurely removed from the bundle (or incorrectly added to the bundle).

13.5.1.3.3 *In-Band/Out-of-Band Policy*

The decision of block criticality and support block policy are known to the bundle source, likely as a function of BPA policy. An implementation decision must be made as to whether the decisions made by the bundle source are presumed to be known to other BPAs as a function of their own local policy (as an out-of-band mechanism), or whether these decisions need to be codified in the bundle itself (as an in-band mechanism).

It is recommended that these decisions be codified in the bundle itself – an in-band mechanism. Presuming that the policy of all nodes in a bundle network maintain the same policy over potentially long periods of time makes the proper operation of the network fragile. Capturing policy within the bundle itself allows for network policy changes to occur without changing the intent of individual bundle handling.

The recommended way to provide in-band policy is through the definition of an extension block that identifies the critical and support blocks in a bundle as well as some flags associated with the policy selections made relating to their processing and disposition.

Specifically, such an extension block could carry the following information.

- A listing of all critical blocks in the bundle, identified by block number. The primary block and payload block are always considered critical to a bundle.
- A listing of all supporting blocks in the bundle, identified by block number.

---

2 Unless, of course, routing determined that the bundle destination was one hop away.

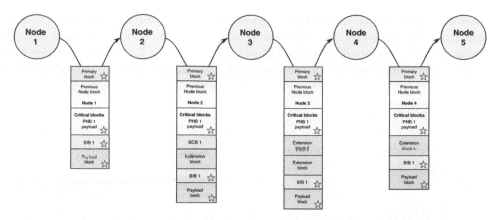

Figure 13.9 In-Band policies are implemented with extension blocks. The in-band identification of critical blocks can be done using a special extension block (noted as *Critical Blocks*). Such an extension block can identify those blocks considered supporting or critical to the bundle. This extension block would not change as the bundle is otherwise changed through the network to its destination.

- A set of processing flags identifying, at a minimum, whether supporting blocks may be added by downstream BPAs.

The concept of such a block is illustrated in Figure 13.9. In this figure, an extension block labeled *CriticalBlocks* is used to identify critical or support blocks. This extension block can be integrity signed (such as with block *BIB*1) when created at *Node*1. While the bundle may undergo multiple other changes along a network path, this extension block – and all other critical and support blocks – can be checked at the bundle destination to ensure that all needed bundles are still present.

13.5.2 Methodology

Detecting harmful bundle composition changes involves representing critical and supporting block identification and policy in an extension block, adding the block to a bundle, and protecting the confidentiality of that extension block. This ensures that only the bundle source and destination know what information is required to be in the bundle.

Since the determination of block changes happen between bundle endpoints, there are processing steps that must be taken at both the bundle source and bundle destination.

13.5.2.1 Bundle Source Processing Steps
The following processing steps can be taken to apply this type of security outcome to a bundle at a bundle source.

1. **Populate the Bundle**: All blocks (other than a manifest block) that will comprise a bundle – to include critical and support blocks – should first be added to the bundle. Attempting to identify blocks prior to this runs the risk that a block number might be changed or that a block intended to be placed in a bundle might experience some error or issue that prevents it from being added to the bundle.
2. **Populate and Add a Manifest Block**: As part of generating some manifest extension block, the bundle source would need to look up what blocks would be considered critical. This lookup likely occurs as a matter of policy associated with the configuration of the manifest block itself. Once critical blocks are identified, supporting blocks can also be calculated and added to a manifest block, as can the capture of policy associated with that block.

3. **Secure the Added Block with a BCB** A manifest block is recommended to be encrypted to ensure that malicious nodes do not review criticality determinations associated with a bundle. In cases where this determination is not considered protected information, a manifest block should, at a minimum, be integrity signed by a BIB. A manifest block in a bundle should never exist without some security service applied to it by the bundle source.

13.5.2.2 Other BPA Processing Steps

A bundle waypoint should act as a security verifier for a manifest block. If the block can be verified and read, then the BPA should process the manifest block. Otherwise, the BPA should ignore the manifest block. Waypoint BPAs should never act as a security acceptor for a manifest block as accepting a security service would remove that security operation from the bundle. Then the manifest block would either be without a security service for the remainder of the journey in the network, or would need to have security reassigned to it by a BPA other than the bundle source. In either case, the security of the manifest block would be compromised.

For this reason, security policy should never assign a security acceptor to a manifest block – this implies that the security acceptor of the block would default to the bundle destination.

A bundle destination should first act as a security acceptor to a manifest block and proceed to process the block. For both a bundle waypoint and a bundle destination, processing of a manifest block should involve the following steps.

1. **Verify/Accept Manifest**: Prior to processing, there must be some determination that the manifest block has remained unchanged since it was created and added to the bundle by a bundle source.
2. **Confirm Required Blocks**: The manifest should be examined and every critical block in the manifest should be confirmed to still exist within the bundle. Similarly, all required supporting blocks should also be confirmed to still exist in the bundle.
3. **Confirm Policy Compliance**: The policy options captured in the manifest should also be checked. If additional support blocks have been prohibited, the bundle should be inspected to ensure that it does not contain them. Similarly, the bundle should be inspected for any additional critical block types that are not listed in the manifest.
4. **Process Success or Failure**: In the event that the bundle has been manipulated such that the manifest is no longer adhered to, it should be processed in accordance with policy. At a minimum, the bundle should be removed from the network. Alternatively, if the manifest has been successfully processed, the bundle may be forwarded (if this check passed at a waypoint BPA) or delivered (if this check passed at the bundle destination).

13.5.3 Potential Issues

Managing the addition and removal of blocks from a bundle is difficult in a networking environment where such additions and removals are seen as part of the normal function of the network. To achieve this security outcome there is a necessary assumption that the bundle source has enough knowledge to create a bundle with everything it needs to arrive at a bundle destination.

There are some circumstances where this assumption might not hold, or where behavior might not be as expected even when the manifest approach does detect issues.

- **Fragmentation**: In cases where a bundle must be fragmented, the manifest block might exist in a single fragment, or in multiple fragments. Care must be taken to not process a manifest block that exists in a bundle fragment, and only process such a block at a bundle destination. This means that waypoint checking of corrupt data can be avoided by early fragmentation of the bundle.

- **No Root Cause**: In cases where a manifest block is encrypted, only the bundle destination can determine whether the bundle composition was improperly changed. In this case, there is no record of which BPA(s) in the network performed the bundle manipulation. As such, it will be difficult to determine root cause. This may be particularly difficult when composition changes as a result of misconfiguration and not malicious intent.
- **Consistent Policy**: When a manifest is used, which prohibits the addition of other supporting blocks to the bundle, and the manifest is otherwise encrypted, then other BPAs in the network must rely on their own local policy configurations to not add supporting blocks to the bundle that would be considered malicious at the bundle destination.

13.6 Summary

Protocol and algorithm specifications focus on individual capabilities with the presumption that these capabilities can be mixed, used, and otherwise enable some trusted communication in a network. The expression of higher-level behaviors related to this kind of secure communication are termed *security outcomes.*

Achieving security outcomes for a network deployment is a complex function of the capabilities of individual security contexts, the types of blocks that appear in a bundle, and the out-of-band policy resident on BPAs in the network. As such, BPSec security blocks are only one piece of a much larger systems engineering approach to securing networks.

Simple expressions of security outcomes such as "encrypt data from a bundle source to a bundle destination" and "make sure a block does not change between a bundle source and a bundle destination" are built into the BPSec specification and only require being enabled to be functioning in the network.

However, there exists a variety of other security outcomes that require much more nuanced behavior. This section presented several such complex outcomes, to include checking for corruption in the network, providing authentication over entire bundles, and determining when important blocks within a bundle have been added or removed.

The number of unique security outcomes will grow with the number of unique BPv7 deployments. The techniques presented in this section are meant to serve as a guide to reasoning about the creation of security contexts, extension blocks, and policy statements to achieve increasingly complex and/or customized security behavior in a network.

Reference

1 Birrane, E. and Soloff, J. [2020]. *Designing Delay-tolerant Applications for Store-and-Forward Networks*, Artech House.

14

Special Considerations

Securing any network is a complex combination of protocols, configurations, policies, and proper software and hardware implementations. High-availability and low-latency networks beneficially leverage their communications infrastructure to support helpful security functions such as just-in-time key negotiation. Networks, particularly those conforming to the Delay Tolerant Networking (DTN) architecture, cannot rely on a communications infrastructure with such assumptions. Instead, these networks must rely on out-of-band mechanisms in cases where end-to-end negotiation is not possible.

Therefore, a Bundle Protocol version 7 (BPv7) network is *at best* as difficult to secure as any other terrestrial network and *at worst* much harder. This chapter discusses some of the unique challenges that must be addressed when securing any type of BPv7 network and those special considerations when securing networks conforming to the DTN architecture.

After reading this chapter you will be able to:

- List the different types of special considerations for securing BPv7 networks.
- Explain special security issues related to bundles.
- Inspect security context behavior for unintended consequences.
- Review security policies for potential issues.

14.1   Scoping Security Concerns

The security of message data in a network must include mechanisms for data confidentiality, integrity, and authentication. BPSec defines the *bcb-confidentiality* service to apply data confidentiality and the *bib-integrity* service for both data integrity and authentication.[1]

The activities necessary to provide individual security services are more encompassing than just the creation of cryptographic materials under the auspices of approved cipher suites. The Internet Engineering Task Force (IETF) documents several additional security considerations for computer networking as Request for Comments (RFC) 3552 "Guidelines for Writing RFC Text on Security Considerations" [1]. These guidelines include considerations such as the following.

- Non-repudiation
- Unauthorized and inappropriate usage
- Denial of service

---

1 The difference between integrity and authentication is a function of the security context used.

Securing Delay-Tolerant Networks with BPSec, First Edition. Edward J. Birrane III, Sarah Heiner, and Ken McKeever.
© 2023 John Wiley & Sons, Inc. Published 2023 by John Wiley & Sons, Inc.
Companion Website: www.wiley.com/go/birrane/securingdelay-tolerantnetworks

- Common issues with shared keys and key distribution centers
- Object versus channel security

There are considerable resources in the literature that address security design for those considerations deemed common to any network user and any network transport. This chapter discusses security considerations unique to elements of BPv7 and BPSec as a function of their structure and behavior. This includes the potential deployment of these tools in the context of the DTN architecture.

Some of the considerations discussed in the following text focus on the implementations of individual network devices and the operational configuration of individual networks. Other considerations apply more directly to the unique characteristics of BPv7 bundles and BPSec mechanisms such as security contexts and security policy statements.

14.2 BPA Resource Considerations

The application of network security – and the systems supporting that security – consumes resources on any network node. However, the features provided by BPv7 and BPSec can consume greater amounts of platform resources than other network security approaches. The decision to deploy bundles and to secure them with BPSec security blocks should consider how this processing affects the performance of nodes in the network.

For example, when end-to-end paths in a network are noncontemporaneous, data may stay at rest on a Bundle Protocol Agent (BPA) for extended periods of time. This time-at-rest allows a BPA to perform additional security operations without impacting the throughput of the network.

14.2.1 Additional Computational Load

The increased fidelity provided by BPSec increases computational load because it requires both deep-bundle inspection and deep-block inspection. Strong evidence of the computational load imposed by bundle processing can be found by examining the relative performance of different bundle protocol implementations. Performance comparisons for BPv6 implementations showed significant performance variances in general [2] and on constrained devices in particular [3].

Variations of BPv6 BPA performance are, of course, a matter of the efficiency of the implementation. Presuming that implementations are not unreasonably inefficient, these results imply that the computational load imposed by bundle (and block) processing requires special attention to ensure that the highest data rates can be achieved. While the performance of a BPv6 BPA will differ from the performance of a BPv7 BPA, both must perform similar levels of bundle and block inspection. Lessons learned from BPv6 processing likely still inform BPv7 processing.

Definition 14.1 *(Deep-Bundle Inspection)* The process of identifying and parsing the contents of a bundle to identify the blocks within the bundle and any potential required processing of those blocks. This is in contrast to shallow-bundle inspection which only looks at a small portion of the bundle – typically the primary block – for route information.

To assess whether security operations are present within a bundle, a BPA must perform deep-bundle inspection looking for the presence of security blocks. Unlike IPv6, BPv7 does not impose a strict ordering of extension blocks; extension blocks may appear in any order and even

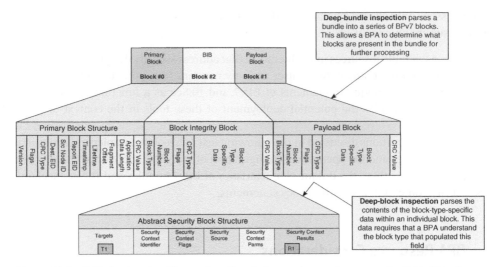

Figure 14.1   Differing levels of inspection. Bundle inspection produces standard BPv7 blocks, whereas block inspection produces the block-type-specific data fields of an individual block.

be re-ordered by a BPA.[2] Isolated networks might attempt to impose certain block orderings – such as placing all security blocks after the primary block – but any imposed ordering would be unsustainable were such a network to be federated with other networks. Therefore, BPAs that are configured to process security must perform deep-bundle inspection on every received bundle.

Definition 14.2   (*Deep-Block Inspection*)   The process of parsing the contexts of a block within a bundle to identify and act on the information within that block. Deep-block inspection implies that a BPA performs deep-bundle inspection first.

The distinction between deep-bundle and deep-block inspection is shown in Figure 14.1. In this figure, a bundle must be deep-bundle inspected to determine that the bundle consists of three blocks: a primary block, a Block Integrity Block (BIB), and a payload block. However, this level of inspection only produces standard BPv7 blocks, which treat any block-type-specific data fields as opaque. To process any individual block, such as the BIB, the block itself must further be inspected to uncover the specific data associated with that block type.

BPSec collects security operations into security blocks. While this aggregation reduces the overall size of the bundle, it also requires that an individual security block be deep-block inspected to ascertain whether any of the security operations in the block must be processed at the BPA. The structure of the BPSec abstract security block attempts to reduce the overall complexity of this inspection by listing the numbers of security target blocks of the encapsulated security operations at the start of the security block's block-type-specific data fields. This allows a BPA to only look at the initial target list rather than decoding the entire security block to identify if further processing is necessary.

This increased level of inspection is in addition to the computational processing already required to extract plaintext and ciphertext from the bundle and pass it through correctly configured cipher suites at the BPA.

---

2 While unlikely that a BPA would reorder extension blocks, a BPA may insert a new extension block between existing extension blocks, thus changing extension block order.

14.2.2 Memory and Storage Requirements

The BIB and BCB extension blocks may appear in any order within a bundle – this includes before or after the blocks that they target. This means that a security block might not be able to be processed when it is decoded from a bundle because the target block of that security block may not have been read from the bundle yet. This is always true when the target block of a security service is the payload block[3] but can happen with any target block and is exacerbated when using security contexts that allow other, non-target blocks to be cryptographically bound to a security operation.[4]

In order for a security block, a target block, or some other cryptographically-bound block to be processed, these blocks must be deserialized from their on-the-wire encoding to identify the various components of the block – particularly the block-type-specific data fields of the blocks. Were the BPA to identify itself as a security verifier or security acceptor of any security operation in the bundle, for every bundle received, the BPA must decode every security block and, possibly, many other blocks in the bundle.

Implementations would likely prefer to keep these decoded forms of related blocks in working memory where they can be accessed and operated on quickly, which would require a larger memory footprint for a BPA, particularly when dealing with a high-rate link where bursts of bundles might be received faster than can be processed in real time.

In some cases, the decoded form of a block might be too large to be practically held in memory. This can certainly be the case for bundles that carry very large payloads, but may also be the case for extension blocks which, themselves, carry large amounts of data. In these cases, spooling decoded block information to disk both increases storage requirements and reduces the practical throughput of the system because disk access is significantly slower than memory access.

Bundle processing may require both large amounts of memory and storage, but also very efficient implementations of storage access and caching mechanisms, otherwise storage quantity and storage speed will become the bottleneck in the system [4].

14.3 Bundle Fragmentation Considerations

The security operations implemented by BPSec are implemented, verified, and accepted as part of processing a bundle at a BPA. This means that BPAs might only provide these services when handling bundles, and not bundle fragments.

> *To fragment a bundle whose payload is of size M is to replace it with two "fragments" – new bundles with the same source node ID and creation timestamp as the original bundle – whose payloads MUST be the first N and the last (M − N) bytes of the original bundle's payload, where $0 < N < M$.*
>
> *BPv7, RFC 9171 §5.8, [5]*

Importantly, a bundle containing a fragment is not the same as the original bundle and, therefore, security services on the original bundle should have no expectation of successful verification and acceptance on a bundle containing the fragment. The BPSec specification is clear that security services on a bundle containing a fragment should never have BPSec security blocks added to them. In cases where a bundle fragment requires its own security, BPSec recommends that either some

---

3 In BPv7 the payload block is always the last block in a bundle.
4 Such as the case when identifying additional authenticated data.

alternative mechanism be used, or the entire bundle fragment be encapsulated in another bundle and security services applied to the encapsulating bundle, similar to the use of Tunnel Mode in IPSec.

> *… a BCB or BIB MUST NOT be added to a bundle if the "Bundle is a fragment" flag is set in the bundle processing control flags field.*
>
> *BPSec, RFC 9172 §5.2, [6]*

The rules associated with bundle fragmentation present multiple difficulties for bundle security. Three important ones being that fragmentation delays certain security processing until bundle reassembly, it results in some extension blocks being duplicated, and the result of security failures is ambiguous in the presence of fragmentation.

14.3.1 Delayed Security Processing

When a bundle is segmented into multiple other bundles any security relating to either the primary block or the payload block cannot be successfully evaluated until the bundle is re-assembled. This reassembly could happen at any BPA in the network and must happen at the bundle destination.

Security services that involve the payload block cannot be processed prior to reassembly because, by definition, fragmentation breaks a payload across multiple new bundles. After fragmentation, the payload (as initially secured) does not exist until the bundle is reassembled. Similarly, security services that involve the primary block cannot be processed after fragmentation, as the original primary block no longer exists, being replaced by the primary blocks of individual fragments.

These changes are illustrated in Figure 14.2. Prior to fragmentation, BIBs can be used to generate integrity signatures over the primary and payload blocks. However, after fragmentation, neither of these target blocks exist. The payload has been replaced by two fragments (Frag 1 and Frag 2) and

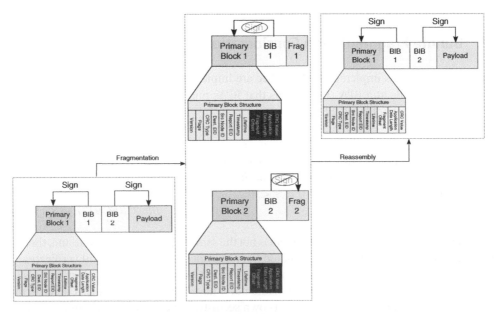

Figure 14.2 Fragmentation delays security processing. Services such as integrity checking cannot happen over the primary and payload blocks, because those blocks do not exist after fragmentation until the bundle is reassembled.

the primary block has been replaced by primary block 1 and primary block 2, each of which have differing fragment offsets, lengths, and CRCs. Regardless of whether BIBs travel with one fragment or another, integrity cannot be checked until the original bundle is later reassembled.

This delayed security processing happens both when the primary or payload blocks are the target blocks of a security operation *as well as* when either of these blocks are otherwise cryptographically bound to a security operation. Consider the case of an integrity mechanism calculated over some extension block E_1 that also is cryptographically bound to the bundle by including the primary block contents as part of the integrity mechanism. Prior to fragmentation, the integrity of E_1 could be verified without issue. However, after fragmentation, even if the BIB and E_1 were located in the same fragment, E_1 could not be verified because the primary block of the fragment is different than the primary block of the original bundle.

Since it is impossible to verify or accept security on a fragment, a malicious node could force the fragmentation of a bundle as a way of circumventing or otherwise delaying security processing of information relating to (or otherwise involving) the primary block or payload block. This is problematic even if security processing of the assembled bundle would, eventually, result in identifying any bundle tampering as the storage and transmission of bundle fragments up to the point of reassembly consume networking resources. Efficient use of networking resources – and the ability to remove bad data from the network early – is an important optimization for the DTN architecture.

14.3.2 Block Duplication

The least significant bit[5] of each extension block's Block Processing Control Flags field is labeled the *Block must be replicated in every fragment* bit. If this bit is set, then the extension block must be replicated in every fragment during the bundle fragmentation process. One use of this flag is when a BCB is used to hold a confidentiality operation over the payload block.

> *BCBs MUST have the "Block must be replicated in every fragment" flag set if one of the targets is the payload block. Having that BCB in each fragment indicates to a receiving node that the payload portion of each fragment represents ciphertext.*
>
> *BPSec, RFC 9172 §3.8, [6]*

The presence of the same extension block in multiple bundle fragments represents unique challenges for interpreting security results. Separate from security services that might involve the primary and payload blocks, other extension blocks *could* have security services applied to them, to include security verification and acceptance. However, the meaning of these functions is made more ambiguous in the presence of fragmentation.

Consider the case where a BCB holds a single security operation over a single extension block E_2 such that neither the primary block nor payload block are involved in the encryption of E_2. Further, consider that a bundle representing this situation is fragmented into three bundle fragments (B_1, B_2, and B_3) at a BPA and that both the BCB and E_2 are duplicated in each fragment. It is possible in a changing topology that these fragments could take three different paths through the network on the way to the bundle destination (or some other point of bundle reassembly). Along this path, several things might happen to these fragments – three outcomes of interest are listed next.

---

5 Meaning the bit represented by the value 0x01.

1. **(Outcome 1)**: The BCB might be unprocessed, having never encountered a BPA that was a security verifier or acceptor for this operation.
2. **(Outcome 2)**: The BCB might be processed and E_2 decrypted by a BPA acting as a security acceptor.
3. **(Outcome 3)**: The BCB might be dropped from the bundle as part of corruption or malicious bundle manipulation.

Given three bundle fragments, any combination of the three outcomes earlier can happen to each bundle fragment – additionally, any bundle fragment might be removed from the network due to a variety of reasons. Assuming that **(Outcome N)** happens to fragment B_n, this implies that the first fragment arrives at the reassembly point with the original BCB and E_2 as it existed at the time of fragmentation. The second fragment arrives without a BCB because its version of E_2 is no longer encrypted, and the last fragment contains no BCB because it was removed, but it's version of E_2 remains encrypted. This situation is illustrated in Figure 14.3.

At this point, the actions taken by the bundle reassembly procedure at a BPA will have a significant impact on the security processing of the *bcb-confidentiality* operation over E_2. If the original BCB and encrypted E_2 are included in the reassembled bundle, then the local BPA or downstream BPA would need to accept the operation. If the decrypted E_2 were included in the assembled bundle, any downstream BPA requiring confidentiality over E_2 might show this lack of encryption as a policy error. If the encrypted E_2 is kept without a BCB, then any downstream BPA attempting to access E_2 would fail block processing as the contents of E_2 would be unintelligible.

14.3.3 Security Block Affinity

When a bundle containing security blocks is fragmented, a decision must be made as to what security blocks would be included in what fragments. The mapping of security blocks to fragments is called the *affinity* of the security block for one (or more) fragments. BPSec requires that security blocks carrying encryption over the payload must be included in every fragment, but is silent on how to make similar determinations regarding other security blocks on other target blocks.

Replicating every security block in every fragment increases the chance of ambiguity at a downstream reassembling BPA. Failing to keep a security block with its target block(s) might cause a BPA with a requirement to verify certain services over certain blocks to issue an error if a required

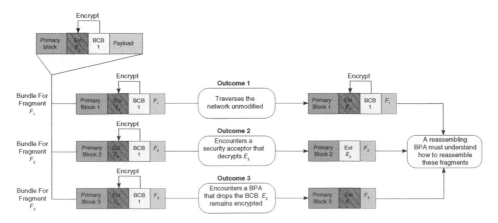

Figure 14.3 **Replicating security blocks can lead to confusion.** If security blocks are processed differently while a bundle is fragmented, it may create ambiguity when attempting reassembly.

security service is missing from a bundle. There are some general recommendations relating to security block affinity, though this must be carefully considered as part of the policy tuning of any BPA.

- Security operations should be included in the same fragment as their target blocks.
- If a target block is replicated in every fragment, its security operation should be replicated in every fragment.
- If a security block contains multiple security operations, the block should, itself, be split to allow operations to stay with their target blocks.

If a BPA chooses to encapsulate a bundle fragment such that only the encapsulating bundle is transmitted to a reassembly point (instead of allowing the bundle fragment to be sent through the network itself) then the aforementioned rules are not needed, as the security blocks within fragments are hidden from BPAs until final reassembly.

14.4 Security Context Considerations

Security contexts within the BPSec are meant to apply processing, parameterization, and other important elements that customize cipher suite handling for a variety of networking conditions. Proper matching of a context to a networking environment helps security processing which implies that *improper* matching of a context to a networking environment impedes security processing.

The selection and configuration of security context information must be carefully considered as a function of the bundle being secured, the block types being secured, and the networking environment. Context definition and selection is an important part of BPv7 network engineering.

It is expected that the security policy configuration at BPAs within a network define the security role of that BPA with respect to types of security operations within the system. This policy configuration includes, in part, the security context to use and the parameters that accompany the implementation of that context.

If a security context is selected that does not match the capabilities of the network, the security source, or the anticipated security verifiers and acceptors, then the security operation may fail within the network. The following are examples of potential security context mismatches.

- The selection of security contexts that use asymmetric keys without access to a private key distribution and authentication mechanism.
- The use of symmetric long-term key security contexts in networks with very large numbers[6] of different source-acceptor pairs.
- The use of a security context that uses a key negotiation algorithm across a network without end-to-end path connectivity.
- The parameterization of security contexts with smaller key sizes or other measures of cryptographic strength when bundles might be stored for very long periods of time.
- The use of security contexts that employ cipher suites that may, over the lifetime of the network, be deprecated.
- The use of security contexts that generate ciphertext larger than the plaintext, which may result in undesirable bundle fragmentation when encrypting the payload.

---

6 For example, there are almost 500,000 source-acceptor pairs in a network of 1000 nodes.

14.5   Policy Considerations

The BPSec policy framework is flexible to allow the configuration of both default security procedures and specific security requirements for individual bundles. The ability to handle general security policy statements is provided by the use of wildcards in the security policy statements. The use of wildcards or a range of values for filter criteria fields such as the bundle source or target block type allows a network operator to define a general security processing approach that applies to many bundles. The support of this ability can lead to security policy conflicts.

A security policy conflict can occur if a BPA identifies multiple security policy rules that are applicable to the same bundle requiring the same security operation be performed by the BPA. It is permissible for the BPA to apply multiple security policy rules to the same bundle if different security roles are defined for these rules. For instance, the BPA may be identified as the security source for a BCB and a security acceptor for an existing BIB in the bundle. However, the BPA could not serve as the security source for a BCB targeting the payload block and a security verifier for that same BCB.

Security policy conflicts are often caused by the presence of both a general and more specific security policy rule at the same BPA. A network operator may define a default security policy rule, then add specific rules later providing additional information such as security policy parameters or changing fields with wildcards to instead identify a specific endpoint or block.

There are many methods of deconflicting security policy, with the selection of the method being up to the implementer. These methods may include determining a specificity score for the security policy rules using their filter criteria, applying security policy rules based on age, or use of the most secure rule when a conflict is detected, with confidentiality being more secure than an integrity service.

This section discusses some common security policy considerations, their impacts, and strategies for mitigating these issues.

14.5.1   Key Management

Security contexts define the algorithms used for key wrapping and, where appropriate, key negotiation. The security policy which determines the use of what security contexts in what circumstances – and the parameters for security contexts – are defined as part of the security policy of the BPA. While security policy configurations are not, themselves, responsible for the generation of key material, these polices are responsible for what key material might be provided to security contexts.

There are several areas where policy misconfiguration may lead to improper key handling, to include using independent keys, exhausting keys, and not planning for key expiration.

14.5.1.1   Key Independence

When using either long-term or session keys, the same keys should not be used by multiple cipher suites, or for multiple types of security operations. The reuse of the same key through different cryptographic algorithms may result in the discovery of the original key material and, thus, the loss of security for related security services. Security policies must ensure that security contexts do not re-use the same key for different purposes – especially avoiding using the same key for both confidentiality and integrity mechanisms.

Similarly, key generation functions often accept one or more nonces – such as an Initialization Vector (IV) – used to introduce additional randomness in the generation of session key material. These nonces must, themselves, be unique. Security policy should not hard-code any nonce value but must, also, ensure that generated values are always unique (for some reasonable period).

Security context designers should take care to ensure that they are following all recommendations from the cipher suite reference material which might include restrictions on key use, nonce generation, and parameter selection, among other topics.

14.5.1.2 Key Exhaustion

Security policy must not specify the use of any key that has been exhausted. Different cipher suite algorithms will have different definitions of exhaustion, but exhaustion is usually defined as either a maximum number of bytes processed or number of invocations of operations using that key.

Security policies that attempt to define and use session keys across multiple security operations and security blocks must track how much each key is being used to avoid exhausting that key. Security policies that specify long-term keys that, themselves generate session keys must also determine similar metrics for the use of these keys both at the security source and also predict this same usage at the security verifier and acceptor.

When symmetric keys are used as long-term keys for the generation of symmetric session keys, then a unique pair of long-term keys should be calculated for every supported combination of (source, acceptor) pairs in the network. For small networks this may be a feasible function of network management or other operation control.

When calculating when a key might be exhausted, the use of the key at the security source, verifier, and acceptor must all be considered. The use of a key at a security source and security acceptor should match – a network cannot accept more security operations than were sourced. However, the use of a key at a security verifier BPA may be greater than the use of that symmetric key at the security source or security acceptor because of bundle retransmissions. While a bundle only exists at a source once and at a destination once, it may be sent and re-sent through waypoint nodes more than once.

Consider the case illustrated in Figure 14.4. In this figure, some security operation is created at a security source, causing some key at that source to be used to generate the appropriate security result. Later, that bundle is sent to Security Verifier 1 (SV1) which uses a local key to verify the security result. SV1, in this instance, is also the point of retransmission for the bundle[7] SV1 sends

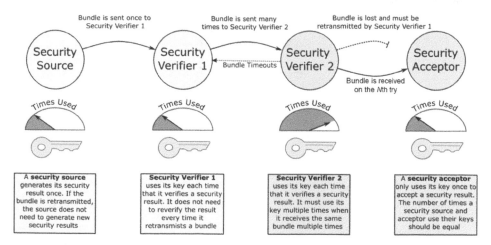

Figure 14.4 Bundle retransmission causes uneven key use for a single security operation. Every time a retransmitted bundle is received, security operations in that bundle must be verified. If Security Verifier 1 retransmits a bundle through Security Verifier 2 multiple times, then keys at Security Verifier 2 will exhaust more quickly.

---

7 This can happen if other nodes – such as Security Verifier 2 – do not have the resources to store bundles.

the bundle to Security Verifier 2 (SV2) which also uses a local key to verify the security result. If SV2 fails to send the bundle along, it will be retransmitted by SV1, which will cause SV2 to re-verify the security result of the retransmitted bundle. As this may happen multiple times, the key at SV2 will be used much more to verify the security result than the key for that security result at any of the other nodes.

Therefore, key exhaustion should be predicted as a function of the overall number of uses of the key at a security verifier given expected numbers of retransmissions in a system.

14.5.1.3 Planning for Key Expiration

Keys that have not yet become exhausted may still be deprecated as a function of how long they have been active in the network. Key expiration is in some ways simpler and in some ways more difficult to plan for than key exhaustion. Expiration is simpler to plan for because the expiration time of a key is known in advance. It is harder to plan for because planning for expiration requires understanding the time at which a bundle will be at a source, a verifier, and an acceptor.

One example of the difficulty of planning for key expiration is shown in Figure 14.5. This figure illustrates three BPAs fulfilling the roles of security source, verifier, and acceptor all using symmetric keys. Each BPA has the same, shared key timing information that Key_1 is valid in the time range $T_0 - T_1$ and that Key_2 is valid in the time range $T_1 - T_2$. This example presumes that the bundle arrives at the security verifier prior to time T_1 and arrives at the security acceptor after time T_1. Selecting the appropriate key to use at each of these nodes is important to ensuring proper security processing.

If a security source uses an about-to-expire long-term key, then that same long-term key may no longer be appropriate when the bundle arrives at a security verifier or security acceptor. There are a few sources of information that a security source can examine to try and determine when this situation might occur.

- Route computations in a BPv7 network might provide estimates for when a bundle is expected to be delivered. However, these are only estimates and may change as a function of long-term topological changes.

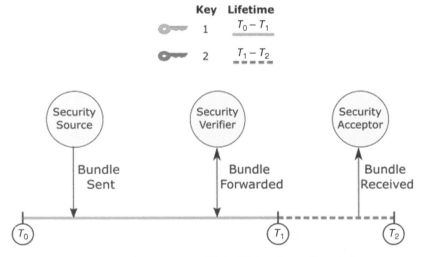

Figure 14.5 Key expiration must be considered for long lived bundles. Assuming all nodes have the same key rotation schedule, were a security source to use Key_1 to create a security result, that key would be considered expired at the security acceptor.

- Special blocks such as the Bundle Age block may assert the lifetime over which a bundle may exist in a network. For some bundles, this overall time to live function may be a large value – such as greater than the lifetime of a given key.

For either long-term or session keys, a security source must select values that can be understood by a security verifier or a security acceptor.

14.5.1.4 Mitigations

Similar to nonce values, key material is not a static element. As such, security policy configurations on a BPA *should not* hard-code a specific key value in a specific circumstance. BPAs are recommended to implement a key manager that provides key generation material to security contexts as a function of the characteristics of the key, the expected time until a security operation reaches a security acceptor, and the amount of data and operations already processed by existing keys.

A key manager function should provide, at a minimum, the following basic services.

- Use different keys for every security context and security service in the system. Keys must not be reused across cipher suites or security services.
- Enforce nonce (such as IV) uniqueness such as the use of a monotonically increasing value with a large overflow value or a pseudo-random number generator with a large period prior to repeating.
- Use asymmetric keys to generate symmetric keys to avoid having to manage combinations of long-term symmetric key exhaustion.
- When long-term symmetric keys are required, generate and track pairs of keys for every combination of sources and acceptors in the network.
- Apply margin to exhaustion estimates for any key to take into consideration the likelihood of bundle retransmission and, thus, multiple key uses at a verifier BPA. For example, a very conservative estimate that every bundle would be retransmitted five times through a single security verifier would cause a key to be rotated when it would otherwise be considered 20% exhausted.
- Use long-term keys that are expected to be active at the time of bundle receipt and not the current key value at the security source. Alternatively, wrap a session key more than once using a variety of long-term keys to help security verifiers and security acceptors unwrap the session key when they receive the bundle in the (perhaps distant) future.
- Defining security contexts that use multiple session keys wrapped by multiple long-term keys to handle cases where a long-term key expires or becomes exhausted while the bundle is in the network.

14.5.2 Cryptographic Binding

A security context may be defined such that it affects both the target block of the security operation and other non-target blocks as well. The common examples of this binding are to include other blocks as additional authenticated data when calculating either integrity or confidentiality services. While the use of such binding can provide some protection against certain types of replay attacks, this binding can also cause problems as these other blocks are modified as the bundle traverses the network.

Definition 14.3 *(Bound Block)* A block in a bundle to which a security operation is cryptographically bound. This means that the content of the bound block is in some way included in, and necessary for the verification and acceptance of, a security operation. One example of a bound block is a block included as additional authenticated data when applying the *bcb-confidentiality* service.

Another example of a bound block is one that is prepended to a target block when applying the *bib-integrity* service.

Since the cryptographic binding capabilities of a security context are a function of the security policy at a BPA, the misuse of binding is a security policy consideration. There are two circumstances related to cryptographic binding that can lead to an inability to process security operations: encrypting bound blocks and forensic analysis when something goes wrong.

14.5.2.1 Bound Block Changes

When a security context binds a security operation to other blocks in a bundle, any changes to those other blocks would cause the security operation to fail to verify or fail to be accepted. For security operations that bind to items such as the primary block this is not an issue, as that block is treated as immutable in the system – changes to the primary block may be seen as an error condition where related security operations can be expected to fail.

Security policy should not attempt to encrypt a bound block. Encrypting a bound block would break the verification and acceptance of all security operations bound to that block until the block itself is decrypted. While it might be possible for a BPA to determine the set of bound blocks by examining every security operation in a bundle, this process would be time consuming. Since security policy would be the mechanism for identifying bound blocks security policy can be the mechanism to avoid encrypting bound blocks.

Figure 14.6 illustrates what can happen when a bound block is encrypted. The bundle at the top of this figure represents one in which an integrity signature is calculated over the payload block *and* the extension block E_2. One way of performing such a binding is by catenating the contents of E_2 and the payload block together into a single plaintext input and calculating a signature over it. If, at a later time, E_2 is encrypted, as shown by the lower bundle, then the integrity signature held by BIB 1 cannot be verified. This is difficult to detect in a bundle, as there are no explicit indications

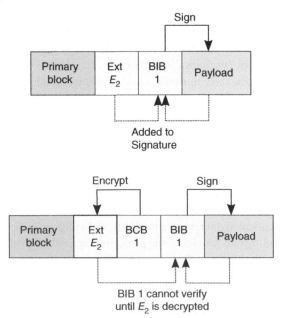

Figure 14.6 Manipulating bound blocks can prevent security processing. If an integrity signature over the payload binds itself to extension block (E_2), then the signature can only be verified if both the payload and E_2 remain unchanged. Encrypting E_2 causes the integrity signature over the payload to no longer be verifiable.

that the BIB whose target block is the payload block would be impeded by the BCB whose target block is E_2.

Security policy must not specify bound blocks that are known to change between a security source and a security acceptor. Blocks such as the Previous Node and Bundle Age block will change at every hop. Making these bound blocks would require that every security operation be accepted at every hop in the network, which is unlikely to be the expected behavior when a bundle must travel multiple hops to its destination.

14.5.2.2 Forensic Analysis

Security policy must determine the actions to take when a security operation fails. Policy actions involve changes to the bundle itself, such as removing operations, removing blocks, or removing the bundle. These actions can also involve reporting and other metrics collection to help with forensic analysis by operators at a later time, such as producing logging messages, quarantining bad data, and generating reports sent to a bundle's Report-To Endpoint Identifier (EID).

Security policy that specifies the use of cryptographic binding can make it difficult to determine why a security operation fails to verify or be accepted by a BPA. In cases where bound blocks are included only as additional authenticated data or otherwise catenated to a target block the cipher suite algorithms that process these data only see the unified set of data input to it, not how it was originally generated.

This means that, upon security operation failure, a BPA will not know if the error was caused by an issue with the security block, the target block, or a bound block. The decision to use bound blocks must consider the extent to which logging and support analysis is important in the event of error.

Some potential ways to mitigate these affects include the following.

- When failing a security operation, include the list of bound blocks as part of logging the associated security operation failure.
- Depending on the criticality of the failure, quarantine the security block, target block, and all bound blocks associated with the failed security operation.
- When using more expansive logging and quarantining policies, make sure to not duplicate information. If multiple security operations in a bundle fail at a BPA, there is no reason to save the same bound block once for each failure.
- When verifying or accepting streams of bundles, consider summarizing and sampling data rather than logging for each operation in each bundle. One example of a threshold would be to log the security, target, and bound blocks of a failed security operation only if a similar failure had not already been logged in the last 30 seconds.

14.5.3 Role Misconfiguration

The use of security policy to determine the role of a BPA for a given security operation is fundamental to the proper application of security in BPSec. Any instance where a BPA is configured to have an incorrect or unintended role in the processing of a security operation is considered a role misconfiguration. When used in this sense, misconfiguration does not only refer to syntax errors, but also to semantic errors that cause the BPA to apply more or less security to a bundle.

Two special considerations relating to role misconfiguration are those that result in an error indicating that a required security block is not present and those that cause a BPA to attempt to duplicate an existing security operation.

14.5.3.1 Missing Security Operations

A missing security operation is any such operation that is required to be in a bundle at a security verifier or a security acceptor. The lack of such an operation can be interpreted as a security error.

Except conditions where a security operation is removed from a bundle as a function of malicious bundle manipulation or block corruption, there are two kinds of policy misconfigurations that can result in a missing operation – failure to set a source and having too many acceptors.

14.5.3.1.1 Missing Security Source Suppose a network operator intends to write a security policy statement requiring a security operation to be added to the bundle at a certain BPA. The operator must associate that policy statement with the Security Source role, as the Security Source is responsible for applying and adding a security operation to the bundle. However, if the security verifier or security acceptor role is instead associated with that policy statement, the BPA will not add the security operation to the bundle and instead will expect that it is already represented in the bundle it receives. When the security verifier or security acceptor BPA then tries to process the non-existent security operation, it will be identified as missing.

14.5.3.1.2 Multiple Security Acceptors If an operator configures multiple BPAs in the network to function as a required security acceptor for an operation, this can lead to a security operation being accepted at one BPA and considered missing at another BPA. This type of error is easy to make in cases where operators attempt to re-use security policy statements across various BPAs in the network. This example of misconfiguration leading to a missing security operation is illustrated in Figure 14.7.

14.5.3.2 Duplicated Security Operations

The implementation of any BPSec-compliant BPA enforces the requirement that security operations are unique within a bundle. Attempts to duplicate an existing security operation will result in an error condition at the Security Source that attempted to add the duplicate operation. A simple way in which this error can be made is through a copy-and-paste error where the security policy rule from a source is copied at a verifier or an acceptor without updating the role from source. Such policy configuration typos can be mitigated through manual or automated checks of policy configurations across a network.

A more complex way in which this same error can manifest is when there is some ambiguity as to the path a bundle might take through a network, such that multiple BPAs can be the legitimate

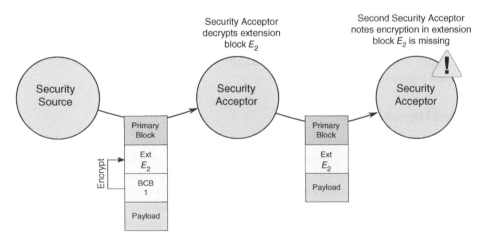

Figure 14.7 Multiple security acceptors. Having multiple security acceptors for a required security service creates problems. The first encountered security acceptor removes the security operation from the bundle. This will cause any subsequent security acceptors to see a required security operation as missing.

source of a security operation. If there are cases where topological change justifies the presence of multiple security sources for the same security operation, then it is possible – also in a changing topology – that a single bundle would pass through two or more of these sources.

14.5.3.3 Mitigations

These types of configuration errors can be avoided by adhering to certain guidelines the network operators can follow when configuring a BPv7 network. Suggestions for guidelines include the following.

- Analyze the security configurations of every security operation at every BPA in the network and ensure that every operation has expected source, verifier, and acceptor nodes specified.
- Define special security BPAs in the network as security processing centers. These may be at proxy, gateway, or other border nodes, at well-resourced nodes in the network, or nodes with special connectivity.
- Make security verification at all non-special security BPAs optional, meaning that operations should only be verified if they are present in the bundle. The lack of a security operation in a bundle should not be considered an error if the verifier rule is optional at a BPA.
- Only specify security acceptors at special security BPAs.
- Only make security acceptor rules mandatory (meaning that the lack of a security operation is treated as an error) at gateway/border nodes and bundle destinations.
- Security policy should differentiate when a BPA must be the source of a security operation and when a BPA can be a source of a security operation. If a BPA can be the source of an operation, but the operation is already present in the bundle, that condition should not be seen as an error condition.
- As a general rule, a BPA should only be the security source for blocks that it adds to the bundle. If BPAs only secure blocks that they add to the bundle, the chance of encountering a duplicate security source is removed.

Guidelines such as these must be analyzed for appropriateness as part of the networking architecture of any BPv7 network. However, a careful consideration of topology, security, and policy can result in configurations that are expressive, secure, and less likely to result in misconfigurations.

14.5.4 Security Context Misuse

A security policy configuration identifies the security context and optional parameters necessary to apply the required security operation to the bundle. If an incorrect security context or parameter is provided in the security policy statement, the security operation will consistently be identified as misconfigured at each BPA it encounters that attempts to process it.

To illustrate, a network operator may intend to create a security policy configuration which adds a *bib-integrity* operation for the primary block to each bundle in the network and processes that operation to verify primary block integrity at the acceptor. If however, the network operator associates the *bib-integrity* operation with a *bcb-confidentiality* security context, the generated cryptographic material will not support the intended integrity service.

If a security source uses the *bcb-confidentiality* context to generate the expected integrity signature for the required *bib-integrity* service, the security operation will be identified as misconfigured by that BPA, as shown in Figure 14.8.

If a security verifier or security acceptor attempts to verify integrity of the target block using a *bcb-confidentiality* security context, the BPA will identify the discrepancy between the generated and provided security results. If the security policy statement at that BPA identifies the wrong

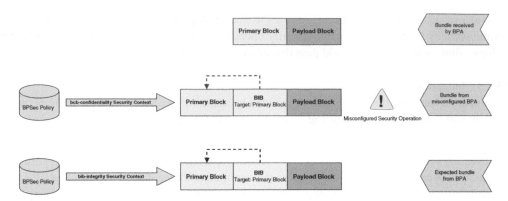

Figure 14.8   Security contexts must match their security usage. A BPA configured to use a *bcb-confidentiality* security context to generate a security result for a BIB will generate a misconfigured security operation. Only a security context providing a *bib-integrity* service can be used to populate BIBs. The same holds when attempting to use a *bcb-confidentiality* security context to populate a BCB.

security context to use for verification, the security operation is identified as misconfigured as the security context associated with the security block will not match.

To resolve this security policy misconfiguration, ensure that the following steps are met:

1. All *bib-integrity* operations are associated with *bib-integrity* security contexts.
2. All *bcb-confidentiality* operations are associated with *bcb-confidentiality* security contexts.
3. The security context parameters provided are appropriate for that context and are consistent across the BPAs in the network.
4. Each BPA in the bundle's path uses the same security context and parameters to process the security operation.

14.5.5   Bundle Matching

Bundle matching refers to the way in which a BPA determines how policy statements apply to bundles. Policy statements often accomplish this matching by applying some type of filtering criteria to identify bundles and blocks within bundles. Some criteria are associated with the block being secured, other criteria may examine the bundle source and bundle destination.

When security policy is based on bundle endpoints, policy must understand how these endpoints are named and how policy rules can be made to match these endpoints. Failure to correctly specify the bundles to which a policy statement might apply can result in unintended matches and mismatches, either of which may lead to unintended security consequences.

Misconfigurations related to bundle matching can occur when confusing bundle EID concepts and when dealing with multiple EID naming schemes.

14.5.5.1   Nodes versus EIDs

BPv7 distinguishes between the *node* on which a BPA resides and the *endpoints* available on that node. Endpoints are identified by an Endpoint Identifier (EID) and represent messaging source and destinations associated with a bundle. BPv7 does not define a separate naming convention for a node, and the Node Id is formatted as an EID. This dual use of the EID syntax for both endpoints and nodes can cause some confusion when specifying security policies.

At any BPv7 node in a network, there is only one service guaranteed to be both running on a node and not running on any other node – the administrative endpoint of that node. Since this endpoint has a unique 1-1 relationship with the node on which it resides, this EID can be unambiguously used as the Node Id. Generally, BPv7 allows any singleton EID associated with the node to be used as a Node Id.

> *The EID of any singleton endpoint is allowed to serve as a "node ID" identifying the node that is the sole member of that endpoint.*
>
>

BPv7, RFC 9171 §4.2.5.2, [5]

This distinction is important because, within the bundle primary block, the bundle source must conform to the definition of a Node Id whereas the bundle destination may reference any EID. For that reason, the criteria for matching based on a bundle source should match as a function of the node that sources the bundle, independent of any additional non-node identifying information in the bundle source EID.

14.5.5.2 Multiple Naming Schemes

A BPv7 EID is defined as a Uniform Resource Identifier (URI) which, itself, may use different naming schemes as a function of the naming and addressing practices of the network in which the bundle is deployed. RFC 9171 [5] defines the EID structure as shown in Expression (14.1).

$$< scheme - name >:< scheme - specific - part > \tag{14.1}$$

RFC 9171 also identifies two schemes that can be used to identify bundles (Interplanetary Network [IPN] and DTN) and it is possible that additional schemes are defined over time. To the extent that security policy uses EIDs to match bundles to policy, a policy engine must understand the naming schemes in use by the network. This is particularly important when a node or endpoint might have different EIDs, such as one for each naming scheme.

If a security policy only uses one naming scheme to perform bundle matching, then bundles applying an alternative naming scheme would not match the policy. Consider the case where a node has two identities (an IPN scheme identity and a DTN scheme identity) and sometimes produces bundles with IPN scheme EIDs and sometimes produces bundles with DTN scheme EIDs. Were security policy to only match based on a single naming scheme, such as the IPN naming scheme, then bundles produced using the DTN naming scheme would never match. This situation is illustrated in Figure 14.9.

The use of a particular naming scheme *might* be an important consideration in the application of security, if naming scheme selection carries some semantic meaning within a network. Consider a case where a policy engine only wants security applied to bundles using an IPN naming scheme rather than a DTN naming scheme. However, when a BPA wants to process security regardless of the naming scheme using in that block's bundle, it must be able to match the bundle regardless of the naming scheme used by the bundle source.

There are a few ways in which security policy statements can handle multiple naming schemes, such as the following.

- In cases where security is processed solely by block type, without concern for bundle source and destination, EIDs can be ignored in policy statements.
- Policy may determine that the security source EID present in a security block always use a specific naming scheme, and policy can be specified based on the security source EID, and not EIDs present in the bundle's primary block.
- Security policy may specify matching criteria for all supporting naming schemes.

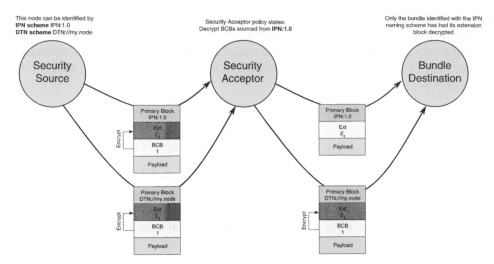

Figure 14.9 Security policies using EIDs. Policies that use EIDs must be aware that there are multiple EID naming schemes. Defining a policy using only one naming scheme might match some EIDs and not others, even when all EIDs refer to the same node or endpoint.

- Security policy may specify matching criteria for a single, preferred naming scheme, and separately attempt to transcode all incoming bundle EIDs to that preferred naming scheme.
- Treat different naming schemes as different security criteria, and define different policy rules as a function of the naming scheme used.

Security Policy Consideration

When a node can be identified by multiple EIDs or multiple naming schemes, care should be taken when constructing policy conditioned based on EID or naming scheme. If security policy is not defined for all possible EIDs and naming schemes, an attacker may be able to craft malicious bundles which take advantage of those undefined cases to bypass security mechanisms.

14.5.6 Rule Specificity

Security policy statements grow with the number of nodes and block types secured. This exponential rise in combinations of bundle sources, bundle destinations, security sources, and block types means that attempting to build individual policy statements for every combination of security operation is not feasible. Maintaining such a large catalog of security operations would be computationally intensive and likely require significant storage resources on what could be very resource-constrained devices. Furthermore, as the number of rules increase, there is increased opportunity for policy misconfigurations, omissions, or conflicts which may create vulnerability in the system.

A common approach to addressing this situation is to incorporate the use of pattern matching and *wildcards* when specifying policy matching criteria. A single set of patterns can, otherwise, replace hundreds or thousands of individual policy statements. This allows the number of policy statements to be driven by the number of unique security actions rather than the number of unique security target blocks. For example, if a BPA should accept every BIB and BCB in every bundle it receives, then those two actions can be codified using two security policy matching criteria, even if the BPA ultimately applies those two criteria to thousands of different combinations of bundles and blocks.

Pattern matching is an important tool in creating policy statements, but can be a source of misconfiguration when attempting to set up default behavior that is *overridden* in specific circumstances. Consider the following four policy statements.

1. Encrypt unencrypted payloads using security context 1.
2. Encrypt unencrypted payloads going to *EID* 2 using security context 2.
3. Never encrypt a payload block going to *EID* 3.
4. Encrypt unencrypted payloads from *EID* 4 using security context 3.

Policy statements 1–3 can co-exist and represent a coherent security policy if the evaluation of security policies can be made unambiguously hierarchical. Of the first three, statement 1 is most generic, and statements 2 and 3 are more specific and not overlapping. Therefore, a BPA can always determine the correct course of action by evaluating statements from most-specific to least-specific.

Policy statements 2 and 4, however, represent a misconfiguration. Neither of these statements are more or less specific than the other, and they are overlapping. A bundle sourced at *EID* 4 and destined for *EID* 2 would match both of these statements. Since these statements require different, mutually exclusive responses and one is not necessarily more specific than another, different BPAs might respond differently causing unexpected security behavior. An example of this type of collision is illustrated in Figure 14.10.

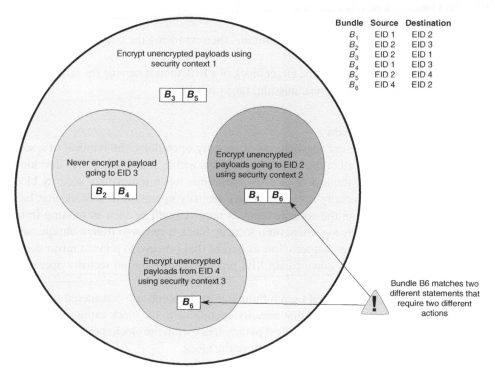

Figure 14.10 Rule Collisions. Nesting rule statements from general to specific does not cause problems unless multiple statements with the same level of specificity require two mutually exclusive actions. Given a set of policy statements and six bundles (B_1–B_6), bundle B_6 matches two different statements that have the same specificity.

Policy statements must be evaluated to ensure that they do not introduce ambiguous behaviors. In cases where ambiguity might exist, some deconfliction rules must be established, such as introducing a priority or other ordering for statements.

14.5.7 Cascading Events

Policy statements must not only determine what security processing should happen at a BPA, but also the actions to take in the event that some security processing fails at the BPA. For example, policy may specify certain actions if a required security operation is not present in a bundle or if a security operation cannot be verified or accepted due to a change in bundle contents. These actions can range from the relatively simple action of generating reports to more complex actions such as removing blocks from the bundle or the bundle from the network.

The actions of one policy statement might impact the evaluation and actions of another policy statement at a BPA – especially when the policy action is to remove information from a bundle. Three types of actions that are most likely to cause cascading events involve removing a target block and removing a security block.

14.5.7.1 Removing Target Blocks

Security policy might require that a target block be removed from the bundle in the event that the block cannot be decrypted or have its integrity verified. Removing target blocks can have impacts on other security processing – even at the same BPA – for a few reasons, as listed next.

- If the target block of a BCB is a BIB, then removing the BIB removes an integrity service over the BIB's target block.
- If the target block is a critical block in the bundle, then removing the target block might cause the BPA to drop the bundle.
- If the target block of a BCB is also the target block of a BIB, then removing the target block will cause the BIB to fail to verify its (now missing) target block.

14.5.7.2 Removing Security Blocks

Because security blocks represent aggregations of security operations, the removal of a security block implies the removal of all of the security operations within that block. Policy development must consider the impact of the loss of security operations when removing a security block in response to a single failed security operation. Every security operation in a block that has not yet been processed at the time the security block is removed will be seen as missing from the local BPA. Policy should clearly log whenever a security block is removed from a bundle, and the unprocessed security operations that were lost as part of that process, to prevent misunderstandings when the local BPA or any downstream BPA processes any required security operations as missing.

A security block might be removed from a bundle if the contents are considered to have been corrupted or otherwise modified such that security operations in the block cannot be considered for evaluation. Similarly, if the security context parameters within the block specify outdated information, then a BPA may similarly choose to remove the block.

BCB blocks should never be removed from a bundle without also removing the target block of each operation within the block. Removing just a BCB block removes from the bundle evidence that other blocks in the bundle have been encrypted.

14.6 Summary

The implementation of a comprehensive security ecosystem for BPv7 bundles requires careful consideration of special environments and use cases. The unique capabilities that make bundles useful in certain environments also represent potential opportunities for human error through misconfiguration. In particular, the application of security must consider the resource impact of security at the BPA, the way in which bundles may be fragmented, and how security policy can be specified across a BPv7 network.

The specification of security policy, in particular, includes broad topics such as key management, cryptographic binding, role misconfiguration, and matching policy statements to bundles. Each of these topics represents their own unique set of concerns and opportunities for misconfiguration.

While the application of any security policy is specific to the nature of the network being secured, there are some common techniques to mitigate negative outcomes. When identified, common techniques have been presented in this chapter to help inform the construction of deterministic, coherent policies for specific network deployments.

References

1 Rescorla, E. and Korver, B. [2003]. Guidelines for Writing RFC Text on Security Considerations, RFC 3552. URL: https://www.rfc-editor.org/info/rfc3552.

2 Pöttner, W.-B., Morgenroth, J., Schildt, S. and Wolf, L. [2011]. Performance comparison of DTN bundle protocol implementations.

3 Johnson Williams, G. M., Walker, B. and Hennessy, A. [2012]. A performance comparison of DTN bundle protocol implementations on resource constrained nodes, *Proceedings of the 7th ACM International Workshop on Challenged Networks*, CHANTS '12, Association for Computing Machinery, New York, NY, USA, pp. 75–78. URL: https://doi.org/10.1145/2348616.2348634.

4 Pöttner, W.-B., Morgenroth, J., Schildt, S. and Wolf, L. [2011]. Performance comparison of DTN bundle protocol implementations, *Proceedings of the 6th ACM Workshop on Challenged Networks*, pp. 61–64.

5 Burleigh, S., Fall, K. and Birrane, E. J. [2022]. Bundle Protocol Version 7, RFC 9171. URL: https://www.rfc-editor.org/info/rfc9171.

6 Birrane, E. J. and McKeever, K. [2022]. Bundle Protocol Security (BPSec), RFC 9172.URL: https://www.rfc-editor.org/info/rfc9172.

Appendix A

Example Security Contexts

Bundle Protocol Security (BPSec) defines the mechanisms necessary to implement security in Bundle Protocol version 7 (BPv7) networks. Some of those mechanisms, such as extension block definitions for the Block Integrity Block (BIB) and Block Confidentiality Block (BCB), are common across all BPv7 networks. Other mechanisms, such as security policies and security contexts are expected to be customized for particular implementations.

Two exemplar security contexts are documented in the "Default Security Contexts for BPSec" specification, Request for Comments (RFC) 9173 [1]. These security contexts provide a minimal security capability appropriate for use over the Internet. One security context is given for the *bib-integrity* service and another is given for the *bcb-confidentiality* service.

The security contexts of RFC 9173 serve two purposes. First, they provide an example for BPv7 network security designers to expand upon for particular use cases, requirements, or other constraints. Second, they provide a basic level of protection that while optional to use, is mandatory to implement in a BPA.

> To ensure interoperability among various implementations, all BPSec implementations MUST support at least the current, mandatory security context(s) defined in IETF Standards Track RFCs. As of this writing, that BP mandatory security context is specified in [RFC9173], but the mandatory security context(s) might change over time in accordance with usual IETF processes.
>
> BPSec, RFC 9172 §9.1, [2]

While additional security contexts may be defined and preferred for communication within a given BPv7 network, the mandatory support for the default security contexts ensures interoperability across implementations.

This appendix provides a summary and discussion of both the integrity and confidentiality security contexts defined in RFC 9173. Basics of these security contexts are covered here, including cipher suite selection, inputs, parameters, outputs, and input canonicalization.

How to use Appendix

This appendix is intended to be a companion to the full text of RFC 9173, bridging the gap between the concepts discussed in this book and a full security context definition, which also includes the necessary implementation and processing requirements. The topics here are presented at a level to convey key aspects of the design of these security contexts. The full text of RFC 9173 contains sufficient information to the implementer, including processing definitions, ties to local security policies, as well as test vectors and usage examples.

Securing Delay-Tolerant Networks with BPSec, First Edition. Edward J. Birrane III, Sarah Heiner, and Ken McKeever.
© 2023 John Wiley & Sons, Inc. Published 2023 by John Wiley & Sons, Inc.
Companion website: www.wiley.com/go/birrane/securingdelay-tolerantnetworks

A.1   Integrity Security Context

The default integrity security context, BIB-HMAC-SHA2, is a single target, single result security context that provides the *bib-integrity* service using a keyed authentication code.

As implied by the name, the BIB-HMAC-SHA2 context uses the Secure Hash Algorithm 2 (SHA-2) [3] combined with the Hashed Message Authentication Code (HMAC) keyed hash [4]. These algorithms were selected based on their wide adoption in existing systems and the diversity of available implementations.

The selection of HMAC means the use of symmetric keys. This choice was made in order to create a security context that can be used in all networking environments. Rather than requiring a complicated asymmetric key infrastructure, implementations may choose to communicate symmetric keys in whatever manner they see fit. This can include the use of asymmetric key algorithms to synchronize symmetric keys, the preplacement of symmetric keys, or other key generation methods.

As the default security contexts are defined to be used in the broadest set of use cases, the BIB-HMAC-SHA2 security context only requires the support for symmetric keying.

A.1.1   Security Context Scope

As an integrity service, BIB-HMAC-SHA2 constructs a set of plaintext over which a keyed hash can be created. The set of plaintext information input into this context is termed the *Integrity-Protected Plaintext (IPPT)*. The process of identifying and assembling the IPPT is the process of determining what information in a block and/or in the bundle will be included in the resultant digest.

The minimal set of information comprising the IPPT is the block-type-specific data field of the security operation's target block. However, the BIB-HMAC-SHA2 context provides additional options for extra data that can be included in the construction of the IPPT, thus allowing more information to be cryptographically bound to the successful verification of that security operation.

> **Integrity-Protected Plaintext**
>
> The scope of the *bib-integrity* security service. The plaintext over which the keyed hash is calculated.

The IPPT field can include up to five sources of data, some of which are mandatory and some of which are optionally included as a function of security policy. The overall IPPT field is constructed as the catenation of the following fields, in the order that they are presented next. When an optional field is not present, it is not represented in the catenation.

1) **Integrity Scope Flags**.
2) **Primary Block**. *Optional.*
3) **Target Block Headers**. *Optional.*
4) **BIB Block Headers**. *Optional.*
5) **Target Block-Type-Specific Data Field**.

The following sections describe each of these fields in more detail, to include when and why optional fields might be included in a keyed digest.

A.1.1.1   Integrity Scope Flags

The Integrity Scope Flags field identifies what optional information must be included when constructing the IPPT. It is expected that the determination of IPPT contents will vary as a function of evolving security policy over the lifetime of the network. A BPA may determine these scope flags as a function of local policy, default values specified in RFC 9173, or as encoded in the security block as a security parameter.

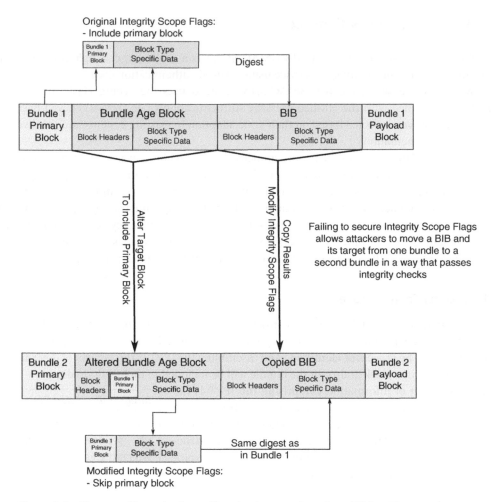

Figure A.1 Unsecured Integrity Scope Flags lead to security vulnerabilities. If an attacker changes scope flags, they can manipulate data in such a way as to cause replicated or fabricated data to be verified without needing to alter signed digest. Protecting the Integrity Scope Flags value prevents this attack.

Verifying the proper selection of these flags at a BPSec verifier or acceptor is important, as flag manipulation could otherwise be a threat vector into potentially passing manipulated data. Consider the case illustrated in Figure A.1, where an attacker replays a BIB and an altered Bundle Age Block from Bundle 1 into Bundle 2. If the integrity Scope Flags in the BIB were to be unsecured, then this type of attack would not be detectable by a BPSec verifier or acceptor of the copied BIB in Bundle 2.

To avoid integrity scope attacks, the Integrity Scope Flags value must always be included in the signed digest. Doing so would cause any alteration of the scope flags to invalidate the resultant keyed digest and more effectively prevent replay attacks.

A.1.1.2 Primary Block

The BPv7 primary block contains information necessary to identify a bundle. Once created, this block is immutable for the lifetime of the bundle. Because this block both uniquely identifies a bundle and cannot change over the life of the bundle, it represents the perfect information needed to cryptographically bind information to a bundle.

When the primary block contents are included in the IPPT, then the integrity operation can only be verified if they are contained in the same bundle as when the keyed digest was first calculated. This cryptographic binding protects the integrity of the association of the integrity security operation and its target block with the bundle.

Bundle Association

If a BIB and its target block are not cryptographically bound to a bundle by including the primary block in the IPPT, both the BIB and its target block can be copied to a different bundle and integrity can still be successfully verified. This cryptographic binding associates the security operation with the bundle it is added to at the security source, and allows for detection of modification to the bundle contents.

A.1.1.3 Target Block Headers

Some other fields of the target block (other than its block-type-specific data) may be included in the IPPT in order to protect the integrity of the identity of the target block's identity and policy information.

The other fields of the security target block are used by BPAs in the bundle's path to determine how its block-type-specific data should be handled. If this information is modified, the way in which BPAs process the otherwise unchanged block-type-specific data of that target block may be different than intended.

Modification of Target Block Policy

The target block processing control flags should be included in the integrity protection of the target block, as these flags affect the overall processing of the block and its bundle. For example, an attacker could change block behavior to cause a BPA to drop the bundle if a block cannot be processed, which would cause data loss in BPv7 networks where not every BPA can process every known block type. Failing to include the target block processing control flags would make detecting this condition impossible.

A.1.1.4 Security Block Headers

Some other fields of the BIB include information used by BPAs to determine how to identify and process the information it carries. Modification of this data can result in downstream BPAs handling the BIB in a different way than was configured at the security source. Therefore, security operations carried in a BIB should also include identifying and processing information associated with the BIB in their own signed digests.

A.1.1.5 Target Block-Type-Specific Data

The block-type-specific data field of the target block, along with the Integrity Scope Flags, represents the minimum input to construct the IPPT. This data is included to provide integrity for the target block's contents.

A.1.2 Security Context Parameters

BIB-HMAC-SHA2 has three optional parameters that instruct the receiving BPA how to process the BIB and protected blocks, as listed in Table A.1. These parameters define how the HMAC-SHA2 algorithm should be configured, passes wrapped key material if it is provided, and identifies the scope of the integrity service.

A.1.2.1 SHA Variant

The SHA Variant is an optional parameter used to select the variant of the SHA-2 algorithm that is used to generate the authentication code. The options for this parameter are as defined in Table A.2. The HMAC algorithm values provided in the table are as defined in Table 7 of [5].

A default value of 6, corresponding to HMAC 384/384, should be used when the SHA Variant parameter value is not provided, unless a different default value is established by local security policy.

A.1.2.2 Wrapped Key

The optional wrapped key parameter is used to hold the output of the AES key wrap function, defined in [6].

The value of this parameter is the ciphertext that is generated when the symmetric HMAC key used to generate the associated BIB's security results is provided as input to the AES key wrap algorithm. Using this value, security verifier and security acceptor BPAs can calculate the symmetric HMAC key necessary for the validation and/or processing of the security results for the BIB.

If a value is not provided for this optional parameter, security verifier and security acceptor BPAs instead determine the symmetric HMAC key to use based off of local security policy.

A.1.2.3 Integrity Scope Flags

The optional integrity scope flags parameter identifies the data that must be included in addition to the security target block's block-type-specific data when the IPPT is generated.

Table A.1 Parameters for BIB-HMAC-SHA2.

Parm Id	Parm name	CBOR encoding type	Default value
1	SHA variant	Unsigned integer	6
2	Wrapped key	Byte string	
3	Integrity scope flags	Unsigned integer	7

Source: Birrane et al. [1]/IETF Trust.

Table A.2 SHA variant parameter values.

Value	SHA-2 variant
5	HMAC 256/256
6	HMAC 384/384
7	HMAC 512/512

Source: Data from Birrane et al. [1].

Table A.3 Integrity scope flag values.

Bit(s)	Bit position(s)	Flag meaning
0	0x0001	Include primary block flag
1	0x0002	Include target header flag
2	0x0004	Include security header flag
3-7	0x00F8	Reserved
8-15	0xFF00	Unassigned

Source: Adapted from Birrane et al. [1].

Table A.4 BIB-HMAC-SHA2 security results, from RFC 9173.

Result Id	Result name	CBOR encoding type
1	Expected HMAC	Byte string

Source: Data from Birrane et al. [1].

A default value of 7, corresponding to all assigned fields, should be used when the integrity scope flags parameter value is not provided, unless a different default value is established by local security policy. Table A.3 defines the integrity scope flags used within this parameter.

A.1.3 Security Results

The BIB-HMAC-SHA2 security context generates a single security result, shown in Table A.4, for the single target block it accepts. This security result is the *expected HMAC*, which is the output from the HMAC calculation performed at the security source BPA.

A.1.4 Input Canonicalization

To prepare the IPPT to generate the keyed hash, the following data fields are concatenated, each in their canonical forms, in the order listed here as described in RFC 9173 [1]:

1) **Integrity Scope Flags**: The canonical form of the IPPT starts as the CBOR encoding of the integrity scope flags in which all unset flags, reserved bits, and unassigned bits have been set to 0. For example, if the primary block flag, target header flag, and security header flag are each set, then the initial value of the canonical form of the IPPT will be 0x07.
2) **Bundle Primary Block**: If the primary block flag of the integrity scope flags is set to 1 and the security target is not the bundle's primary block, then a canonical form of the bundle's primary block MUST be calculated and the result appended to the IPPT.
3) **Target Header**: If the target header flag of the integrity scope flags is set to 1 and the security target is not the bundle's primary block, then the canonical form of the block type code, block number, and block processing control flags associated with the security target MUST be calculated and, in that order, appended to the IPPT.
4) **Security Header**: If the security header flag of the integrity scope flags is set to 1, then the canonical form of the block type code, block number, and block processing control flags associated with the BIB MUST be calculated and, in that order, appended to the IPPT.

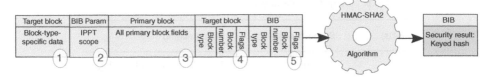

Figure A.2 Input canonicalization of IPPT. This shows the ordering of the canonical forms of input data to generate the keyed hash.

5) **Security Target**: The canonical form of the security target MUST be calculated and appended to the IPPT. If the security target is the primary block, this is the canonical form of the primary block. Otherwise, this is the canonical form of the block-type-specific data of the security target.

Figure A.2 shows the ordering of these fields as they are presented to the SHA2-HMAC algorithm. This data is processed, and the output is the keyed hash, which must be included in the BIB as a security result.

A.2 Confidentiality Security Context

The default confidentiality security context, BCB-AES-GCM, is a single target, multiple result security context that provides *bcb-confidentiality* using an Authenticated Encryption with Associated Data (AEAD) cipher suite.

As implied by the name, BCB-AES-GCM uses the Advanced Encryption Standard (AES) using the Galois/Counter Mode (GCM) [7] to provide authenticated encryption functions to protect the confidentiality of data contained in target blocks. This is a symmetric key algorithm which, within this security context, allows for key sizes of 128 or 256 bits.

A symmetric key cipher suite was selected for this security context for reasons similar to the selection of a symmetric key system with the BIB-HMAC-SHA2 security context. Namely, that the use of symmetric keys allows for multiple key management approaches, thus allowing this context to have the widest adaptability in potential BPv7 network deployments.

While this security context is useable as defined, it can also be extended through other policy or implementation-specific mechanisms. For example, to avoid overuse of symmetric keys, key rotation or other key distribution algorithms can be implemented outside the scope of the security context. This subsection provides an overview of some of the key aspects of BCB-AES-GCM, but is not intended to be a substitute for the full specification found in RFC 9173. There are many requirements and considerations outlined there as well as several in BPSec from RFC 9172 that are important to follow when implementing this security context.

A.2.1 Cipher Suite Selection

The GCM operation of AES provides *authenticated* encryption, which is a requirement of BPSec in RFC 9172. Through this process, the block-type-specific data field of a target block is replaced with the ciphertext generated using the plaintext data and encryption key. Additionally, an authentication tag is generated using the plaintext block-type-specific data and optionally other control fields within the bundle. This ensures that at the destination, the receiving BPA verifies the integrity of the data and the handling of the bundle in transit through the BPv7 network.

A.2.2   Security Context Scope

As an *authenticated* encryption service, BCB-AES-GCM defines two separate scopes in determining the data to which is applies. The first is the scope of the confidentiality service, and the second is the scope of the authentication service.

A.2.2.1   Confidentiality Scope

BCB-AES-GCM encrypts the block-type-specific data field of the target block. The encryption operation generates ciphertext from the plaintext located in that field and replaces the plaintext contents with the ciphertext.

Security Target Limitations

The BCB-AES-GCM security context adheres to all BPSec limitations on allowable target blocks, to include not allowing a *bcb-confidentiality* operation over a primary block, another BCB, or a BIB that does not share the same target blocks as the BCB.

A.2.2.2   Authentication Scope

The authentication scope of the BCB-AES-GCM security context largely mirrors that of the BIB-HMAC-SHA2 scope as described in Section A.1.1. This scope identifies the set of information used to generate the authentication tag in the security results field of the BCB relating to the security operation over the given target block. This includes two mandatory elements and up to three optional elements.

A.2.2.2.1   AAD Scope Flags   This optional element, described in Section A.2.3, identifies the optional data that may be added as additional authenticated data. By requiring this parameter to be in the scope of the authentication service, it can be assured that the receiving node is able to correctly process and verify the authentication tag. Similar to the Integrity Scope Flags defined for the BIB-HMAC-SHA2 service, failure to protect the AAD Scope Flags could lead to security vulnerabilities.

A.2.2.2.2   Primary Block   This optional element refers to the entire primary block of the bundle containing the BCB. Bundle primary blocks both uniquely identify a bundle and are immutable once constructed. Therefore, including the primary block in the authentication scope connects a BCB to the bundle in which it is added. This prevents attacks where a BCB (and associated target blocks) might be incorrectly copied into other bundles.

A.2.2.2.3   Target Block Headers   This optional element includes fields within the target block relating to block identity (type and number) and processing control flags.

A.2.2.2.4   BCB Block headers   Similar to the target block headers, this optional element includes fields relating to the block identity (type and number) and processing control flags of the BCB security block holding this security operation.

A.2.2.2.5   Target Block-Type-Specific Data Field   This is a mandatory element of the scope and is the *plaintext* content of the block-type-specific data field, which is the scope of the confidentiality service.

Table A.5   Parameters for BCB-AES-GCM.

Parm Id	Parm name	CBOR encoding type	Default value
1	Initialization vector	Byte string	
2	AES variant	Unsigned integer	3
3	Wrapped key	Byte string	
4	AAD scope flags	Unsigned integer	7

Source: Birrane et al. [1]/IETF Trust.

A.2.3   Security Context Parameters

BCB-AES-GCM defines four optional parameters that instruct the receiving BPA how to process *bcb-confidentiality* security operations. These parameters define how the AES algorithm should be configured, passes wrapped key material if that is used, and identifies the scope of the authentication service. Table A.5 lists the parameters for this security context.

A.2.3.1   Initialization Vector (IV)

This is an optional parameter that is used to initialize the AES-GCM cipher. It must be between 8 and 16 bytes and SHOULD be 12 bytes unless directed otherwise by local security policy.

IV Usage Requirements

The Initialization Vector can have any value, but an IV MUST NOT be reused for multiple encryption operations using the same encryption key.

A.2.3.2   AES Variant

This is an optional parameter that is used to indicate which variant of the AES cipher is used. The 256-bit AES-GCM variant is the default variant, but the 128-bit variant is also allowed, appropriate for implementations with lesser security requirements. The two allowable parameter values are shown in Table A.6. The variant selected here does not change the length of the generated authentication tag – that MUST always be 128 bits.

A.2.3.3   Wrapped Key

This is an optional parameter that contains the output of the AES key wrap function as defined in RFC 3394 [6]. This parameter value is an input to the AES key wrap authenticated decryption function for security acceptors and security verifiers to be able to determine the symmetric key to decrypt the ciphertext in the block-type-specific data fields of the security targets. If not present, then a processing node must determine the proper key based on its local BPSec policy and configuration.

Table A.6   Parameter values for the optional AES variant parameter.

Value	Description	Default
1	128-bit AES-GCM from RFC 8152	No
3	256-bit AES-GCM from RFC 8152	Yes

Source: Adapted from Birrane et al. [1].

Table A.7 Bit field definitions of the AAD scope flags parameter.

Bit(s)	Bit position(s)	Flag meaning
0	0x0001	Include primary block flag
1	0x0002	Include target header flag
2	0x0004	Include security header flag
3–7	0x00F8	Reserved
8–15	0xFF00	Unassigned

Source: Adapted from Birrane et al. [1].

Table A.8 BCB-AES-GCM security results.

Result Id	Result name	CBOR encoding type
1	Authentication tag	Byte string

Source: Birrane et al. [1]/IETF Trust.

A.2.3.4 AAD Scope Flags

This is an optional parameter that defines any additional data to be included along with the block-type-specific data and the value of this parameter to be used when generating the authentication tag. When not included, implementations SHOULD assume a value of 7, meaning that all additional data should be included. Table A.7 defines the AAD scope flags used within this parameter.

A.2.4 Security Results

The only security result of this security context is the authentication tag, which is 128 bits. Table A.8 lists the result along with its result identifier and CBOR encoding type.

A.2.5 Input Canonicalization

To generate the ciphertext using this security context, the input is simply the plaintext block-type-specific data field of the security target excluding the Bundle Protocol CBOR encoding of that field. The resulting ciphertext is encoded as a definite-length CBOR byte string.

In order to prepare the input AAD to generate the authentication tag, the following data fields are concatenated, each in their canonical forms, in the order listed here as described in RFC 9173 [1]:

1) **AAD Scope Flags**: The canonical form of the AAD starts as the CBOR encoding of the AAD scope flags in which all unset flags, reserved bits, and unassigned bits have been set to 0.
2) **Bundle Primary Block**: If the primary block flag of the AAD scope flags is set to 1, then a canonical form of the bundle's primary block MUST be calculated and the result appended to the AAD.
3) **Target Header**: If the target header flag of the AAD scope flags is set to 1, then the canonical form of the block type code, block number, and block processing control flags associated with the security target MUST be calculated and, in that order, appended to the AAD.
4) **Security Header**: If the security header flag of the AAD scope flags is set to 1, then the canonical form of the block type code, block number, and block processing control flags associated with the BIB MUST be calculated and, in that order, appended to the AAD.

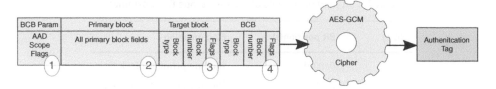

Figure A.3   Input Canonicalization of AAD. This shows the ordering of the canonical forms of input data to generate the authentication tag.

Figure A.3 shows the ordering of these fields as they are presented to the AES-GCM cipher. This data is processed, and the output is the authentication tag, which can either be included in the BCB in the security result flag or included with the ciphertext in the block-type-specific data field of the security target.

References

1 Birrane, E. J., White, A. and Heiner, S. [2022]. Default Security Contexts for Bundle Protocol Security (BPSec), RFC 9173. **URL**: https://www.rfc-editor.org/info/rfc9173.

2 Birrane, E. J. and McKeever, K. [2022]. Bundle Protocol Security (BPSec), RFC 9172. **URL**: https://www.rfc-editor.org/info/rfc9172.

3 National Institute of Standards and Technology [2015]. Secure Hash Standard (SHS), FIPS PUB 180-4. **URL**: https://csrc.nist.gov/publications/detail/fips/180/4/final.

4 Krawczyk, D. H., Bellare, M. and Canetti, R. [1997]. HMAC: Keyed-Hashing for Message Authentication, RFC 2104. **URL**: https://www.rfc-editor.org/info/rfc2104.

5 Schaad, J. [2017]. CBOR Object Signing and Encryption (COSE), RFC 8152. **URL**: https://www.rfc-editor.org/info/rfc8152.

6 Housley, R. and Schaad, J. [2002]. Advanced Encryption Standard (AES) Key Wrap Algorithm, RFC 3394. **URL**: https://www.rfc-editor.org/info/rfc3394.

7 Dworkin, M. J. [2007]. SP 800-38D. Recommendation for block cipher modes of operation: Galois/counter mode (GCM) and GMAC.

Appendix B

Security Block Processing

B.1 Overview

This appendix details some of the processing steps taken by Bundle Protocol Agents (BPAs) as they generate and receive security blocks. Understanding these steps is beneficial for BPA software developers, security context designers, and those building configurations and policies as part of operating Bundle Protocol version 7 (BPv7) networks.

This appendix is segmented into three subsections, defined as follows.

1. Processing single-target, single-result security contexts.
2. Processing single-target, multiple-result security contexts.
3. Processing multiple security sources (with multiple security contexts).

These subsections each provide reference scenarios relating to the use of the *bcb-confidentiality* security service, *bib-integrity* security service, or both. They also enumerate processing steps associated with actions taken at security sources, security verifiers, and security acceptors as required by Bundle Protocol Security (BPSec). Finally, common error conditions that may arise during those steps are listed and these error conditions are mapped to recommended responses.

B.2 Single-Target Single-Result Security Contexts

Single-Target, Single-Result (STSR) security contexts represent a 1-1 mapping of target blocks and cipher suite invocations. This does not mean that such a security context only produces a single cipher suite output value, it simply means that the security service is applied exactly once per target block. A single call to a cipher suite might generate more than one set of output materials (such as both ciphertext and an integrity check value).

Security contexts with this property can be used for both the *bcb-confidentiality* and *bib-integrity* security services.

B.2.1 BCB-Confidentiality

An STSR context providing the *bcb-confidentiality* service based on an Authenticated Encryption with Associated Data (AEAD) cipher suite is a good example of a security context that both:

- Maintains a 1-1 mapping between a target block and a cipher suite invocation.
- Generates two values from that single cipher suite invocation.

Namely, a single invocation of a cipher suite generates both ciphertext and a signature over that ciphertext (and any additional associated data). These two values are still considered a *single result* because they are the resultant outputs of a single invocation of the cipher suite.

Securing Delay-Tolerant Networks with BPSec, First Edition. Edward J. Birrane III, Sarah Heiner, and Ken McKeever.
© 2023 John Wiley & Sons, Inc. Published 2023 by John Wiley & Sons, Inc.
Companion website: www.wiley.com/go/birrane/securingdelay-tolerantnetworks

B.2.1.1 Scenario

A BPA configured as a security source may be required by its security policy to apply the *bcb-confidentiality* service to a target block. This security policy presumes that downstream BPAs in the bundle's path possess the necessary security context implementation, configuration, and policy to authenticate and/or decrypt that target block's data.

B.2.1.2 Processing Steps

Processing associated with an STSR security context applying *bcb-confidentiality* differs based on the security role of the processing BPA. The overall processing of this scenario is illustrated in Figure B.1. The enumerated elements in this figure are described next in greater detail.

1. Actions at the security source.
   1.1 A BPA is assembling a bundle to be transmitted. The bundle contains a block that is required by local security policy to be confidential until it reaches the bundle's destination. This makes the BPA the security source for a *bcb-confidentiality* operation on that block.
   1.2 To protect the block from unauthorized reads, the security source generates a Block Confidentiality Block (BCB) with the security operation $OP(bcb\text{-}confidentiality, Target_{Block})$. This results in the encryption of the block-type-specific data for the target block, which has its plaintext replaced by ciphertext. The BCB is added to the bundle before the bundle is forwarded.
   1.3 Bundle transmission continues from the security source.
2. Actions at the security acceptor.
   2.1 The bundle arrives at its destination which local security policy identifies as the security acceptor that must process the BCB present in the bundle.
   2.2 Each security operation in the BCB is processed, to include the operation $OP(bcb\text{-}confidentiality, Target_{Block})$. The target block's block-type-specific data, which is received as ciphertext, is replaced with the recovered plaintext resulting from this processing.
   2.3 When the security operation has been processed, it is removed from the BCB. Since this was the only operation present in the BCB it is removed from the bundle.

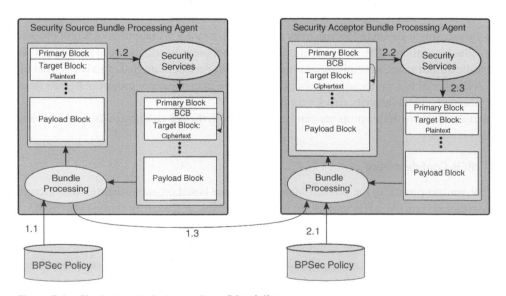

Figure B.1 Single-target single-result confidentiality.

B.2.2 BIB-Integrity

The 1-1 nature of target blocks and results is simpler when applying the *bib-integrity* service, since a single invocation of an integrity mechanism typically produces a single integrity result – such as a keyed digest.

Some STSR security contexts may implement more complex behaviors allowing other data to be associated with the target block when generating this single signature. Again, this construction still qualifies as an STSR security context because a single set of plaintext is sent through a single invocation of an integrity mechanism. Even if that single inputted plaintext is the catenation of more than just information from the target block.

For example, the BIB-HMAC-SHA2 security context defined as part of the Default Security Contexts for BPSec (RFC 9173 [1]) allows for the catenation of other information, such as the primary block of the bundle, when constructing the plaintext input to the HMAC-SHA2 mechanism. This context is still considered an STSR context.

B.2.2.1 Scenario

A BPA configured as a security source may be required by its security policy to apply the *bib-integrity* service to a target block. This security policy presumes that downstream security verifiers and the security acceptor can each verify that the target block's block-type-specific data has not been modified since it was first transmitted by the security source.

B.2.2.2 Processing Steps

Processing steps associated with an STSR security context applying *bib-integrity* differ based on the security role of the processing BPA. The overall processing of this scenario is illustrated in Figure B.2. The enumerated elements in this figure are described next in greater detail.

1. Actions at the security source.
   1.1 A BPA is preparing a bundle to be transmitted. The bundle contains a block that is required by local security policy to have integrity protection. This makes the BPA the security source for a *bib-integrity* operation on that block.

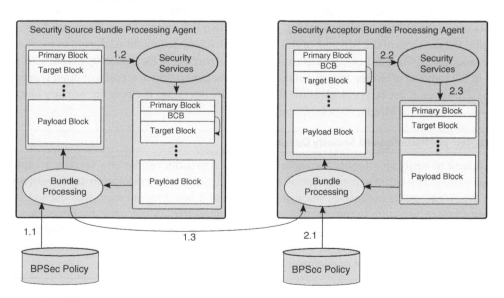

Figure B.2   Single-target single-result integrity.

1.2 To protect the block from tampering, the security source generates a BIB with the security operation $OP(bib\text{-}integrity, Target_{Block})$. This calculates a keyed digest over the target block whose result is stored in the BIB. The BIB is added to the bundle.

1.3 The bundle is transmitted from the security source.

2. Actions at the security acceptor.[1]

2.1 The bundle reaches its destination which is required to function as the security acceptor for the BIB.

2.2 The $OP(bib\text{-}integrity, Target_{Block})$ operation within the BIB is processed to verify the security target's keyed digest.

2.3 When the security operation has been processed, it is removed from the BIB. Since this was the only operation present in the BIB, it is removed from the bundle.

B.2.3 Common Error Conditions

Potential issues with use of a STSR security context include the following.

B.2.3.1 Failed Generation of Cryptographic Material

In the event that the generation of cryptographic material fails, the security result required to populate the security block is not created. This means that the security block is incomplete and cannot be added to the bundle. Failure to generate a required security result is a violation of security policy and must be handled in accordance with some corrective action related to insufficient security at a security source.

Misconfigured Security Operation

The failure to generate cryptographic material is most likely caused by some misconfiguration of the system, such as the specification of an inappropriately sized key. Therefore, this error would most likely be associated with the Security Operation Misconfigured at Source event in the security operation lifecyle.

B.2.3.2 Integrity Verification Failure

This error occurs when the BPA attempting to process the integrity signature associated with the security operation $OP(bib\text{-}integrity, Target_{Block})$ resident in a BIB cannot verify the result. This indicates that the contents of the BIB and/or the target block have been modified during transmission.

The processing BPA handles this verification failure according to local security policy. This may result in the removal of blocks, removal of the bundle, the generation of reports, or other corrective actions. However, if the target block of the failed security operation is either the primary block or the payload block, the bundle must be discarded.

Corrupted or Misconfigured Security Operation

The failure to verify an integrity mechanism at either a security verifier or a security acceptor is likely caused by either some corruption of the bundle, or a misconfiguration at the verifying BPA. When caused by misconfiguration, a BPA may use a key different from that used at the security source, resulting in the creation of mismatching integrity mechanisms. When caused by corruption, some relevant aspect of the bundle has changed since the last time integrity was verified. This error can be associated with either the Security Operation Corrupted or Security Operation Misconfigured set of events in the security operation lifecycle.

---

1 The actions at a security verifier are the same, with the omission of step 2.3, which is unique to the security acceptor.

B.2.3.3 Decryption Failure at the Security Acceptor

This error occurs when the BPA attempting to process the ciphertext associated with the security operation $OP(bcb\text{-}confidentiality,Target_{Block})$ resident in a BCB. This results in a failed decryption and/or authentication of the target block.

The processing BPA handles this decryption failure according to local security policy. This may result in the removal of blocks, removal of the bundle, the generation of reports, or other corrective actions. However, if the target block of the failed security operation is the payload block, the bundle must be discarded as the plaintext contents of the payload is considered unrecoverable.

Any non-payload target block that fails to decrypt must be removed from the bundle and any BIB over that same target block must also be removed.

A BCB does not have the proper security context and/or necessary security context parameters such as the key, to decrypt and/or authenticate the BCB's security target block. This will result in failed decryption and/or authentication of the BCB's security target block at that node.

Corrupted or Misconfigured Security Operation

The failure to decrypt a target block is caused by either some corruption of the bundle or a misconfiguration at the verifying BPA. When caused by misconfiguration, a BPA may use a key different from that used at the security source. When caused by corruption, some relevant aspect of the bundle has changed since the target block was encrypted at the security source. This error can be associated with either the Security Operation Corrupted at Acceptor or Security Operation Misconfigured at Acceptor events in the security operation lifecycle.

B.3 Single-Target Multiple-Result Security Contexts

Single-Target, Multiple-Result (STMR) contexts represent a $1\text{-}N$ mapping of target blocks and cipher suite invocations. In these contexts, the contents of a target block may be input to multiple invocations of a cipher suite, generating multiple sets of results (one set for each cipher suite invocation).

Security contexts with this property can be used to provide both the *bcb-confidentiality* and *bib-integrity* security services.

B.3.1 BCB-Confidentiality

An STMR context providing the *bcb-confidentiality* service may produce multiple versions of ciphertext for a target block, which can be useful when there is ambiguity regarding the identification or configuration of the eventual security acceptor for the associated security operation.

B.3.1.1 Scenario

In network deployments with sufficient resources for security result generation and processing, there are several scenarios which may benefit from the use of a STMR security context for *bcb-confidentiality*. A sample scenario for which an STMR security context should be used is as follows.

A BPA configured to serve as a security source for a *bcb-confidentiality* security operation may be required by its security policy to use a STMR security context. This security context requires that the security source performs the following actions:

1. The calculation of ciphertext n times where n is set by the security context definition.

2. The encryption of the target block, requiring the replacement of the target block's plaintext with one instance of the calculated ciphertext.

3. The storage of the remaining $n - 1$ instances of ciphertext as security results belonging to the BCB.

The actions taken by the security source can accommodate different parameters or sets of parameters when calculating the ciphertext. This may be used in cases where the key used by the intended security acceptor may not be known at the time of encryption. While it may be assumed that the security acceptor will use the agreed upon long-term key to perform decryption, it may also be the case that said key is exhausted before the bundle reaches the security acceptor. For this use case, an STMR security context is used so that the ciphertext generated by the security source can be calculated multiple times using different key inputs.

B.3.1.2 Processing Steps

The processing steps associated with the use of a STMR security context for *bcb-confidentiality* are illustrated in Figure B.3 and described as follows.

1. Actions at the security source.
   1.1 At the security source, BPSec policy is present which identifies a block which must have a *bcb-confidentiality* service applied before transmission begins. The security policy also requires that a STMR security context is used.
   1.2 The security source generates a BCB targeting the specified block using the STMR security context given by policy.
   1.3 The security source generates multiple instances of ciphertext using various input parameters identified by the STMR security context and local security policy.
   1.4 The security source encrypts the security target block's contents, replacing the plaintext block-type-specific data with ciphertext.
   1.5 The remaining instances of ciphertext are stored as security results in the BCB.
   1.6 The bundle is transmitted by the security source.
2. Actions at a security verifier.
   2.1 At each security verifier in the bundle's path, BPSec policy is present which indicates that the BPA must use any of the applicable security results that the BCB holds to verify the authenticity of its security target block.
   2.2 The security verifier checks the authenticity of the security target block using one or more of the security results provided by the BCB.
   2.3 Bundle transmission continues from the security verifier.

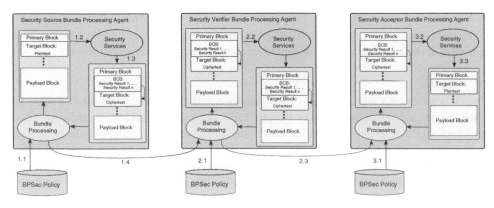

Figure B.3   Single-Target Multiple-Result Confidentiality.

3. Actions at a security acceptor.
    3.1 The bundle arrives at its destination, which local security policy identifies as the security acceptor for the BCB.
    3.2 The BCB is processed. The security target block is decrypted and its authenticity is verified.
    3.3 When the BCB has been successfully processed by the security acceptor, it is removed from the bundle.

B.3.2 BIB-Integrity

B.3.2.1 Scenario

In network deployments with sufficient resources for security result generation and processing, there are several scenarios which may benefit from the use of a STMR security context for *bib-integrity*. A sample scenario for which an STMR security context should be used is as follows.

A node configured to serve as a security source may be required by its security policy to use a STMR security context when adding *bib-integrity* services to a bundle. This security policy configuration allows the security source BPA to generate n integrity signatures, with n being defined by the security context, using different security context parameters as input in addition to the target block's block-type-specific data.

Use of a STMR security context allows security verifiers and the security acceptor to verify the target block's integrity using all of the security results carried by the BIB. If any of the integrity signatures carried by the BIB cannot be verified, the integrity of the *target block* cannot be verified.

B.3.2.2 Processing Steps

The processing steps associated with the use of a STMR security context for *bib-integrity* are illustrated in Figure B.4 and described as follows.

1. Actions at the security source.
    1.1 At the security source, BPSec policy is present which identifies a block which must have a *bib-integrity* service applied before transmission begins. The security policy also requires that a STMR security context is used.
    1.2 The security source generates a BIB targeting the specified block using the STMR security context given by policy. The security source generates multiple security results using multiple keys and/or other security context parameters, as defined by the STMR security context.
    1.3 The bundle is transmitted from the security source.

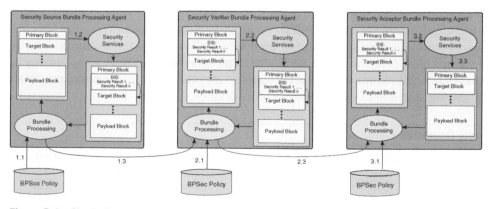

Figure B.4   Single-Target Multiple-Result Integrity.

2. Actions at a security verifier.

    2.1 At each security verifier, policy is present which indicates that the node must process the BIB to verify the integrity of its security target block.

    2.2 Each security verifier processes the BIB, verifying the integrity of the target block using each of the security results provided by the BIB.

    2.3 Bundle transmission continues from the security verifier.

3. Actions at the security acceptor.

    3.1 The bundle arrives at its destination, which policy identifies as the security acceptor for the BIB.

    3.2 The BIB is processed to verify the integrity of its target block. Each of the security results carried by the BIB must match the integrity signatures generated by the security acceptor for verification to be successful.

    3.3 When the BIB has been successfully processed it is removed from the bundle.

B.3.3 Common Error Conditions

When a STMR security context is used, issues may be encountered when additional security results are verified.

Some potential issues with the use of a STMR security context include the following.

B.3.3.1 Failed Generation of Cryptographic Material: Integrity Signature at Security Source

This error occurs when a security source is required by its security policy to generate multiple integrity signatures for a BIB target block, but a security result is not able to be generated and/or added to the BIB. Local security policy determines how the bundle is handled after the failure.

Security Operation Event

This error corresponds to the Security Operation Misconfigured at Source security operation event.

B.3.3.2 Integrity Verification Failure at a Security Verifier

When a BIB that was generated using a STMR security context is processed by a security verifier in the sample scenario, validation of multiple security results may be required for integrity verification of the target block.

This error occurs when one or more of the BIB's security results cannot be verified, leading to the failed integrity verification of the BIB's security target block. The security verifier must handle the security block and its target block according to local security policy. It is recommended that the security target block is removed from the bundle. If the security target block is the primary block or payload block, it is recommended that the bundle is also discarded. In all cases, requested status reports may be generated to reflect bundle or block deletion.

Security Operation Event

This error corresponds to the Security Operation Corrupted at Verifier security operation event.

B.3.3.3   Integrity Verification Failure at the Security Acceptor

When a BIB that was generated using a STMR security context is processed by a security acceptor, validation of multiple security results may be required for integrity verification of the target block.

This error occurs when one or more of the BIB's security results cannot be verified, leading to the failed integrity verification of the BIB's security target block. The security acceptor must handle the BIB and its target block according to local security policy. It is recommended that the security target block is removed from the bundle.

Security Operation Event

This error corresponds to the Security Operation Corrupted at Acceptor security operation event.

B.3.3.4   Failed Generation of Cryptographic Material: Ciphertext at Security Source

This error occurs when a security source is required by its security policy to generate multiple instances of ciphertext for a BCB target block, but the security result is not able to be generated and/or added to the BCB. Local security policy determines how the bundle is handled after the failure.

Security Operation Event

This error corresponds to the Security Operation Misconfigured at Source security operation event.

B.3.3.5   Confidentiality Verification Failed at a Security Verifier

When a BCB that was generated using a STMR security context is processed by its security verifier in the sample scenario, the validation of all of the BCB's security results must be performed successfully to be sufficient for verification of the authenticity of the target block.

This error occurs when one or more of the BCB's authentication tags cannot be verified by the node, leading to the failed verification of the security target block's authenticity. Note that the security verifier does *not* verify the ciphertext as decryption of the security target block is not permitted at a security verifier. The security verifier must handle the BCB and its corrupted target block according to local security policy.

It is recommended that the security target block is removed from the bundle. If the security target block is the primary block or payload block, it is recommended that the bundle is also discarded. In all cases, requested status reports may be generated to reflect bundle or block deletion.

Security Operation Event

This error corresponds to the Security Operation Corrupted at Verifier security operation event.

B.3.3.6   Confidentiality Processing Failed at the Security Acceptor

When a BCB that was generated using a STMR security context is processed by a security acceptor, the decryption of the security target block is unsuccessful only if none of the ciphertext security results provided by the BCB can be decrypted.

This error occurs when each of the BCB's security results have been tried as input to the STMR security context and none have resulted in the successful decryption of the security target block, leading to the failed processing of the BCB's target block. The security acceptor must handle the BCB and its target block according to local security policy. It is recommended that the security target block is removed from the bundle.

Security Operation Event

This error corresponds to the Security Operation Corrupted at Acceptor security operation event.

B.4 Multiple Security Sources

The BPSec provides an augmentation before encapsulation approach to security, meaning that modifications to the bundle to provide the required security operations are preferred to the approach of whole-bundle encapsulation. Following from this approach, the configuration of multiple security source nodes for a single bundle is used when different security services are necessary for different portions of the network.

Multiple security contexts can be used to represent these different security operations in the bundle. Using this method, the BPSec can provide integrity or confidentiality tunnels in the network without the need for bundle encapsulation.

B.4.1 Scenario

Multiple security contexts may need to be used in network deployments composed of segments which require different security services for bundle traffic that must pass through.

A sample scenario for which multiple security contexts may be used to provide different security services as necessary through segments in the network is illustrated in Figure B.5 and described as follows.

A bundle is received by a waypoint node – neither the bundle source nor bundle destination – with a *bib-integrity* security operation present. However, this level of security is insufficient for the next network segment that the bundle will pass through. The waypoint node is configured to serve as a security source for *bcb-confidentiality* for blocks in the bundle which require additional security before passing through the network segment. This node acts as the security gateway for this portion of the network, providing additional security services to the bundle as needed so that its contents are secure as transmission continues.

With this configuration, end-to-end *bib-integrity* is maintained while the bundle is augmented at a network boundary to create a *bcb-confidentiality* tunnel. Two security source nodes are used to add the two security operations to the bundle.

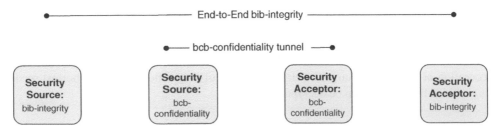

Figure B.5 Multiple Security Sources Create Tunnels.

B.4.2 Processing Steps

The use of multiple security contexts to fulfil the scenario from the Section B.4.1 Scenario is illustrated in Figure B.6 and detailed as follows.

1. Actions at the *bib-integrity* security source.
   1.1 A bundle source assembles a bundle for transmission. The security policy at this Bundle Protocol Agent designates the node as a security source, and requires that a BIB is added to the bundle to provide integrity over a target block
   1.2 A BIB is added to the bundle by the security source targeting the block identified by security policy. An integrity signature is generated over the security target block's block-type-specific data.
   1.3 The bundle is forwarded by the security source.
2. Actions at the *bcb-confidentiality* security source.
   2.1 The bundle is received by another node identified as a security source. Security policy at this security source requires that a *bcb-confidentiality* security operation sharing the *bib-integrity* security target block is added to the bundle before it can be transmitted further. This node serves as the gateway to the network segment requiring *bcb-confidentiality* for that block.
   2.2 The security source adds a BCB to the bundle, encrypting the security target block's block-type-specific data.
   2.3 Bundle transmission continues through the network segment requiring *bcb-confidentiality*.
3. Actions at the *bcb-confidentiality* security acceptor.
   3.1 The bundle arrives at a node with security policy which designates it as the security acceptor for the *bcb-confidentiality* operation in the bundle. This node terminates the confidentiality tunnel provided by that BCB.
   3.2 The security acceptor processes the BCB that was added earlier, authenticating and decrypting the target block's block-type-specific data.
   3.3 The security acceptor gateway node removes the BCB from the bundle after it has been successfully processed.
   3.4 Bundle transmission continues from the security acceptor.
4. Actions at the *bib-integrity* security acceptor.
   4.1 The bundle arrives at its destination, which policy identifies as the security acceptor for the BIB.
   4.2 The BIB is processed to verify the integrity signature for its security target block.
   4.3 The BIB is removed from the bundle when the security acceptor has successfully processed it and verified the integrity of the target block.

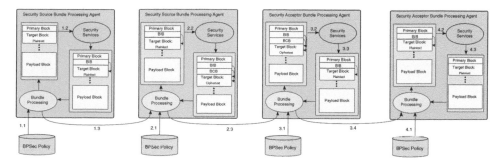

Figure B.6   Multiple Security Contexts Provide Different Security Services.

B.4.3   Common Error Conditions

When adding multiple security operations to a bundle, making use of multiple nodes configured to serve as security sources, issues may be encountered when those additional services are populated in the bundle.

Some potential issues that may occur when multiple security sources are configured for the same bundle include the following.

B.4.3.1   Failed Generation of BIB at Security Source

This error occurs if a security source node is unable to add a required BIB to the bundle. This error can occur because a security result was not generated properly, the BIB was not constructed due to insufficient resources, or a misconfiguration in security policy prevented the BIB from being added to the bundle.

In the event that a required BIB is not added to the bundle, local security policy must be consulted to determine how the bundle is handled.

Security Operation Event

This error corresponds to the Security Operation Misconfigured at Source security operation event.

B.4.3.2   Failed Generation of BCB at Security Source

When a source cannot add a BCB required by security policy at that node, bundle transmission cannot be continued because sufficient security is not present. Local security policy determines how the bundle is handled after the failure to add the BCB.

Security Operation Event

This error corresponds to the Security Operation Misconfigured at Source security operation event.

Reference

1 Birrane, E. J., White, A. and Heiner, S. [2022]. Default Security Contexts for Bundle Protocol Security (BPSec), RFC 9173. URL: https://www.rfc-editor.org/info/rfc9173.

Appendix C

Bundle Protocol Data Representation

C.1   Bundle Protocol Data Objects

Within the set of the Request for Comments (RFCs) which define the Bundle Protocol version 7 (BPv7) and Bundle Protocol Security (BPSec) mechanisms, there are several data objects that define the blocks used to encode and protect information carried in a bundle. This appendix provides a definition of those structures using the Concise Data Definition Language (CDDL) as an aid to the developer to understand the specific fields to be included in each of these structures.

BPSec, defined in RFC 9172 [1], extends BPv7 as defined in RFC 9171 [2] and is the required specification to provide security services for BPv7. Whereas BPv7 defines the basic structure for bundles and blocks, BPSec defines the basic blocks for security infrastructure – the Block Integrity Block (BIB) and the Block Confidentiality Block (BCB).

The BIB and BCB share a common block-type-specific data field structure to carry security-related information. Security related information is generated by a security context defining the cipher suites and cryptographic processing that, when combined with a robust security policy, provide security for a BPv7 implementation. A pair of example security contexts, the Default Security Contexts for BPSec, are defined in RFC 9173 [3].

This appendix details the following objects to better describe expected messages used in BPSec implementation:

- BPv7 (RFC 9171)
  - Bundle
  - Primary Block
  - Canonical Block
- BPSec (RFC 9172)
  - Abstract Security Block (ASB)
  - Block Integrity Block (BIB)
  - Block Confidentiality Block (BCB)
- Default Security Context (RFC 9173)
  - BIB-HMAC-SHA2
  - BCB-AES-GCM

There are other structures defined in BPv7 which are not described here, including specific bundle extension blocks and administrative records. These are important for BPv7 operation, but not directly related to BPSec. The representations for those objects can be derived relatively easily from the BPSec-related representations here in this appendix.

Figure C.1 shows the relationships between these structures and the hierarchy by which the various blocks are derived. To better describe the structures here to motivate the development of the

Securing Delay-Tolerant Networks with BPSec, First Edition. Edward J. Birrane III, Sarah Heiner, and Ken McKeever.
© 2023 John Wiley & Sons, Inc. Published 2023 by John Wiley & Sons, Inc.
Companion website: www.wiley.com/go/birrane/securingdelay-tolerantnetworks

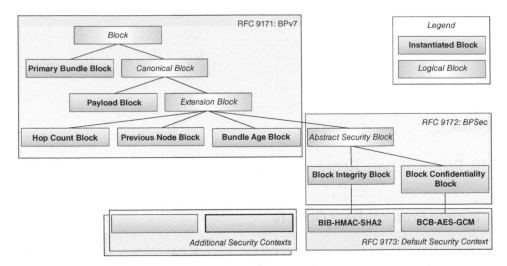

Figure C.1   Block definitions and relationships. BPSec-related blocks are defined hierarchically.

CDDL representation, the structures are divided into two classes: instantiated blocks and logical blocks. Instantiated blocks are those which are actually populated with field values and can be present as part of a bundle. In contrast, logical blocks introduce fields, but are not intended to be populated; instead, these blocks provide structure for other blocks to further define.

For example, in Figure C.1, the BIB-HMAC-SHA2 block defined in RFC 9173 is built upon the BIB, which is in turn built upon the ASB. The BIB and ASB are defined in BPSec and are derived from the Canonical Block which is based on a BPv7 Block.

C.2   Data Representation

This appendix provides a description of block formats from BPv7, BPSec, and the default security context which may be helpful to the reader implementing BP functionality. There are two specifications useful here. Concise Binary Object Representation (CBOR) is a binary serialization data format which is designed with efficiency of implementation in mind. The Concise Data Definition Language is a machine-readable data structure representation format which is used to represent CBOR data objects to facilitate design communication. In this appendix, BPv7 blocks are described using CDDL notation.

C.2.1   CBOR Basics

CBOR is defined in RFC 8949[1] and provides a binary serialization format that attempts to allow for implementation with a relatively small code base and message size [4]. There are several predefined CBOR types which have a specified common encoding scheme. The structures defined by BPv7, BPSec, and the Default Security Contexts all make use of these types to construct and define blocks.

C.2.1.1   CBOR Objectives
From RFC 8949, the objectives of CBOR are:

1. The representation must be able to unambiguously encode most common data formats used in Internet standards.

---
1 RFC 8949 is the latest CBOR specification and obsoletes RFC 7049, which originally defined CBOR.

2. The code for an encoder or decoder must be able to be compact in order to support systems with very limited memory, processor power, and instruction sets.
3. Data must be able to be decoded without a schema description.
4. The serialization must be reasonably compact, but data compactness is secondary to code compactness for the encoder and decoder.
5. The format must be applicable to both constrained nodes and high-volume applications.
6. The format must support all JavaScript Object Notation (JSON) data types for conversion to and from JSON.
7. The format must be extensible, and the extended data must be decodable by earlier decoders.

C.2.1.2 CBOR Encoding

The CBOR encoding scheme relies on header bytes at the beginning of the binary object to define the specific data type encoded in the remaining bytes of the binary object. There can be anywhere from one to five header bytes used to describe how the decoder should interpret the binary object.

Figure C.2 shows a generalized CBOR binary object with initial and additional bytes which are referred to as the head of the binary object. There are eight major types of data structures indicated by the high-order 3 bits of the first header byte. Some examples of major types include unsigned integers, byte strings, arrays, and other common data structures. Additionally, there is a "tag" major type which allows for other data types and encoding structures to be defined. The remaining low-order 5 bits of the initial byte are used to encode the additional information to indicate either the encoded data or that additional bytes follow to define the argument of the specified major type.

For a full description of the CBOR encoding process, see Section 3 of RFC 8949.

C.2.2 CDDL Basics

CDDL is defined in RFC 8610 and is used to provide an unambiguous description of a CBOR data structure that is able to both be read by humans and parsed by computers [5]. This is particularly useful for describing message constructions in protocols that make use of CBOR encoding, such as BPv7, BPSec, and the default security contexts.

CDDL uses a syntax inspired by Augmented Backus–Naur Form grammar notation which is defined in RFC 5234 [6] and is intended to be both machine- and human-readable. This section describes some basic elements of CDDL including terminology, syntax, and control to help understand the examples in Section C.3. For a thorough description of CDDL syntax and usage, see RFC 8610.

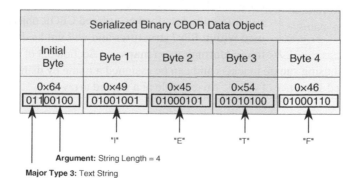

Figure C.2   CBOR Text Encoding. This is an example from RFC 8949 of the encoding of the 4-character string "IETF" [4].

C.2.2.1 Groups

> **Group**
>
> A group is a list of **entries** where each entry is itself an object. Groups are inherently unordered lists, meaning that the sequence of entries is not part of the definition of a group.

CDDL has a single primary structure used for object composition called a **group**. A group is defined using parentheses and can be assigned a name by using it in a rule. This group becomes a named group and can be used in other group definitions. An example of a named group called "filledshape" with an arbitrary number of sides and a color which is red, blue, yellow is defined as:

```
1    filledshape = (
2        sides: uint,
3        color: "red" / "blue" / "yellow"
4    )
```

C.2.2.2 Entries

> **Entries**
>
> Entries are the elements that make up groups using **name/value pairs**. Depending on how the group is used, the name and value or only the value will be stored in the resulting serialized data.

An entry can be another group or a **name/value** pair. If the entry is another group, that is indicated by using the name assigned to that group. This then incorporates the entries from that group into the group being defined. If the entry is a name/value pair, that is indicated using the syntax "*name*: *value.*"

Values of entries in CDDL are one of several types. There is a list of predefined types which correspond to CBOR data types, found in Appendix D of RFC 8610. The CDDL representations for BPv7-related data only make use of several of these, primarily the following:

- uint: unsigned integer
- tstr: text string
- bstr: byte string

C.2.2.3 Group Contexts: Arrays and Maps

> **Array versus Map**
>
> An array is an ordered list of name/value pairs within the group. A map is an unordered list of key/value pairs that come from the name/value pairs of the group.

There are two other object compositions that derive from a group: an **array** and a **map**. An array enforces ordering of the entries within a group such that two arrays will match only if they have the same entries and in the same order. Arrays are central to the description of bundles and blocks within BPv7 as those structures are described as ordered CBOR objects. Maps are not used within BPv7 structures and will not be discussed here. For more information on maps, see Section 3.5 of RFC 8610.

Arrays are defined using square brackets and can be assigned a name by using it in a rule. This will define an ordered sequence of values. An example of a named array is:

```
1    header = [
2        messagetype: uint,
3        encoding: tstr,
4        errorchecking: bool
5    ]
```

Table C.1 Occurrence indicators are used to build array a from predefined groups b and c.

Indicator	Example	Occurrences of Group c
?	$a = [b, ?c]$	c is optional and may occur zero or one times
*	$a = [b, *c]$	c can occur any number of times, including zero
+	$a = [b, +c]$	c can occur one or more times
$n * m$	$a = [b, 2 * 5c]$	c can occur 2 through 5 times, including 2 and 5
$*m$	$a = [b, *5c]$	c can occur up to 5 times, including 0 and 5
$n*$	$a = [b, 2 * c]$	c can occur 2 or more times

C.2.2.4 Entry Occurrence Indicators

In CDDL, **occurrence** is shown through an occurrence indicator which is placed in front of a group entry. This specifies how many times the indicated entry can be present in the group. CDDL allows each of the following occurrence indicators, each defined through example in Table C.1.

C.2.2.5 Choices

Entries can also be specified as choices, either allowing multiple types or restricting possible values. In the definition of a data object, there are cases where it is necessary to define an entry as one of several options. In BPv7 there are objects with fields that can take only one of several options. Within a group definition, a choice between types is written with a "/" operator. In the following example, a group "icecream" is defined with an entry for flavor that can be either "chocolate" or "vanilla":

```
1    icecream = (
2        flavor: "chocolate" / "vanilla"
3    )
```

C.2.2.6 Building Objects: Sockets, Plugs, and Within

Another important way that CDDL can be used is through **sockets** and **plugs**. Sockets are a type or a group which is intended to be extended through subsequent rules. In the BPv7-related structures, type sockets are primarily used to extend arrays defined as higher-level objects. A type socket is assigned a name with a preceding "$" operator and is then extended through the "/=" operator.

Another control operator used for the BPv7-related structure definitions is the ".within" operator. This indicates that the left-hand side is meant to be a subset of the right-hand side. This is used in the examples within this appendix to indicate when blocks derive from other previously defined structures. Just as in Figure C.1, the *BIB-AES-GCM* block defined in the Default Security Context would be a subset of the Block Confidentiality Block defined in BPSec.

> **Sockets and Plugs**
>
> Sockets are an empty type that are intended to be extended through subsequent definition, referred to as plugs.

C.3   CDDL Representations

C.3.1   Bundle Protocol v7

The CDDL here was adapted from Appendix B of RFC 9171 [2].

Common Groups

```
1    ; BPv7 - RFC 9171
2    ; Endpoint ID
3    endpoint-id = $endpoint-id.within endpoint-id-structure
4    endpoint-id-structure = [
5      uri-scheme-code-number: uint,
6      scheme-specific-part: any
7    ]
8    ; DTN URI Scheme Type
9    $endpoint-id /= [
10     uri-scheme-code-number: 1,
11     scheme-specific-part: (tstr / 0)
12   ]
13   ; IPN URI Scheme Type
14   $endpoint-id /= [
15     uri-scheme-code-number: 2,
16     scheme-specific-part: [
17       endpoint-id-node-number: uint,
18       endpoint-id-service-number: uint
19     ]
20   ]
21   ; CRC
22   crc-type = &(
23     crc-none: 0,
24     crc-16bit: 1,
25     crc-32bit: 2
26   )
27   ; Either 16-bit or 32-bit
28   crc-value = (bstr.size 2) / (bstr.size 4)
29
30   block-control-flags = uint.bits blockflagbits
31   blockflagbits = &(
32     reserved: 7,
33     reserved: 6,
34     reserved: 5,
35     block-must-be-removed-from-bundle-if-it-cannot-be-processed: 4,
36     reserved: 3,
37     bundle-must-be-deleted-if-block-cannot-be-processed: 2,
38     status-report-must-be-transmitted-if-block-cannot-be-processed: 1,
39     block-must-be-replicated-in-every-fragment: 0
40   )
41
42   bundle-control-flags = uint.bits bundleflagbits
43   bundleflagbits = &(
44     reserved: 20,
45     reserved: 19,
46     bundle-deletion-status-reports-are-requested: 18,
47     bundle-delivery-status-reports-are-requested: 17,
48     bundle-forwarding-status-reports-are-requested: 16,
49     reserved: 15,
50     bundle-reception-status-reports-are-requested: 14,
51     reserved: 13,
52     reserved: 12,
53     reserved: 11,
54     reserved: 10,
55     reserved: 9,
```

```
56    reserved: 8,
57    reserved: 7,
58    status-time-is-requested-in-all-status-reports: 6,
59    user-application-acknowledgement-is-requested: 5,
60    reserved: 4,
61    reserved: 3,
62    bundle-must-not-be-fragmented: 2,
63    payload-is-an-administrative-record: 1,
64    bundle-is-a-fragment: 0
65   )
```

BPv7 Groups

```
1    ; BPv7   RFC 9171
2    ; Bundle structure
3    bundle = [primary-block, *extension-block, payload-block]
4
5    ; Primary Block
6    primary-block = [
7      version: 7,
8      bundle-processing-control-flags,
9      crc-type,
10     destination: endpoint-id,
11     source-node: endpoint-id,
12     report-to: endpoint-id,
13     creation-timestamp: [
14       bundle-creation-time: uint,
15       sequence-number: unit
16     ],
17     lifetime: uint,
18     ? (
19       fragment-offset: uint,
20       total-application-data-unit-length: uint
21     ),
22     ? crc-value
23   ]
24
25   ; Canonical Block
26   canonical-block = [
27     block-type-code: uint,
28     block-number: uint,
29     block-control-flags,
30     crc-type,
31     block-type-specific-data: bstr,
32     ? crc-value
33   ]
34
35   ; Extension Block, derived from Canonical Block
36   extension-block = [
37     block-type-code: uint,
38     block-number: (uint.gt 1),
39     block-control-flags,
40     crc-type,
41     block-type-specific-data: bstr,
42     ? crc-value
43   ]
```

C.3.2 BPSec

```
1    ; BPSec - RFC 9712
2    ; Abstract Security Block, derived from Extension Block
3    abstract-security-block = [
4      block-type-code: uint,
5      block-number: uint,
6      block-processing-control-flags,
7      crc-type,
8      block-type-specific-data: [
9        security-targets: [+ uint],
10       security-context-id: uint,
11       security-context-flags: uint,
12       security-source: endpoint-id,
13       ? security-context-parameters: [ + [parameter-id: uint,
14           parameter-value: any]],
15       security-results: [ * [result-id: uint, result-value: any]]
16     ],
17     ? crc-value
18   ]
19
20   ; Block Integrity Block, derived from Abstract Security Block
21   block-integrity-block = [
22     block-type-code: 11,
23     block-number: uint,'
24     block-processing-control-flags,
25     crc-type,
26     block-type-specific-data: [
27       security-targets: [+ uint],
28       security-context-id: uint,
29       security-context-flags: uint,
30       security-source: endpoint-id,
31       ? security-context-parameters: [ + [parameter-id: uint,
32           parameter-value: any]],
33       security-results: [ * [result-id: uint, result-value: any]]
34     ],
35     ? crc-value
36   ]
37
38   ; Block Confidentiality Block, derived from Abstract Security Block
39   block-confidentiality-block = [
40     block-type-code: 12,
41     block-number: uint,
42     block-processing-control-flags,
43       ;per RFC 9172, these flags must be set:
44       ;block-must-be-replicated-in-every-fragment = true
45       ;block-must-be-removed-from-bundle-if-it-cant-be-processed = false
46     crc-type,
47     block-type-specific-data: [
48       security-targets: [+ uint],
49       security-context-id: uint,
50       security-context-flags: uint,
51       security-source: endpoint-id,
52       ? security-context-parameters: [ + [parameter-id: uint,
53           parameter-value: any]],
54       security-results: [ * [result-id: uint, result-value: any]]
55     ],
56     ? crc-value
57   ]
```

C.3.3 Default Security Context

```
1   ; Default Security Context - RFC 9173
2   ; Derived from Block Integrity Block
3   bib-hmac-sha2 = [
4     block-type-code: 11,
5     block-number: uint,
6     block-processing-control-flags: uint,
7     crc-type,
8     block-type-specific-data: [
9       security-targets: [+ uint],
10      security-context-id: uint,
11      security-context-flags: uint,
12      security-source: endpoint-id,
13      ? security-context-parameters: [
14        ? [1, 5 / 6 / 7.default 6],
15        ? [2, bstr],
16        ? [3, uint.default 7]
17      ],
18      security-results: [
19        [1, bstr]
20      ]
21    ],
22    ? crc-value
23  ]
24
25  ; Derived from Block Confidentiality Block
26  bcb-aes-gcm = [
27    block-type-code: 12,
28    block-number: uint,
29    block-processing-control-flags: uint,
30     ;per RFC 9172, these flags must be set:
31     ;block-must-be-replicated-in-every-fragment = true
32     ;block-must-be-removed-from-bundle-if-it-cant-be-processed = false
33    crc-type,
34    block-type-specific-data: [
35      security-targets: [+ uint],
36      security-context-id: uint,
37      security-context-flags: uint,
38      security-source: endpoint-id,
39      ? security-context-parameters: [
40        ? [1, bstr.size 8..16],
41        ? [2, 1 / 3.default 3],
42        ? [3, bstr],
43        ? [4, uint.default 7]
44      ],
45      security-results: [
46        [1, bstr.size 16]
47      ]
48    ],
49    ? crc-value
50  ]
```

References

1 Birrane, E. J. and McKeever, K. [2022]. Bundle Protocol Security (BPSec), RFC 9172.
URL: https://www.rfc-editor.org/info/rfc9172.

2 Burleigh, S., Fall, K. and Birrane, E. J. [2022]. Bundle Protocol Version 7, RFC 9171.
URL: https://www.rfc-editor.org/info/rfc9171.

3 Birrane, E. J., White, A. and Heiner, S. [2022]. Default Security Contexts for Bundle Protocol Security (BPSec), RFC 9173. URL: https://www.rfc-editor.org/info/rfc9173.

4 Bormann, C. and Hoffman, P. E. [2020]. Concise Binary Object Representation (CBOR), RFC 8949. URL: https://www.rfc-editor.org/info/rfc8949.

5 Birkholz, H., Vigano, C. and Bormann, C. [2019]. Concise Data Definition Language (CDDL): A Notational Convention to Express Concise Binary Object Representation (CBOR) and JSON Data Structures, RFC 8610. URL: https://www.rfc-editor.org/info/rfc8610.

6 Crocker, D. and Overell, P. [2008]. Augmented BNF for Syntax Specifications: ABNF, RFC 5234. URL: https://www.rfc-editor.org/info/rfc5234.

Index

Securing Delay-Tolerant Networks with BPSec, First Edition. Edward J. Birrane III, Sarah Heiner, and Ken McKeever.
© 2023 John Wiley & Sons, Inc. Published 2023 by John Wiley & Sons, Inc.
Companion website: www.wiley.com/go/birrane/securingdelay-tolerantnetworks

Printed and bound by CPI Group (UK) Ltd, Croydon, CR0 4YY

27/10/2024

14580678-0001